*Nanotechnologies for the
Life Sciences*
Volume 5

**Nanomaterials –
Toxicity, Health and
Environmental Issues**

Edited by
Challa S. S. R. Kumar

Nanotechnologies for the Life Sciences
Volume 5

Nanomaterials – Toxicity, Health and Environmental Issues

Edited by
Challa S. S. R. Kumar

1st Edition

WILEY-VCH Verlag GmbH & Co. KGaA

The Editor of this Book

Dr. Challa S. S. R. Kumar
The Center for Advanced
Microstructures and Devices
(CAMD)
Louisiana State University
6980 Jefferson Highway
Baton Rouge, LA 70806
USA

Library of Congress Card No.: applied for
British Library Cataloguing-in-Publication Data:
A catalogue record for this book is available from the British Library.

Bibliographic information published by Die Deutsche Bibliothek
Die Deutsche Bibliothek lists this publication in the Deutsche Nationalbibliografie; detailed bibliographic data is available in the Internet at ⟨http://dnb.ddb.de⟩.

© 2006 WILEY-VCH Verlag GmbH & Co. KGaA, Weinheim

Printed in the Federal Republic of Germany.
Printed on acid-free paper.

Typesetting   Asco Typesetter, Hong Kong
Printing   betz-druck GmbH, Darmstadt
Binding   J. Schäffer GmbH, Grünstadt
Cover Design   Grafik-Design Schulz, Fußgönheim

ISBN-13 978-3-527-31385-3
ISBN-10 3-527-31385-0

Contents

Nanotechnologies for the Life Sciences Vol. 5
Nanomaterials – Toxicity, Health and Environmental Issues. Edited by Challa S. S. R. Kumar
Copyright © 2006 WILEY-VCH Verlag GmbH & Co. KGaA, Weinheim
ISBN: 3-527-31385-0

12 Toxicology of Nanoparticles in Environmental Air Pollution   294

Ken Donaldson, Nicholas Mills, David E. Newby, William MacNee, and Vicki Stone

Preface

It is my pleasure to welcome the readers back to *Nanotechnologies for the Life Sciences* with the first volume published in 2006. I am presenting to you, on behalf of yet another dedicated team of contributors and supporters, the fifth volume, *Nanomaterials – Toxicity, Health and Environmental Issues*, of the ten volume series. We are bringing the fifth volume while the fourth is still in print for a number of reasons. The most important being the fact that a potential $1 trillion nanotechnology market hinges on understanding the toxic effects of nanomaterials on our health and environment. With continuous world-wide increase in both government and private funding in nanoscience and nanotechnology touching close to $35 billion, the stakes are even higher. With increase in stakes, there is a worldwide awakening to understand the toxic effect of nanomaterials and the scholarly chapters presented in this book are testimony to the efforts of several research groups to understand these effects. While the current knowledge base is small compared to what needs to be understood, it certainly provides a scaffold for this knowledge base to take definite shape.

Some of the critical risk assessment issues that are currently being investigated by the health & environmental nano researchers are toxicology, exposure assessment, environmental and biological fate, transport, persistence, transformation, recyclables and overall sustainability of manufactured nanomaterials. I am aware that the scientific data generated so far is very scanty and requires more worldwide concerted effort in this direction. Nevertheless, the amount of information presented by the authors covers almost everything of what is currently available in the literature. The book is divided into three distinct sections in an attempt to emphasize the three major issues related to nanomaterials, which are toxicology, health and environment. The boundaries are only artificial and have been created for the sake of clarity. I am aware that the three issues are interrelated, yet unique in their own way. I am also aware the field is very nascent and hence there could be some amount of overlap in terms of information that is presented in the chapters. However, the USP of the book is that all the chapters provide very unique and intellectually stimulating perspectives on the most important topics in the field of nanoscience and nanotechnology.

The first section of the book deals, in general, with issues around the toxicity of nanomaterials and begins with a scholarly report on the toxic effects of metal oxide

Nanotechnologies for the Life Sciences Vol. 5
Nanomaterials – Toxicity, Health and Environmental Issues. Edited by Challa S. S. R. Kumar
Copyright © 2006 WILEY-VCH Verlag GmbH & Co. KGaA, Weinheim
ISBN: 3-527-31385-0

nanoparticles which are by far, commercially, the most significant materials as they find applications in cosmetics, sunscreens, fillers in dental materials, water filtration processes, catalysis, glare-reducing coating for glasses, and so on. Amanda M. Fond and Gerald J. Meyer from the Department of Chemistry, Johns Hopkins University, USA have reviewed the literature, in addition to capturing their own findings, on biotoxicity of metal oxide nanoparticles keeping the emphasis more on *in vitro* rather than *in vivo* studies. In this chapter entitled, *Biotoxicity of Metal Oxide Nanoparticles*, their critical analysis provides to the reader possible mechanisms by which the metal oxide nanoparticles enter the environment and the body, and the potential health impacts that might be expected. Eva Oberdörster from Southern Methodist University, Patricia McClellan-Green from NC State University and Mary Haasch from University of Mississippi have collaborated to present their critical evaluation in the second chapter, *Ecotoxicity of Engineered Nanomaterials*, impact of nanomaterials on the environment, and more specifically on air, water and soil. In addition, readers will find very useful the authors' insight into how the activity of nanomaterials is effected by extraneous factors such as abiotic factors, microbial degradation/activation and identification of biomarkers associated with nanoparticle exposure.

In the second section of the book, illuminating perspectives on the effect of nanomaterials on health are presented. Relative to the increased use of nanomaterials in a variety of industrial applications, the amount of information regarding their health effects is limited. Peter Hoet from Katholieke Universiteit Leuven, Belgium, Irene Brüske-Hohlfeld from GSF-Forschungszentrum für Umwelt und Gesundheit, Germany, and Oleg V. Salata from Sir William Dunn School of Pathology, University of Oxford, UK, teamed up in order to review the epidemiological studies of the technogenic nanoparticles and to highlight the apparent health effects associated with the inhalation of ultrafine particulate matter. The third chapter by them, aptly entitled *Possible Health Impact of Nanomaterials*, provides information on likely pathways for nanoparticulates in general and nanofibers in particular inside the body, the effects associated with their interactions on the cellular level, and analysis of the origins of bioactivity of nanomaterials. Continuing on the same theme, chapter number four, *Dosimetry, Epidemiology and Toxicology of Nanoparticles*, describes the dosimetry, epidemiology and toxicology of nanoparticles with reference to generally well established principles and paradigms. The chapter is contributed by Wolfgang G. Kreyling, Manuela Semmler and Winfried Möller from GSF-National Research Centre for Environment & Health, Institute for Inhalation Biology, Focus-Network Aerosols and Health, and Clinical Research Group 'Inflammatory Lung Diseases' respectively, from Germany. The highlight of the chapter, in my view, is described best by the authors themselves: "extrapolating findings and principles observed in particle inhalation toxicology into recommendations for an integrated concept of risk assessment of nanoparticles for a broad range of use in science, technology and medicine." Focusing more specifically on ceramic and metallic nanoparticles, the team lead by Kirsten Peters from Institute of Pathology, Johannes Gutenberg University, Germany, discusses in chapter five their effects on primary human endothelial cells which are highly relevant for nanoparticle transmigration from the blood into tissues. The chapter, *Impact of Ceramic and Metallic Nano-Scaled Particles on Endothelial Cell Functions in vitro*, clearly helps readers to understand, with an example of pro-inflammatory stimulation of endothelial cells by nanoparticles, that even though it is clear that the nanoparticles exert effects that are relevant in vitro, these cannot be easily interpreted and may not be of relevance in vivo. The sixth chapter, *Toxicity of Carbon Nanotubes and its Implications for Occupational and Environmental Health*, written by the team lead by Chiu-Wing Lam from the Division of Space Life Sciences, NASA Johnson Space Center, and Wyle Laboratories, Houston, USA, is a comprehensive review on the toxicological risk of carbon nanotubes (CNT) and the impurities present in them due to inhalation exposures using both rodent and in vitro cell culture studies. In addition, the authors discuss the mechanisms of CNT pathogenesis in the lung and other toxicological manifestations. In view of the growing expectations that CNTs will find extraordinary applications in the field of not only life sciences but also in electronics, computer, and aerospace industries, the chapter is timely and will be a single source of information for the readers. The final chapter in this section is the seventh chapter, wherein the authors review the latest results from various studies on the biological effects of nanoparticles that may be the basis for adverse effects, especially in humans. The chapter, *Toxicity of Nanomaterials – New Carbon Conformations and Metal Oxides*, provides a comparative study on two most important classes of nanomaterials, viz carbon and metal oxide based nanomaterials, with respect to their cellular uptake and possible influence on important cellular mechanisms in vitro. The chapter is a testimony to the intensive analysis on the topic carried out by the authors Harald F. Krug, Katrin Kern, Jörg M. Wörle-Knirsch, and Silvia Diabate from the Institute of Toxicology and Genetics at Forschungszentrum Karlsruhe, Germany.

The final section and the most important one, in my view, is dedicated to the investigations related to impact of nanomaterials on environment. While the chapters 1–7 in the previous sections dealt with possible negative effects of nanomaterials, this sections portrays the positive aspects of nanomaterials. The first chapter in this section (8th in the book) is contributed by Glen E. Fryxell and Shas V. Mattigod of Materials Chemistry and Surface Research Group, Pacific Northwest National Laboratory, USA. In this chapter, *Nanomaterials for Environmental Remediation*, the authors address one of the key global political and economic issues of the 21st century – how does one ensure that the majority of the world population has clean environment in general and air & water in particular in future? An analysis of nanoparticle-based remediation technologies for air and water treatment including field tests on actual waste streams is presented. Moving into the ninth chapter, readers will find more specific information regarding the variety of approaches being utilized for treatment of water with nanomaterials. In this chapter, entitled *Nanomaterials for Water Treatment*, Peter Majewski of Ian Wark Research Institute, University of South Australia, Australia, is upbeat about various technologies currently under development and more specifically about the approach using magnetic iron exchange resin (MIEX) which is already commercially applied in water treatment. It is heartening to read the next chapter, chapter ten, wherein

Heather Coleman from the Centre for Particle and Catalyst Technologies of the University of New South Wales, Sydney, Australia, elaborates on how nanotechnologies are proving to be playing a major role in alleviating the concerns about the release into the aquatic environment of natural and synthetic oestrogens and compounds that have the ability to mimic oestrogens. In this chapter, *Nanoparticles for the Photocatalytic Removal of Endocrine Disrupting Chemicals in Water*, the author describes nanoscale titanium dioxide photocatalysis for the degradation of the natural and synthetic oestrogens in water. Chapter eleven by Wan Y. Shih and Wei-Heng Shih, Department of Materials Science and Engineering, Drexel University, Philadelphia, USA, is very unique in the sense that the authors describe their own investigations into the development of piezoelectric microcantilever sensors of different sizes and types that can perform rapid, in-situ, in-water pathogen detection with sensitivities well above that of the current techniques. The chapter describes both theoretical and experimental studies that were carried out to characterize the sensors. While the information provided in the chapter, *Nanosensors for Environmental Applications*, clearly demonstrates that we have a long way to go before realizing the dream of fabricating truly nanosize sensors, it is hoped that the chapter will form a strong basis for readers in designing their own nanosensors for environmental applications. The final chapter, *Toxicology of Nanoparticles in Environmental Air Pollution* by Ken Donaldson and his collaborators, puts forward the evidence that nano-sized air pollutants play adverse role on our health. I confess that this chapter could have been included in the previous section. However, since the chapter describes nanosized partriculate matter present in the natural environment, I have decided to include it in this section. As a final chapter, I also wanted the reader to take home the message that while certainly nanomaterials can be utilized to clean up our environment and treat variety of diseases, one needs to be aware of the deleterious effects of nano-sized particulate matter in the environment.

In the end, I would like to state that I am indeed very grateful to all the authors for their contribution of quality manuscripts on time. I am thankful to my employer, family, friends and Wiley-VCH publishers for making this book a reality. I am always indebted to you, the reader, who is an integral part of this journey into brining nanotechnologies to life sciences and life sciences into nanotechnologies. I am eagerly waiting to receive your comments, suggestions and constructive criticism to make this journey even more enjoyable and a learning experience for all.

March 2006, Baton Rouge   *Challa S. S. R. Kumar*

List of Authors

Irene Brüske-Hohlfeld
GSF-Forschungszentrum für Umwelt und
Gesundheit GmbH
Ingolstädter Landstraße 1
85764 Neuherberg
Germany

Heather M. Coleman
Centre for Particle and Catalyst Technologies
School of Chemical Engineering and
Industrial Chemistry
The University of New South Wales
Sydney
New South Wales 2052
Australia

Silvia Diabaté
Forschungszentrum Karlsruhe
Institute of Toxicology and Genetics
Department of Molecular Environmental
Toxicology
Postfach 3640
76021 Karlsruhe
Germany

Ken Donaldson
University of Edinburgh
ELEGI Colt Laboratory
Queens Medical Research Institute
47 Little France Crescent
Edinburgh EH16 4SU
UK

Amanda M. Fond
Johns Hopkins University
Department of Chemistry
3400 North Charles Street
Baltimore MD 21218
USA

Glen E. Fryxell
Materials Chemistry and Surface Research
Group
Pacific Northwest National Laboratory
PO Box 999
Richland, WA 99352
USA

Andrea Gambarelli
University of Modena and Reggio Emilia
Laboratory of Biomaterials
Department of Neurosciences
Via Campi 213A
41100 Modena
Italy

Antonietta M. Gatti
University of Modena and Reggio Emilia
Laboratory of Biomaterials
Department of Neurosciences
Via Campi 213A
41100 Modena
Italy

Mary Haasch
The University of Mississippi
School of Pharmacy
National Center for Natural Products
Research, ETRP
347 Faser Hall
University, MS 38677
USA

Andrij Holian
University of Montana
Center for Environmental Health Sciences
Missoula, MT 59812
USA

Peter H. M. Hoet
Katholieke Universiteit Leuven
Pneumologie, Longtoxicologie
Campus GHB
Herestraat 49
Leuven B-3000
Belgium

Robert L. Hunter
University of Texas Medical School
Department of Pathology and Laboratory
Medicine
Houston, TX 77030
USA

John T. James
JSC Toxicology Group
Space Life Sciences
NASA Johnson Space Center
Houston, TX 77058
USA

Katrin Kern
Forschungszentrum Karlsruhe
Institute of Toxicology and Genetics
Department of Molecular Environmental
Toxicology
Postfach 3640
76021 Karlsruhe
Germany

C. James Kirkpatrick
Johannes Gutenberg University
Institute of Pathology
Langenbeckstr. 1
55101 Mainz
Germany

Wolfgang G. Kreyling
GSF – National Research Center
for Environment and Health
GSF-Focus: Aerosols and Health
Institute for Inhalation Biology
P.O. Box 1129
85758 Neuherberg
Munich
Germany

Harald F. Krug
Forschungszentrum Karlsruhe
Institute of Toxicology and Genetics
Department of Molecular Environmental
Toxicology
Postfach 3640
76021 Karlsruhe
Germany

Chiu-wing Lam
JSC Toxicology Group
Space Life Sciences and Wyle Laboratories
NASA Johnson Space Center
Houston, TX 77058
USA

William MacNee
University of Edinburgh
ELEGI Colt Laboratory
Queens Medical Research Institute
47 Little France Crescent
Edinburgh EH16 4SU
UK

Peter Majewski
University of South Australia
Ian Wark Research Institute
Mawson Lakes Blvd.
Mawson Lakes, SA 5095
Australia

Shas V. Mattigod
Materials Chemistry and Surface Research
Group
Pacific Northwest National Laboratory
PO Box 999
Richland, WA 99352
USA

Patricia McClellan-Green
North Carolina State University
Department of Environmental and Molecular
Toxicology
NCSU-Center for Marine Sciences and
Technology
303 College Circle
Morehead City, NC 28557
USA

Richard McCluskey
Medical Operation Branch
Space Life Sciences
NASA Johnson Space Center
Houston, TX 77058
USA

Gerald J. Meyer
Johns Hopkins University
Department of Chemistry
3400 North Charles Street
Baltimore, MD 21218
USA

Nicholas Mills
University of Edinburgh
Cardiovascular Research
49 Little France Crescent
Edinburgh EH16 4SU
UK

Winfried Möller
GSF-National Research Center
for Environment and Health
GSF-Clinical Research Group "Inflammatory
Lung Diseases"
Institute for Inhalation Biology
P.O. Box 1129
85758 Neuherberg
Munich
Germany

David E. Newby
University of Edinburgh
Cardiovascular Research
49 Little France Crescent
Edinburgh EH16 4SU
UK

Eva Oberdörster
Southern Methodist University
Department of Biology
6501 Airline Road
Dallas, TX 75275-0376
USA

Kirsten Peters
Johannes Gutenberg University
Institute of Pathology
Langenbeckstr. 1
55101 Mainz
Germany

Enrico Sabbioni
European Commission
Institute for Health and Consumer Protection
(IHCP)
ECVAM Unit, Joint Research Centre (JRC)
Via Fermi 1, JRC-Ispra, TP. 580
21020 Ispra (Varese)
Italy

Oleg V. Salata
Sir William Dunn School of Pathology
University of Oxford
South Parks Road
Oxford OX1 3RE
UK

Manuela Semmler-Behnke
GSF-National Research Center
for Environment and Health
GSF-Focus: Aerosols and Health
Institute for Inhalation Biology
P.O. Box 1129
85758 Neuherberg
Munich
Germany

Wan Y. Shih
Drexel University
Department of Materials Science
and Engineering
Philadelphia, PA 19104
USA

Wei-Heng Shih
Drexel University
Department of Materials Science and
Engineering
Philadelphia, PA 19104
USA

Vicki Stone
Napier University
School of Life Sciences
10 Colinton Road
Edinburgh EH10 5DT
UK

Ronald E. Unger
Johannes Gutenberg University
Institute of Pathology
Langenbeckstr. 1
55101 Mainz
Germany

Jörg M. Wörle-Knirsch
Forschungszentrum Karlsruhe
Institute of Toxicology and Genetics
Department of Molecular Environmental
Toxicology
Postfach 3640
76021 Karlsruhe
Germany

I
Toxicity

1
Biotoxicity of Metal Oxide Nanoparticles

Amanda M. Fond and Gerald J. Meyer

1.1
Introduction

Nanotechnology is a relatively new and evolving field. Although the uses and technological advances in nanotechnology are endless, very little is known about its future consequences or impacts. This leaves some a little skeptical about current research and advances. Concerns range from the health and economic impacts that the once popular material asbestos had on society [1] to nanotechnology careening out of control [2]. However, even though nanotechnology is a fairly new field, nanomaterials are not. Some nanomaterials stem back to the 10th century, such as nanometer-diameter particles of gold and silver, which were used in stained glass and ceramics to generate different hues [3]. In addition, Egyptians were thought to have consumed colloidal gold, believing that it would raise vitality [4]. Nowadays, nanoparticles are frequently found in such commercial products as cosmetics and sunscreens (TiO_2, Fe_2O_3, and ZnO), fillers in dental fillings (SiO_2), water filtration processes, catalysis, and glare-reducing coating for glasses. In addition, they are currently being used in the development of stain and wrinkle-free fabrics and to make longer-lasting tennis balls [5].

Metal oxide nanoparticles have a rich history with applications in food, materials, and chemical and biological studies. The thermodynamically stable form of most metals are their oxides. In many cases metal oxides, e.g., SiO_2, TiO_2, ZnO, have been approved by the Food and Drug Administration for decades [6]. It is, therefore, tempting to assume that metal oxide nanoparticles will also be non-toxic. However, as this chapter demonstrates, and asbestos toxicity has taught us, the shape, size and morphology can also play a significant role in biotoxicity [5].

For such a rapidly growing field, surprisingly little is known about either nanotoxicology or the toxicity of nanoparticles. Funding for nanotoxicology is necessary because nanomaterials often behave differently than their bulk counterparts. At the nanoscale, the surface area of particles greatly increases and can result in a higher reactivity of the material, since the surface atoms now dominate the particle's physical and chemical properties. The material's electrical, optical and thermal properties change and quantum effects become significant [1]. For example, gold

Nanotechnologies for the Life Sciences Vol. 5
Nanomaterials – Toxicity, Health and Environmental Issues. Edited by Challa S. S. R. Kumar
Copyright © 2006 WILEY-VCH Verlag GmbH & Co. KGaA, Weinheim
ISBN: 3-527-31385-0

particles are inert when in bulk material; however, gold nanoparticles are highly reactive and are used in catalysis.

Donaldson et al. concluded that ultrafine particles cause more inflammation than larger respirable particles of the same material when delivered at the same mass dose. Although the exact role of ultrafine particle toxicity remains unknown, experimental evidence showed that ultrafine particles inhibit phagocytosis more than fine particles of the same mass. In addition, even when composed of low toxicity materials, ultrafine particles caused inflammation in the lungs. Many believe it is because of the large surface area of ultrafine nanoparticles [7, 8].

Both *in vivo* and *in vitro* studies are currently underway around the world to evaluate the biotoxicity of metal oxide nanoparticles. However, difficulty arises in marrying these two sets of experiments. Hart states that the main reason *in vivo* and *in vitro* studies are not complementary is due to biopersistence, which relies on particle dissolution rate and the capability of the particles to be translocated out of the lung [9]. In addition, *in vitro* studies are used to measure more short-term toxicity effects and fail to look at how a specific cell type will interact when incorporated with other cell types within an animal. However, *in vivo* studies can prove to be very time consuming and costly.

This chapter will only review *in vitro* biotoxicity literature reports with metal oxide nanoparticles, and broadly overviews mechanisms by which they enter mammalian systems. *In vitro* studies help in the understanding of toxicity mechanisms at a molecular level, information that is difficult if not impossible to gain from *in vivo* studies. In addition, *in vitro* studies make it possible to determine a relationship between toxicity and particle characteristics [9]. Therefore, this chapter will cover literature studies of metal oxide nanoparticles with cells, bacteria and biopolymers, and will not cover the vast breadth of animal studies found in the literature. Additionally, studies on sulfides, selenides, noble metals or organic coatings are not included. However, notably, these particles may prove to have biotoxic effects as well.

First, it is worth defining "nanomaterial" as descriptions in the literature often vary. The National Nanotechnology Initiative defines nanotechnology as: (1) Research and technology development involving structures with at least one dimension on the 1–100 nm range. (2) Creating/using structures, devices, systems that have novel properties and functions because of their nanometer scale dimensions. (3) The ability to control or manipulate on the atomic scale [5]. Here, we have adopted a broader definition and have included all studies of metal oxide materials with length scales less than 1000 nm, as a result of the size relationships of ultrafine particles and cellular structures described by Donaldson (Fig. 1.1) [8].

In the remainder of this chapter we discuss the mechanisms by which nanoparticles enter the environment and the body, and the potential health impacts that might be expected. We then review literature including biotoxicity studies of cells with metal oxide nanoparticles. The literature in this area is conveniently divided into areas based on materials: (a) iron oxide; (b) titanium dioxide; and (c) other oxides. In addition to reviewing the published literature, some background on the materials is also included.

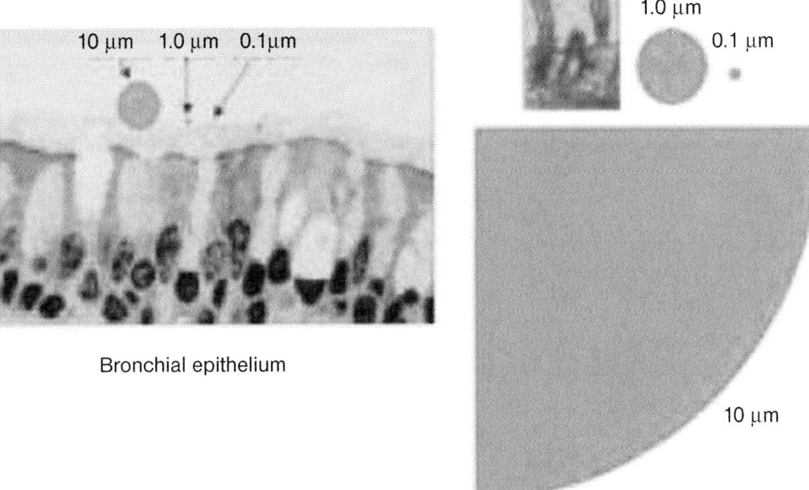

Fig. 1.1. Left-hand side: relationship between ultrafine particles and cellular structures of the lung. Right-hand side: same three particles relative to cilia. (Figure taken from Donaldson [8] with permission from the BMJ Publishing Group.)

1.2
Nanoparticles in the Environment

According to the U.S. Department of Labor, in the U.S. alone, 2 million people work with nanometer-diameter particles regularly in development, production, and use of nanomaterials and products [10]. The National Institute for Occupational Safety and Health (NIOSH) claimed that between 1997 and 2005 the U.S. government investment went from $432 million to $1240 million per year [5], and by 2015 global investment is expected to be $1 trillion [11]. If growth continues as expected, an additional 2 million workers will be required worldwide [5].

Nature has also utilized "nanotechnology." Nanoparticles are found everywhere in the environment. Natural materials such as proteins and colloids, like milk, are composed of nanoparticles. Indeed, most subcellular structures are "nanomaterials." The left-hand side of Fig. 1.2 shows additional examples [12]. Man-made particles produced as a by-product of industry are also a source of nanoparticles in the environment (Fig. 1.2) [12]. However, the lack of information on the environmental impact of nanoparticles has society concerned. Some of these concerns have been brought to the attention of the Department of Health and Human Services. As a result, the National Toxicology Program is assessing the health effects associated with nanoscale materials, such as size and composition dependent biological

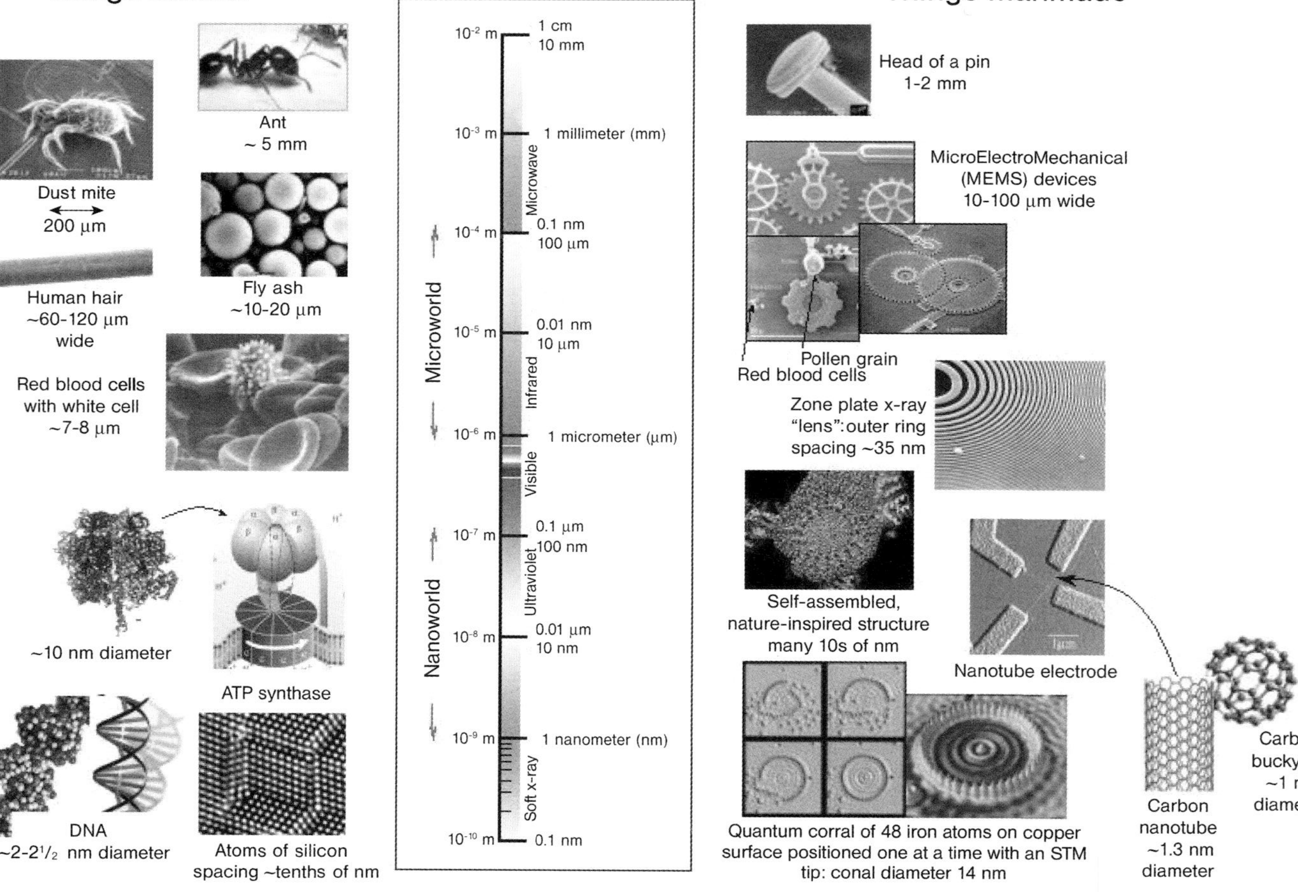

Fig. 1.2. Examples of natural and synthetic nanometer-sized materials. (Figure adapted from www.nano.gov [12].)

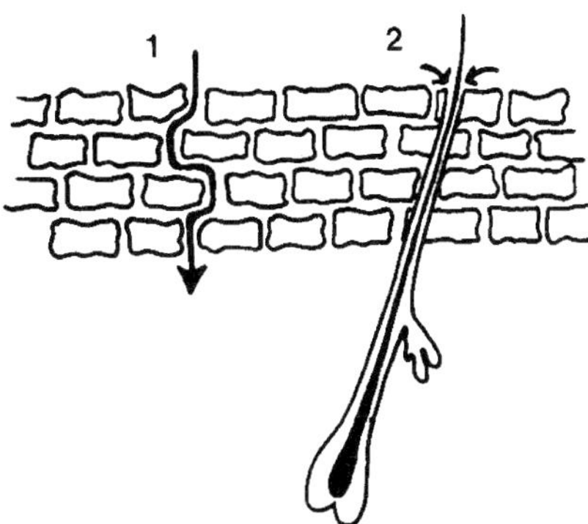

Fig. 1.3. Penetration routes of nanoparticles into human skin. Path 1 shows an intercellular route and path 2 a route through a hair follicle. (Figure taken from Bennat [15] with permission from Blackwell Publishing.)

disposition of fluorescent semiconductor nanomaterials and the phototoxicity of metal oxide nanoparticles [13].

1.3
How Nanoparticles are Introduced into Mammalian Systems

Currently, there is a vague understanding of a nanoparticles' path of entry into one's system, cell uptake, distribution, and health effects. Three main routes of nanoparticle exposure are penetration through the skin, ingestion, and exposure by inhalation – from which the particles may then be able to translocate from the respiratory system to other organs [14].

There can be two routes of entry into the skin, transepidermal intercellular or penetration via pores and hair follicles (Fig. 1.3). Bennat et al. believe that TiO_2 nanoparticles penetrate the skin through the lipids of hair follicles. They found that the more hair follicles in the skin, the deeper the TiO_2 nanoparticle penetration. Furthermore, TiO_2 particles from an oily dispersion penetrated deeper than those from an aqueous solution, possibly because the palmitic acid component of the skin lipids was acting as a penetration enhancer [15]. In contrast, a separate study using pig skin samples showed that the stratum corneum layer of the skin effectively prevented dermal uptake of an oil-in-water emulsion of TiO_2 particles (20–50 nm) [16]. Animal studies, quantified by autoradiography, in which emitted radiation is measured from a tissue specimen that has been treated with a radio-actively labeled isotope, have indicated that ZnO nanoparticles pass through rat and rabbit skin [17–19]. Therefore, some reports raise the idea that it may be possible for ZnO and TiO_2 nanoparticles to pass through human skin. For example, particles of 10–50 nm in diameter would be able to penetrate skin because the intracellular space in the stratum corneum is around 100 nm [20, 21], and the gap in

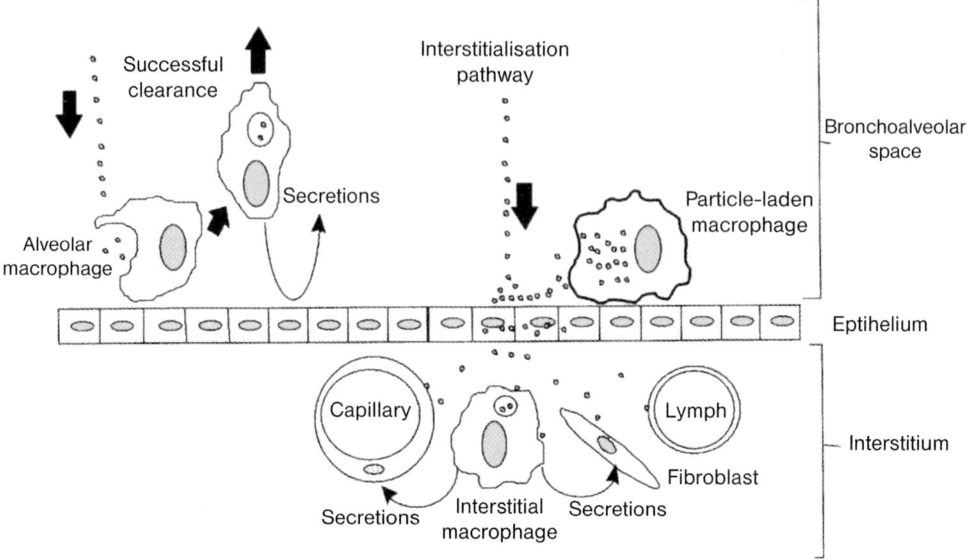

Fig. 1.4. Potential pathways for nanoparticles once they enter the lung; see text for additional details. (Figure taken from Donaldson [25] with permission from Elsevier.)

the lamellar bilayer is 0.5–1 nm. When filled with topically applied materials, the gap has the ability to enlarge [21–23]. Currently, human skin models, such as Skin$^{2\,\mathrm{TM}}$, are being used for *in vitro* penetration studies of UV-irradiated particles [24].

Once inhaled, particles enter the deep lung region where they are engulfed by macrophages and removed before damage to the epithelium occurs. However, with nanoparticles, the burden becomes too large for the macrophages to remove all of the particles. The particles can then interact with the epithelium and cause inflammatory effects, enter the interstitium where they promote chronic effects on cells, or transfer to lymph nodes (Fig. 1.4) [25].

1.4
Health Threats

Some believe that human exposure to most nanoparticles is not large enough to cause significant health effects in healthy individuals [1, 8]. TiO_2 is reported to be harmless when swallowed by man [26]. However, occupational health risks may be significant due to exposure of nanoparticles at levels higher than ambient conditions. In addition, man-made nanomaterials may have novel sizes, and physical and chemical properties, which can lead to biocompatibility problems when intro-

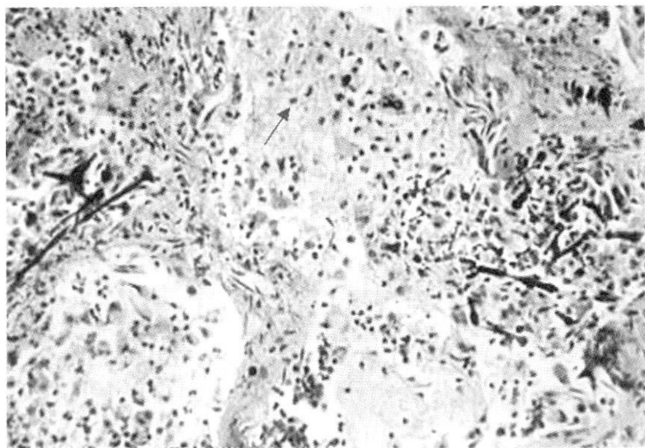

Fig. 1.5. Asbestos fiber surrounded by macrophages in the lung. (Figure taken from www.cdc.gov [5].)

duced into one's system [1, 5, 11, 12]. NIOSH concluded, on the basis of thousands of studies, that it was the shape of the asbestos fibers that caused its high toxicity, not its elemental composition. Indeed, asbestos is a general term used to describe a broad class of alumina silicate minerals (Fig. 1.5). For example, the long, thin dimensions of asbestos fibers enable them to reach the gas-exchanging part of lung when inhaled; however, they are not easily removed by macrophages in the lung, thus leading to inflammation and scarring [1]. Long-term exposure may even lead to cancer. Additional health effects of particulate materials include increased attacks of asthma in asthma patients, silicosis, asbestosis, and "black lung" [1, 8, 27].

1.5
Nanomaterials and Biotoxicity

1.5.1
Iron Oxide

Iron oxide nanoparticles have been used extensively for biological applications and as pigments [28, 29]. The common oxidation states of iron found in the environment are +2 (ferrous) and +3 (ferric). Nanoparticles with a wide degree of morphologies and crystal structures exist. According to Schwertmann, there are fifteen known polymorphs of ferric oxide [28, 29]. Ferric oxide nanoparticles are in fact one of the few classes of nanomaterials approved by the FDA for parenatal (IV) administration to humans [30, 31].

The magnetic properties of mixed valent Fe(II), Fe(III) oxides are finding increased applications for imaging, drug delivery, and separations [32, 33]. The toxicity of these mixed valent materials is far less clear. The ability of many microorgan-

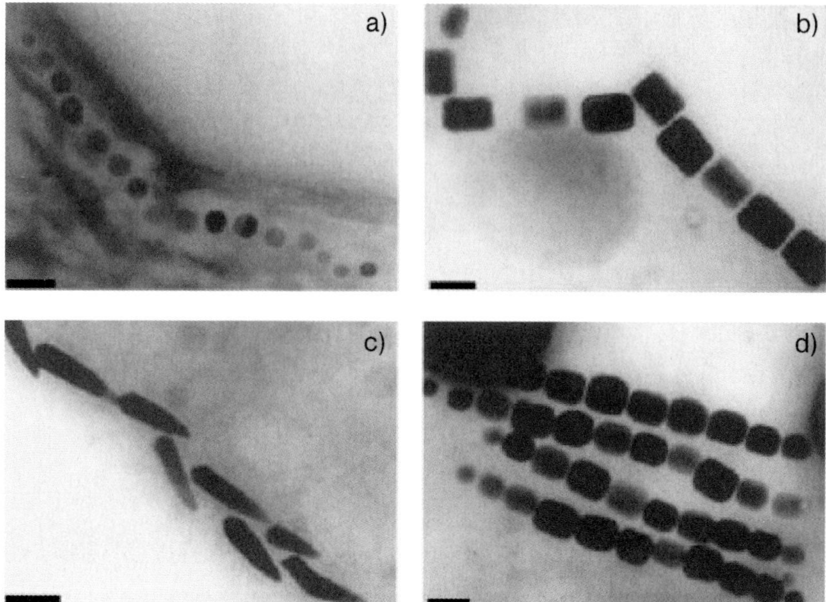

Fig. 1.6. Electron micrographs of magnetosomes found in magnetotatic bacteria. Scale bar: 100 nm. (Figure taken with permission from Safarik [34].)

isms (i.e., magnetosomes), fish and mammals to produce and/or utilize magnetite, Fe_3O_4, demonstrates that they are not toxic under all conditions. Figure 1.6 shows a transmission electron micrograph of Fe_3O_4 particles synthesized by a microorganism [34]. Magnetic nanoparticles are also thought to be exploited by more advanced organisms such as trout, migrating birds, and whales. Conversely, the well-known Fenton reaction of Fe(ii) yields hydroxyl radicals that damage DNA and can oxidize a wide variety of organic and biological reagents [35].

Below we review recent cellular studies of magnetic iron oxide nanoparticles. The vast majority of these studies are focused on superparamagnetic Fe_3O_4 particles that respond rapidly to magnetic fields but retain no residual magnetism when the field is removed. Such materials have long been commercially available as micron-sized magnetic beads, in which the superparamagnetic particles are encapsulated within an organic sphere [36]. The use of nanometer-sized materials presents new opportunities for separations and imaging technologies, where possible biotoxicity is a critical concern.

Goetze et al. prepared biocompatible superparamagnetic nanoparticles of 2–30 nm diameter. The particles were coated with citric acid or carboxymethyl dextran (CMD) [37]. Lacava et al. had previously studied the biological effects of ionic and citrate based magnetic fluids composed of $MnFe_2O_4$ on mice. The citrate and ionic nanoparticles caused diarrhea and ultimately animal death. While citrate alone did not cause toxicity, it was not clear whether the manganese or the iron were responsible for death [38].

Mikhaylova et al. have studied the effects of biocompatible coating layers on superparamagnetic iron nanoparticles. Bovine serum albumin (BSA), poly(L,L-lactic acid), or poly(ε-caprolactone) were coated on 8 nm particles. FTIR spectroscopy was used to characterize the nanoparticles and confirm the presence of the coatings. For uncoated or gold-coated nanoparticles, superparamagnetic behavior was observed. However, Mössbauer and magnetic susceptibility studies indicated significant cluster formation in the case of BSA modified particles, and chain-like structures for the lactic acid and caprolactone modified nanoparticles [39].

Gupta and Gupta reported a cytotoxicity decrease and internalization increase for pullulan-coated superparamagnetic nanoparticles with human fibroblasts. Uncoated, 20 nm iron oxide particles were toxic to human dermal fibroblasts. Internalization of these particles resulted in disruption of the cell cytoskeleton. Pullulan coated particles were non-toxic and had a different effect on the cytoskeleton. TEM data indicated that the internalization mechanisms were different for the two particles – behavior that was attributed to the hydrophilic nature of the pullulan coating [40].

Petri-Fink et al. studied the effects of surface-coated superparamagnetic iron oxide nanoparticles with human cancer cells. Nine-nm iron oxide nanoparticles were coated with poly(vinyl alcohol) (PVA) or PVA with carboxylate, amine or thiol functional groups. The PVA and the carboxyl and thiol functionalized PVA nanoparticles were non-toxic to the melanoma cells. Some cytotoxicity was observed for the amine functionalized PVA nanoparticles, particularly when the polymer concentrations were high. The amine groups increased cellular uptake of the nanoparticles [41].

Stroh et al. reported on studies of rat macrophages incubated with citrate coated iron oxide nanoparticles (9 nm). Atomic absorption and NMR studies showed a large uptake of the nanoparticles that could be easily visualized by confocal microscopy (Fig. 1.7). Rhodamine green-labeled iron oxide nanoparticles were incubated with the cells for 90 min at 37 °C. The cells were then centrifuged, washed with PBS buffer, resuspended in medium, and seeded in six-well plates. The next day the adherent cells were incubated with the lipophilic fluorescent dye ANEPPS, which is a common stain for outer and intracellular membrane structures including vesicles. After 45 min incubation, the cells were washed and studied by confocal microscopy. With 488 nm laser excitation, both the rhodamine green emission and the ANEPPS red emission were simultaneously monitored. Control experiments without the iron oxide nanoparticles are also shown [42].

The confocal results clearly indicate that the iron oxide nanoparticles were taken up by the cells. Even though some cell autofluorescence was seen in the control data, it was much weaker in intensity. The high fluorescence intensity from

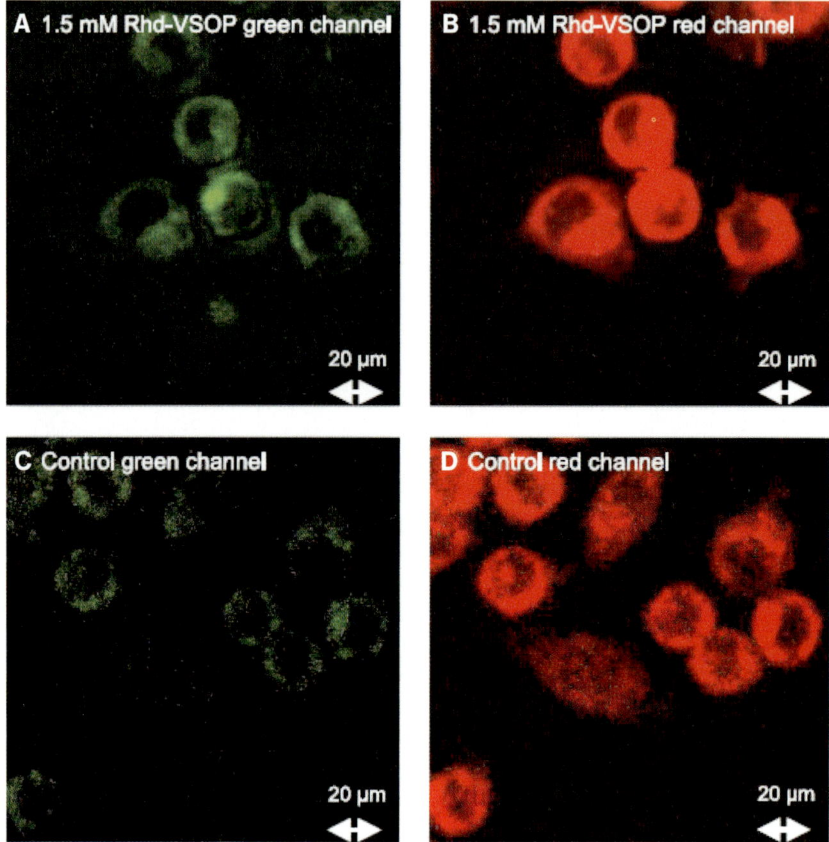

Fig. 1.7. Cell internalization quantified by confocal microscopy; see text for details. (Figure taken from Stroh [42] with permission from Elsevier.)

ANEPPS in the same region as the nanoparticles led the authors to suggest that the iron oxide nanoparticles form intracellular aggregates in membrane coated vesicles [42].

The cells were lysed at various times after nanoparticle exposure and the levels of malonydialdehyde (MDA) and protein carbonyls were measured. At short incubation times, a significant increase in protein oxidation and MDA was observed. Interestingly, the yields diminished with time and by 24 h there was no evidence for oxidative stress. Therefore, the oxidative stress was transient and the cells remained viable and useful for magnetic imaging applications. Iron chelators and spin traps caused a reduction in the concentrations of MDA and oxidized proteins, leading the authors to conclude that free iron present during the incubation procedure caused the transient oxidative stress [42].

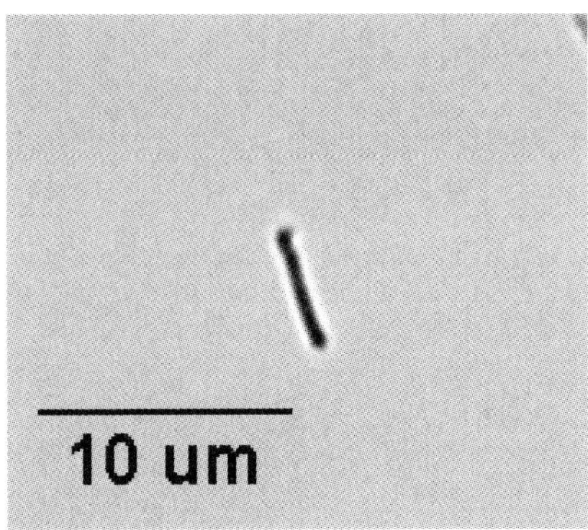

Fig. 1.8. Optical image of a high-aspect ratio TiO_2 nanoparticle or nanowire.

1.5.2
Titanium Dioxide

Titanium dioxide is commonly found in three crystalline forms: rutile, anatase and brookite [43]. In all three crystalline forms, Ti(IV) is in an octahedral coordination environment.

For bulk TiO_2, rutile is the thermodynamically stable form, while for nanoparticles (<14 nm) the anatase form is most stable [44–46]. Therefore, most synthetic routes for the preparation of TiO_2 nanoparticles yield anatase, and rutile is far less common [47–49]. Some preparations yield a mixture of the three phases. The commercially available DeGussa P25 consists of ~25 nm particles of about 80% anatase and 20% rutile. Because of the large effective mass of electrons in TiO_2, quantum size effects, which are well known for other semiconductors, are not observed until the particle size is less than 1 nm [50]. While spherical particles are by far the most common, it is possible to synthesize high-aspect ratio particles like those shown in Fig. 1.8.

Titanium dioxide is classified as a wide band-gap semiconductor, the anatase form having a band gap of 3.2 eV (Fig. 1.9) [43]. Much of the biotoxicity of TiO_2 is attributed to photoeffects wherein the material is illuminated with ultraviolet light. Band gap illumination produces an electron–hole pair excited state that is a much stronger oxidant and reductant than is the ground state. Under standard conditions the excited electron has a reduction potential of about 0.0 V vs. NHE while the hole has an oxidizing power of about +3.2 V [51].

Under many conditions, the initially formed electron–hole pairs trap at specific sites to yield radicals. The nature of these radicals has been the subject of many investigations, particularly because of their possible relevance in splitting water into hydrogen and dioxygen [52]. It is now widely accepted that the electrons trap at localized Ti(IV) sites. Titanium(III) is a reductant that reduces dioxygen to form superoxide ions [53]. Superoxide has long been thought to abstract hydrogen atoms from various biological substrates [53].

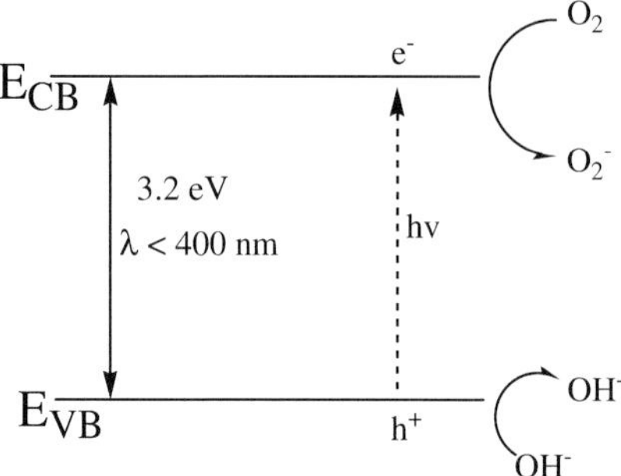

Fig. 1.9. Simplified band structure for anatase TiO$_2$. Band gap light excitation produces an electron in the conduction band and a hole in the valence band. The electron reduces dioxygen to superoxide, a reactive oxygen species that can abstract hydrogen atoms from organic biomolecules. The valence band hole can oxidize water to dioxygen, and can also produce hydroxyl radicals – potent reactive oxygen species that oxidize most organic compounds. Not shown is the trapping of the photogenerated carriers. See text for additional details.

The fate of the photogenerated hole in TiO$_2$ is less certain. Electron pair repulsion spectroscopic measurements indicate that the hole is initially trapped at an oxygen bridged between two Ti(IV) sites [54]. This "hole" is very reactive and is thought to ultimately yield a hydroxyl radical under ambient aqueous conditions. Much of the environmental photocatalysis of TiO$_2$ is best understood by invoking the presence of hydroxyl radicals. Hydroxyl radicals are highly reactive and generally react with the first substrate they encounter [53].

Titanium dioxide nanoparticles are of considerable industrial interest. The high refractive index (2.7) makes it an ideal material for light scattering and it has historically been used in paints, polymers, enamels, and coatings. It is also an ingredient in some suntan lotions and used as a colorant in foods. Growing applications in solar energy conversion and environmental remediation have been envisioned.

The body of literature for TiO$_2$ nanoparticle biotoxicity was by far the largest. We have organized the literature descriptions based on whether the nanoparticles were illuminated or kept in the dark.

1.5.2.1 **Dark Studies**

Donaldson et al. have examined the cytotoxic effects of TiO$_2$ by measuring the DNA strand breakage on a supercoiled DNA band caused by free radical activity. A DNA plasmid (290 ng φ X174 RF) was incubated with either TiO$_2$ (0.5 μm) or ultrafine TiO$_2$ (0.02 μm) particles. The plasmid was separated into the three possible forms, super-coiled, relaxed coil or linear by electrophoresis and quantified by scanning laser densitometry. Findings showed that TiO$_2$ particles had little effect on DNA strand breakage, whereas the ultrafine particles caused complete destruc-

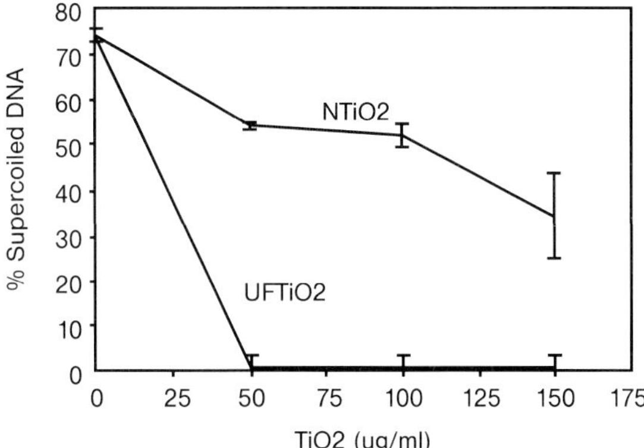

Fig. 1.10. Relationship between TiO_2 ($NTiO_2$) and ultrafine TiO_2 ($UFTiO_2$) particles and DNA strand breakage. (Figure taken from Donaldson [55] with permission from Elsevier.)

tion of the super-coiled DNA plasmid at concentrations greater than 50 µg mL^{-1} (Fig. 1.10). In addition, crocidolite and amosite asbestos caused supercoiled DNA depletion. At lower ultrafine TiO_2 concentrations (5 µg mL^{-1}) and in both asbestos samples the amount of DNA damage was improved by the addition of a radical scavenger, such as mannitol, indicating the role of free radical activity at the surface of the particles. The researchers also suggested that the surface of the particles can react with reductants generated by inflammatory cells and physiological chelators to generate more free radicals to assist in the destruction of DNA [55].

Tan et al. studied human subjects whom applied sunscreen to skin for 2–6 weeks until up to 2 days before excision of the skin lesion. Studies were performed by removing the stratum corneum by using cyanoacrylate ester and elastic plaster and a tissue sample was obtained. The samples were digested by microwave extraction and analyzed using mass spectrometry. The results showed that the levels of TiO_2 in the dermis were higher, yet not statistically different from, the control samples (post mortem cadavers). However, if the outlier in the control samples was excluded, the TiO_2 concentrations were significantly higher than the control. No correlation was found between the concentration of TiO_2 in the samples and the duration of application. A small test group, and concentrations of TiO_2 close to the detection limits, made it difficult to compare the concentrations of TiO_2 in the dermis samples with the control [21].

In a study by Hart et al., CHO-K1 cells were incubated for 2–5 days with dusts: chrysotile (1.4×0.1 µm), crocidolite (1.8×0.2 µm), and TiO_2 (0.6 µm). Viability was determined by an esterase activity viability assay, where cells are treated with 5(6)-carboxyfluorescein diacetate, which is a non-fluorescing ester conjugate. When internalized by cells, carboxyfluorescein is cleaved by cytoplasmic esterases and becomes polar and fluorescent. If the cell is not viable, the ester conjugate would not be retained by the cell. In all samples the loss of cell viability was not significant.

However, micronuclei and polynuclei tests with acridine orange staining expressed nuclear abnormalities. Chrysotile was the most cytotoxic sample, while TiO_2 was the least, indicating a particle size dependence on toxicity. Since viability remained high and only nuclear abnormalities resulted from particle interactions, the proposed mechanism of toxicity involved the interference of the internalized particles with mitosis, which causes a distortion in nuclear morphology and cytostasis, cessation of cell division [9].

In another study by Hart et al., Chinese hamster ovary (CHO) cells were exposed to particles for 3 days and then counted using a Coulter counter. For all fibers tested, ranging from glass, ceramics, and slag wool to asbestos, similar toxic effects were observed. Cultures showed little loss in viability (~90%); however, cell proliferation was almost completely inhibited and a concentration-dependent increase in morphological changes was observed. Unexposed samples retained a viability of 99% and showed a 20-fold increase in cell population [56].

Size comparisons indicated that thinner fibers were more toxic than thicker ones of similar lengths when concentration was a function of fiber mass per unit area. However, when concentration was expressed as a function of number of fibers per unit area, the difference in effects was non-existent [56].

The researchers concluded that cytotoxicity and genotoxicity correlate with fiber length and the mechanism of toxicity was by cytostasis, or the disruption of cell division. Possible explanations include longer fibers being more easily entangled in migrating chromosomes or spindle apparatus, which leads to the formation of micronuclei [56, 57]. In addition, long fibers are more biocompatible because they support cell growth *in vitro* by providing a substratum for attachment and proliferation of fibroblast cell lines [56, 58, 59].

Peters et al. have analyzed the effects of TiO_2 (14 nm) and SiO_2 (70 nm) particles on human dermal microvascular endothelial cells. Viability tests along with Ki67, a protein expressed in the nucleus of proliferating cells, and the cytokine interleukin-8 (IL-8) measurements were performed to determine the cytotoxic effects of the metal oxide particles on cells [60]. The CellTiter AQ$_{ueous}$ non-radioactive assay was used to determine cell viability by measuring the conversion of an enzymatic tetrazolium salt (MTS) via mitochondrial dehydrogenase [61]. Results showed no significant difference in cell viability; however, an increase in IL-8 production for both the SiO_2 and TiO_2 particle (50 µg mL^{-1}) treated samples was taken as evidence for pro-inflammatory effects (Fig. 1.11) [60]. The only sample that induced a decrease in Ki67 expression was SiO_2 (50 µg mL^{-1}), indicating a decrease in the number of cells participating in the active part of the cell cycle. Since biocompatible TiO_2 showed some inflammatory effects, the authors concluded that particles can possess different features when in the nano versus bulk scale [60].

Shanbhag et al. have studied the effects of TiO_2 particles on P388D$_1$ macrophages. Viability was measured as a function of ^{3}H-thymidine (^{3}H-TdR). Cells were incubated with particles for 8 h followed by the addition of ^{3}H-TdR for 16 h. ^{3}H-TdR suppresses DNA synthesis and is used as a way to measure DNA fragmentation. The results showed that TiO_2 decreased ^{3}H-TdR levels in macrophages in a size- and concentration-dependent manner [62].

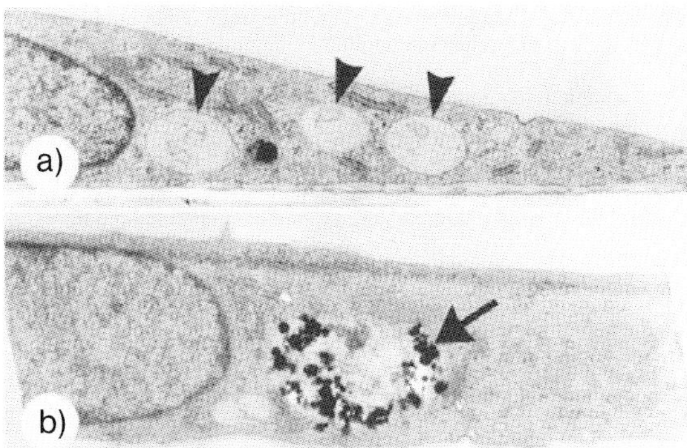

Fig. 1.11. A human dermal microvascular endothelial cell exposed to (a) no particles and (b) TiO_2 particles. (Figure adapted from Peters [60] with kind permission of Springer Science and Business Media.)

Stearns et al. have investigated the effects of TiO_2 particles (50 nm) on a human epithelium cell line (A549). In as little as 1–2 h, minimal internalization of the particles was observed by energy-filter TEM. Short exposure times to TiO_2 yielded particles found in the vacuoles; however, longer incubation times showed TiO_2 in the lamellar bodies. Addition of the inhibitor cytochalasin D (cyto D), which is known to affect actin polymerization and particle internalization, to cells before they were introduced to TiO_2 neither inhibited phagocytosis of the nanoparticles nor decreased cell viability. However, a change in cell morphology was observed in the presence of cyto D and more particles were internalized in membrane-bound vacuoles rather than the lamellar bodies [63].

Westmoreland et al. have used TiO_2 as a reference compound because it is known as a "nuisance dust." In the study, TiO_2 particles were introduced into an epithelial cell line (16HBE14o-) for 24 h [64]. However, after 24 h, there was no toxicity induced as measured by the MTT reduction assay [65], neutral red uptake assay or transepithelial resistance.

Kilgour et al. have modeled inhalation exposure *in vitro* by exposing olfactory and respiratory turbinates of rats to TiO_2 (<1 μm) for short and long durations. They found that TiO_2 exposure had no effect on adenosine triphosphate (ATP) or potassium concentrations in olfactory epithelium for any of the times studied. Conversely, when exposed to respiratory epithelium, a minimal decrease in ATP and potassium concentration was observed for the 4 h particle incubation and 20 h fresh media incubation. An observed decrease in potassium concentration at the 24 h exposure time was attributed to mechanical damage arising from TiO_2 particle precipitation. These studies concluded that TiO_2 is not acutely toxic to the nasal cavity [66].

Miller et al. found TiO_2 to have no effect on CHO cells with or without metabolic activation, which is potentially a result of TiO_2 nanoparticle insolubility. The effect

viability. The decrease in cell viability, and eventually cell death, occurred as a result of intracellular components leaking out of the permeable membrane and additional TiO_2 nanoparticles trafficking into the damaged cells and directly attacking the nucleus and other intracellular components [70].

Cai et al. observed that HeLa cells incubated with TiO_2 particles without UV irradiation had 90% survival, as did cells irradiated with UV light for less than 19 min. However, cultures exposed to both TiO_2 particles and UV light showed a dramatic fall in survival rate that decreased with increasing TiO_2 concentration. Irradiation of HeLa–TiO_2 adducts with visible light did not cause any photodynamic effects, and simply raised the temperature of the culture to around 36 °C. Since this is less than the culture temperature of the cells, thermal death is not very likely and the observed behavior with UV light was attributed to photochemistry, not thermal effects. Irradiation with wavelengths greater than 440 nm produced cell survival of 90%, but when the sample was irradiated with a slightly lower wavelength (300–400 nm), for an equal amount of time, all cells in the culture were killed. This same irradiation wavelength range on cells without TiO_2 yielded an 85% survival rate [71].

Cell death was proposed to take place by two possible mechanisms. In the first, cells were oxidized by photogenerated holes in the valence band. In the second, the holes reacted with water to produce OH• radicals that can attack the cell membrane and intracellular components [71]. Cai et al. and Sakai et al. deduced that the viability of T-24 cells decreased with both variations in UV irradiation intensity and TiO_2 particle concentration [72, 73].

Saito et al. have investigated the polycatalytic bactericidal effects of TiO_2 particles (21 nm) on three strains of bacteria (*Streptococcus mutans, S. rattus,* and *S. cricetus*). TiO_2 particles were introduced to the specimen, irradiated with UV light and incubated for two days. The bactericidal action of TiO_2 increased with TiO_2 particle concentration. In addition, potassium leakage, measured by flame photometry, paralleled the loss of cell viability. In bacteria, a cell wall of peptidoglycan is formed around the cell membrane. When observed by TEM, the TiO_2 particles took over 30 min to reach the cell membrane of the bacteria, yet leakage of intracellular protein and cell death occurred in less than 1 min. Cell death was rapid and the cell wall was not destroyed until after 60–120 min; therefore, the TiO_2 particles could not have been able to attach to the cell membrane directly to cause any damage. Rather, cell death was assumed to result from membrane damage caused by the superoxide and perhydroxy radicals produced from TiO_2 photocatalysis [26].

Nakagawa et al. studied the effects of four sizes of UV irradiated TiO_2 particles on a mouse lymphoma cell line. DNA tail length was measured by means of a SCG assay. The results showed that UV-irradiated Degussa P-25 (anatase, 21 nm), TP-3 (rutile, 420 nm) and WA (anatase, 255 nm) samples all elicited increased DNA damage and a decrease in cell survival (Fig. 1.13) [74, 75, 76].

Warmer et al. determined whether nucleic acids were targets for photoxidative damage caused by UV-irradiated TiO_2 by investigating the effects of UV-irradiated TiO_2 on calf thymus DNA and human skin fibroblasts. A suspension of TiO_2 particles was added to samples of calf thymus DNA and fibroblasts and exposed to

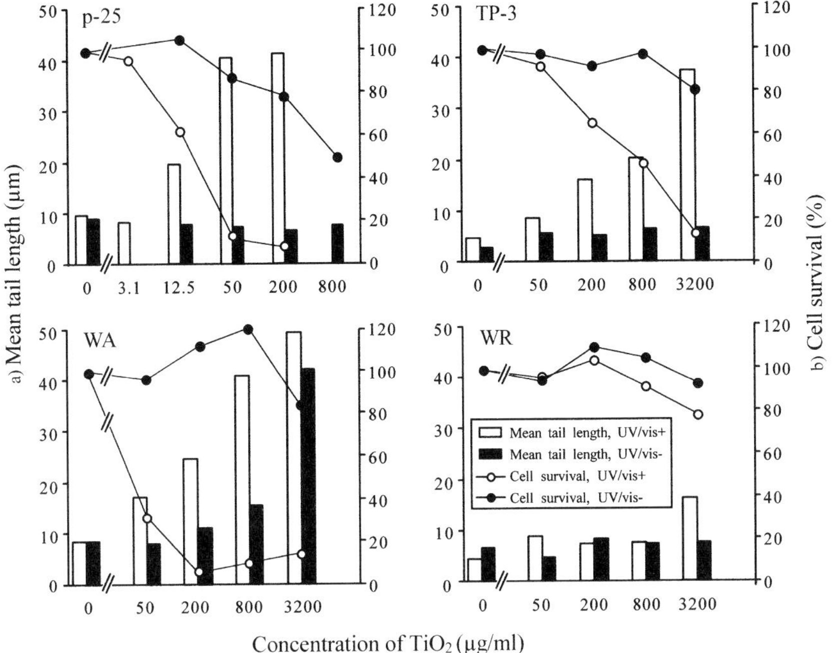

Fig. 1.13. Effects of TiO$_2$ on DNA tail length and cell survival. Abbreviations: p-25, anatase 21 nm; WA, anatase 255 nm; WR, rutile 255 nm; TP-3, rutile 420 nm. (Figure taken from Nakagawa [74] with permission from Elsevier.)

UVA light for 1 h. The calf thymus DNA and cellular DNA samples were then enzymatically hydrolyzed and analyzed using reversed-phase high-performance liquid chromatography (HPLC). HPLC was used to measure the hydroxylation of guanine bases (forms 8-oxodG) and indicated whether any nucleic acids were oxidatively damaged [75].

Results with calf thymus DNA showed that levels of 8-oxodG increased directly with the addition of TiO$_2$ and fluence of UV irradiation, when they were both present in calf thymus DNA experiment. Furthermore, the fibroblast samples showed that both TiO$_2$ particulates and UV light must be present to induce oxidative damage, in which case cytotoxicity was 85%. The phototoxicity was UV fluence dependent. Although oxidative damage did not occur in cellular DNA, there was a 3-fold increase in 8-oxoG in the presence of TiO$_2$ and UV irradiation, indicating the oxidative damage of the guanine bases in cellular RNA. This gave evidence that oxidative stress caused by irradiated TiO$_2$ particles was occurring in the cytoplasmic and nuclear compartments of the cell. The photocytotoxicity of TiO$_2$ is a result of intracellular damage induced by reactive oxygen species. However, the relative importance of the reactive oxygen species formed by photoexcited TiO$_2$ is still unknown.

ESR studies to measure the formation of a DMPO-OH radical adduct were also performed. The formation of adduct increased with time and then leveled off [77]. Since there is evidence of reactive oxygen species, the hydrolysis of guanine in calf thymus DNA can be due to any of the following: hydroxyl radical formation, decomposition of hydrogen peroxide from UV irradiation, or the Fenton reaction [78–80].

Kubota et al. incubated T-24 human bladder cancer cells with TiO_2 particles, cultured them for 24 h and observed the effects. The TiO_2 particles were contained mainly in the cytoplasm and cell membrane of the T-24 cells, as found by transmission electron microscopy (TEM). There was a 90% survival rate amongst cells introduced with up to 300 μg mL^{-1} of TiO_2. However, when exposed to UV light for as little as 5 min there was as much as a 20% decrease in the survival rate even with TiO_2 concentrations as low as 10 μg mL^{-1}. They also found that cell killing was more effective in phosphate buffered saline (PBS) than in F-12 media. This is possibly due to either the absorptive abilities of F-12 or the fact that it contains radical scavengers such as mannitol and tryptophan. These researchers also investigated the role that radical scavengers play in irradiated TiO_2 cytotoxicity. Molecular scavengers of both hydrogen peroxide and hydroxyl radicals, catalase and L-cysteine, respectively, effectively diminished cell death when added to the cell samples. This provided evidence of the role of hydroxyl radicals and hydrogen peroxide in cell death [81].

In addition, they investigated the mechanism of photoexcited TiO_2 biotoxicity by depositing TiO_2 nanoparticles onto conductive tin oxide glass. The cells were subsequently cultured on the conductive glass. In the dark, when a potential was applied to the TiO_2 electrode, the cells remained viable. However, in the presence of UV light, cells were killed when the electrode potential was more positive than -0.5 V. The percent of cells surviving was proportional to the photoinduced current. Due to the strong photocurrent, at anodic potentials, the researchers concluded that photogenerated TiO_2 holes were responsible for cell death [81].

An *in vitro* experiment using T-24 cells was also carried out by the same group. TiO_2 particles (0.03–10 nm) were added to cell cultures and irradiated with a 500-W Hg lamp. The cells were cultured for another 10 days before fixing and staining with Giemsa in order to count. With either TiO_2 or UV light alone, survival was >90%, indicating no cytotoxic effects. However, in the presence of both TiO_2 and UV light a cytotoxic effect was observed that increased with increasing TiO_2 concentrations. Scavenger experiments were conducted to determine the mechanism of cell death. In the presence of the hydrogen peroxide scavenger, catalase, and hydroxyl radical quencher, L-cysteine, cell death of the photoexcited TiO_2 samples decreased, pointing to the participation of H_2O_2 and OH$^{\cdot}$ radical in cell death [81].

Cai et al. set out to determine the mechanism behind the photodamaging of cells with TiO_2. To investigate the mechanism by which TiO_2 can photokill HeLa cells, superoxide dismutase (SOD), which converts the superoxide anion (O_2^{-}) into H_2O_2, was added to a TiO_2 infused cell culture. The TiO_2 nanoparticles were incubated in cultured HeLa cells for 24 h, while SOD was added to the cell culture for the final hours of incubation. After a short irradiation with UV light, the cells were counted to determine viability. Cell survival in the absence of SOD was a meager 55%, which indicated that cell death may be due to H_2O_2 and $OH^\bullet$ radical formed by the irradiation of TiO_2. The addition of SOD caused a decrease in survival rate, which increased with SOD concentration when irradiated for the same amount of time. Controls showed that samples containing SOD in the absence of TiO_2 had no effect on survival rate, therefore indicating that the decrease in cell survival was due to the conversion of O_2^- into H_2O_2 by SOD [82]. Further evidence of the production of H_2O_2 was given by the addition of the fluorophore scopoletin (6-methoxy-7-hydroxy-1,2-benzopyrone), whose fluorescence is quenched by H_2O_2 [83]. A much higher concentration of H_2O_2 was produced in irradiated HeLa–TiO_2 samples that contained SOD than in samples without SOD. With the addition of catalase (EC 1.11.1.6), which converts hydrogen peroxide into water and molecular oxygen, to the TiO_2–SOD sample, the surviving fraction of cells increased, further confirming the production of H_2O_2. In addition, the presence of catalase increased survival even in the absence of SOD, indicating that H_2O_2 can be produced by another method:

$$O_2^- + H^+ \rightarrow HO_2$$

$$HO_2 + e^- \rightarrow HO_2^-$$

$$HO_2^- + H^+ \rightarrow H_2O_2$$

However, O_2^- must also be converted into some other reactive oxygen species because cell death still occurred in the presence of catalase and the absence of SOD [82].

Jang et al. have studied bacterial death by photocatalyzed TiO_2. The TiO_2 was added to cultures of either *Escherichia coli* (*E. coli*) or *Pseudomonas areruginosa* and were then irradiated with UV light. Smaller particles of TiO_2 had a larger effect on the degree of decomposition of the bacteria, while increased anatase mass fraction caused an increase in decomposition (Fig. 1.14) [84].

Sakai et al. have investigated the cytotoxicity of TiO_2 particles (30 nm) on a T-24 human malignant cell line. Cells were incubated with TiO_2 particles for 24 h and irradiated with UV light. Cell viability was determined by a colony forming assay, and the change in Ca^{2+} concentration was monitored by ethidium bromide staining [73]. The Ca^{2+} ions play a role in differentiation, intracellular transport, secretion and metabolism [73, 85]. A change in Ca^{2+} concentration is linked to cytotoxicity [73, 86–91]. The addition of TiO_2 particles to T24 cells yielded a 90% survival rate of the cells. Conversely, when the samples were irradiated with UV light, the Ca^{2+} concentration increased as determined by ratiometric imaging. With increased TiO_2 concentration, less irradiation is required to cause an increase in Ca^{2+} concentration. However, since the stepwise increase in Ca^{2+} concentration remained constant it was concluded that there is a minimum amount of reactive oxygen species needed to trigger their uptake.

The processes for Ca^{2+} mobilization include influx through the plasma membrane and release from Ca^{2+} storage in the endoplasmic reticulum. The change

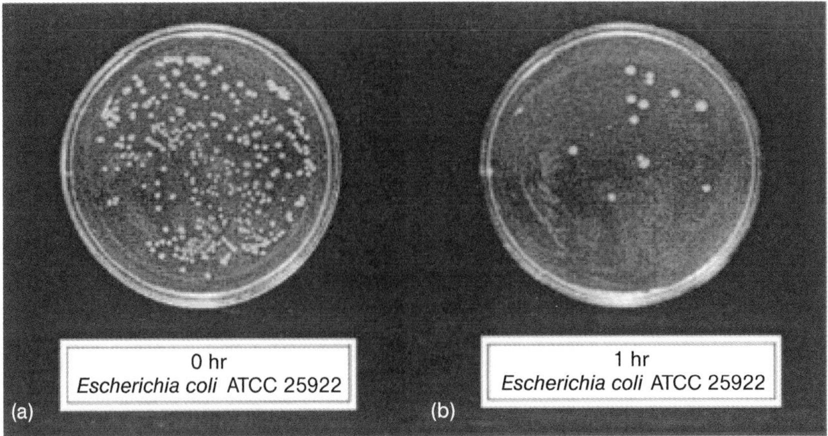

Fig. 1.14. *Escherichia coli* in (a) absence and (b) presence of TiO$_2$ nanoparticles. (Figure taken with permission from Jang [82].)

in Ca^{2+} distribution was monitored to determine the mechanism taking place. When the concentration of Ca^{2+} was monitored with UV and TiO$_2$ present in cell culture, the concentration was greatest near the cell membrane. However, without the addition of TiO$_2$ particles, the concentration was low and evenly distributed. The increase was attributed to an increase in cell membrane permeability, but no loss in cell viability, and the Ca^{2+} was from the buffer. However, in later stages the cell dies. Ca^{2+} cannot directly be responsible for cell death because in buffer without Ca^{2+} there is still loss in cell viability. There is a simply an increase in Ca^{2+} caused by a change in permeability associated with cell death. Instead, the hydroxyl radical and hydrogen peroxide promote cell death [73].

Cai et al. showed that HeLa cells were effectively killed in the presence of photoexcited TiO$_2$ particles. The cytotoxic effects were suppressed in the presence of L-tryptophan and catalase, which quench hydroxyl radicals and scavenge hydrogen peroxide, respectively. This suggests that cell death resulted from the production of reactive oxygen species on the particle surface [72].

Linnainmaa et al. found no effect upon the addition to rat liver epithelial cells of three particle types, uncoated anatase, rutile coated with aluminum hydroxide and stearic acid, and P25 Degussa TiO$_2$. The inhibition of cell growth was measured by the multinuclei assay, in which the addition of cytochalasin B prevents cell division but not division of the nucleus. Cytotoxicity was measured by the increase in the amount of cells that contained only one nucleus after treatment. The micronucleus test assessed the chromosomal damage of cells caused by the addition of TiO$_2$. After 1 h incubation with TiO$_2$ particles, the cells were irradiated with UV light for 5 min. The results indicated no inhibition of cell growth or cytotoxic effects with any of the TiO$_2$ samples. Small cytotoxic effects were seen in the irradiated samples, but they were not of statistical significance [92].

Donaldson et al. have stated that TiO_2 particles are cleared from the lung by phagocytosis of alveolar macrophages. They studied the impact of ultrafine (20 nm) versus fine (200 nm) particles of TiO_2 and carbon black on a macrophage cell line. Fluorescent latex beads were then added to the cell line to indicate phagocytic uptake. The results indicated that the ultrafine particles inhibited phagocytosis by the macrophages when compared to the fine particles, which may lend one possible explanation for their adverse effects. The mechanism is a result of the increased surface area and increased number of ultrafine particles present, due to inhibited phagocytosis, that interact with the epithelium and lead to oxidative stress and thus inflammation [8].

Wilson et al. have stated that the surface of particles may be a source of reactive oxygen species. Ultrafine and fine particles were incubated with a compound that undergoes activation to a fluorescent state when oxidized. Fluorescence intensity measurements revealed that ultrafine particles yielded a dose-dependent increase in fluorescence, whereas fine particles did not. This result is likely a consequence of the high surface area of the particles yielding more reactive oxygen species [93].

Maness et al. set out to determine the mechanism by which photocatalytic TiO_2 kills bacteria. They studied the effects of Degussa P25 TiO_2 particles (surface area 50 m^2 g^{-1}) on a strain of *E. coli* in the presence of UV light. Membrane damage was determined by measuring the production of malondialdehyde (MDA), a product of lipid peroxidation, by the colored adduct it forms with thiobarbituric acid. More MDA is produced when TiO_2 was present than without. Reactive oxygen species are proposed to play a role because they attack the polyunsaturated phospholipids in *E. coli*, causing deterioration of the cell membrane and loss of functions within the cell [94].

1.5.3
Other Metal Oxides

Our literature searches revealed a handful of biotoxicity studies with other metal oxide nanomaterials. These studies are described below.

Yamamoto et al. have looked at the cytotoxic effects of metal oxide particles on murine fibroblasts and murine monocyte macrophages. The particles were added to cells and the relative plating efficiency was obtained after 6–8 days, depending on the cell type. The results showed that the cytotoxicity of Al_2O_3 and ZrO_2 particles ($d = 500$–700 nm) were enhanced relative to TiO_2 particles ($d = 130$–180 nm). TiO_2 particles and Al_2O_3-coated TiO_2 particles both demonstrated similar cytotoxic effects, and showed larger particles to be more toxic than the smaller ones. Shape-dependent cytotoxicity was also determined, and dendritic TiO_2 proved to have the highest cytotoxicity when calculated as a function of number, volume and surface area. When cytotoxicity was determined as a function of volume, the particles ranked as dendritic > spindle > spheric. But when particle toxicity was ranked as a function of surface area the cytotoxic effects changed: dendritic > spheric > spindle. The overall conclusion on cytotoxicity ranked the dendritic particles as the most cytotoxic, followed by spindle and spheric particles. The number of particle edges is important when determining cytotoxicity – the more edges the more of a cytotoxic effect [95]. Cytotoxicity was then compared to that of the parent metal ions. It was concluded that toxicity results because of two processes: chemical toxicity of released metal ions or other soluble components [95–97] or mechanical stimulation caused by sizes and shapes [95].

Cytotoxicity of metal ions and other chemicals differs among cell lines. Larger particles (only if phagocytosed) tended to have higher cytotoxicity than smaller particles. For example, larger TiO_2 particles caused a higher inhibition of ^{3}H-thymidine incorporation of human monocyte macrophages. However, if the particle is too large to be phagocytosed by the cells, then there is no cytotoxic effect. The authors concluded that the cytotoxicity of insoluble particles does not depend on chemical composition. In addition, cytotoxicity was not dependent on chemical species but on particle size and phagocytic properties [95].

Hanawa et al. have studied the toxicity of metal oxide nanoparticles ranging from 500 to 3000 nm in diameter. The particles were incubated in human fibroblasts for 24 h and stained with haematotoxylin and eosin to determine the magnitude of toxicity. With this assay, cells that adhered to the coverglass would stain, while dead cells would detach from the glass during staining. A digitizer was used to assess the area that was stained. The area stained was considered to be proportional to the magnitude of cytotoxicity of the metal oxide particles. Cells incubated with Al_2O_3, TiO_2, Fe_2O_3, Fe_3O_4, Co_2O_3, NiO, Ga_2O_3, SnO, SnO_2, HgO showed no cytotoxic effects. A difference in formal oxidation state of some of these metals yielded different effects, e.g., CoO, Co_3O_4, and Ni_2O_3 appeared to be toxic. In addition, Cr_2O_3, Cu_2O, CuO, ZnO, and Ag_2O proved to be cytotoxic. A potential problem would be that the study was based on particles dissolving into elements/ions, which have a cytotoxic response. Therefore, larger areas of affected cells indicated a more cytotoxic effect, which does not necessarily correlate with the components of the particles being more effective at killing cells. Larger cytotoxic effects could have been a result of a higher particle concentration in the medium [98].

Lison et al. have studied the surface area effects of MnO_2 particles (Fig. 1.15). Mouse peritoneal macrophages were incubated with MnO_2 particles of varying sur-

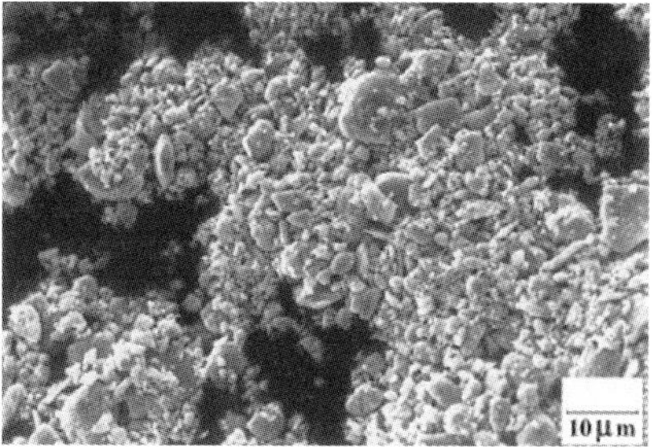

Fig. 1.15. Scanning electron micrograph of MnO_2 particles. (Figure taken from Lison [99] with kind permission of Springer Science and Business Media.)

face areas (0.5, 17, and 62 $m^2\ g^{-1}$) for 6 h. Lactate dehydrogenase (LDH) release was measured to indicate the degree of membrane damage. At the same particle concentrations, the particle with the highest surface induced the greatest amount of cytotoxic activity. However, freshly ground particles (5 $m^2\ g^{-1}$) from the 0.5 $m^2\ g^{-1}$ sample showed the highest toxicity of all. The researchers concluded that the toxicity of MnO_2 particles is surface dependent, indicating the possible effect of surface chemistry on cytotoxicity [99]. Lehnert et al. have claimed that cytotoxic effects are a result of intracellular dissolution of the nanoparticles in the phagolysosomes [100]. Therefore, since dissolution is a function of specific surface area, toxicity should increase with increasing surface area, which is consistent with the results of Lison. The increased toxicity of the freshly ground particle is attributed to additional reactive sites present on the surface [99].

Keceli et al. have studied the cytotoxicity of various metal oxides. Titanium, tantalum, and niobium are biocompatible due to the native oxide layer formed on the surface. Toxicity studies with these particles were preformed with African green monkey kidney cells (Vero fibroblasts). Glass plates were spin-coated with a metal oxide layer (>100 nm) from a sol–gel solution. Cells were then cultured on the metal oxide for 7 days before they were counted. At the end of 7 days, no visible morphology change or apoptosis was observed, indicating that the samples were not cytotoxic. However, there was an effect on cell proliferation, which indicated an effect on biocompatibility. Al_2O_3 and Nb_2O_5 showed a 30% decrease in cell proliferation, while Ta_2O_5 and ZrO_2 showed a 45% and 58% decrease, respectively. The TiO_2 sample did not differ from the control. The observed decrease in proliferation was consistent with the dielectric constants of the metal oxides, showing that metal oxides with a higher dielectric constant have more isolating effects and, as a result, are more biocompatible [101].

Chiu et al. have treated CHO cells with GeO_2 particles of varying concentrations for 12 h to determine their impact on cytotoxicity. Cytochalasin B, a proliferation inhibitor, was then added and an additional 24 h incubation applied. The cells were stained with Giemsa solution and counted to determine the number of binuclear cells. The number of binuclear cells decreased with the addition of GeO_2 particles, indicating that the particles induce G2/M block [102]. G2/M block is indicative of the cells not entering the mitosis stage of the cell cycle (Fig. 1.16). The G2, or Gap 2, phase is an intermittent stage that occurs after the synthesis of DNA, but before nuclear and cytoplasmic division of the cell. Prolonged periods in the G2 phase inhibit cell proliferation (M phase) without inhibiting the formation of daughter cells (S phase), thus yielding binuclear cells [103].

Viability was determined by the sulforhodamine B viability assay [31] in which cells are fixed and stained with sulforhodamine B followed by dye extraction and analysis with an ELISA plate reader at 540 nm. Cell survival decreased with an increase in GeO_2 particle concentration. However, with the clonogenic survival assay, in which cells were cultured for an additional 7 days after treatment with the GeO_2 particles before they were assayed, survival rates were above 80%. This discrepancy is possibly due to delayed cell growth after treatment. Studies were also conducted to determine the phase of the cell cycle that the cells were in after treatment with GeO_2 particles. Treated samples showed a dose-dependent increase in the number

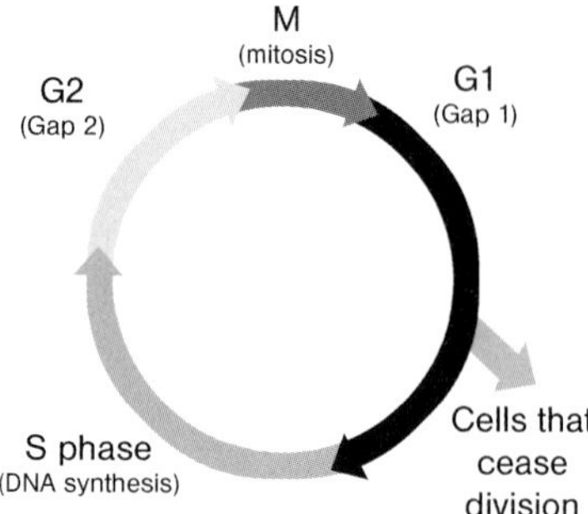

Fig. 1.16. Cell cycle; see text for details. (Figure taken from Ref. [101].)

of cells in the G2/M phase of the cell cycle. Therefore, GeO_2 particles slow cell proliferation, but do not play a major role in cytotoxicity [102].

Gaudenzi et al. have demonstrated the effects of CdO on a lymphocyte cell line (Jurkat cells). Cells were treated with CdO particles and a survival index, the ratio of the average number of viable cells in the treated and control samples, was calculated. The CdO particles induced a decrease in the survival index of Jurkat cells over time until 24 h when a zero factor was obtained. The mechanism of cell death was probed by FT-IR studies, which showed a decrease in intensity of the amide bands, suggesting a decrease in H-bonding energy and protein degradation. In addition, evidence of additional carbonyl groups was present, which is indicative of protein oxidation [104].

Pigott et al. have found that amorphous SiO_2 (100 nm) has a biphasic response when exposed to Chinese hamster lung cells. At a dosage of less than 30 μg mL^{-1}, there were little or no cytotoxic effects, based on cloning efficiency; however, there was a progressive increase in cytotoxicity at higher concentrations [105]. Amorphous SiO_2 was also found to be cytotoxic in other studies, and disrupts cell membrane functions when studied in cell culture [106–108]. However, this should not be of great concern for exposure of airborne amounts of SiO_2 by inhalation.

1.6
Conclusions

Currently, the National Institute of Occupational Safety and Health (NIOSH) is trying to answer the questions: In what ways might employees be exposed to nanomaterials in manufacture and use? How do nanoparticles enter the body? Once in the body, where would they travel? How would they interact physiologically and chemically with the body's systems [5]? The cellular and subcellular studies described herein provide some clues that address these questions.

In general, the reviewed studies showed that ferric oxide and titanium dioxide nanomaterials are not biotoxic in the dark and that TiO_2 illuminated with ultraviolet light has a high cytotoxicity to both bacteria and mammalian cells. The toxicity of ferric oxides under illumination has not been previously studied to our knowledge. There is an environmental photochemistry of ferric oxides with environmental pollutants [28], the products of which often include ferrous ions that could undergo Fenton-like chemistry and produce reactive hydroxyl radicals. Additional studies are warranted in this area.

However, there exist some important exceptions to the generalized statement above. For example, Linnainmaa et al. found no cytotoxicity with illuminated TiO_2 toward rat liver epithelial cells. Since the experimental conditions of light source, TiO_2 materials, and cell lines were different, it is difficult to pinpoint why this study differs. Likewise, Donaldson reported conditions where 20 nm TiO_2 particles linearized plasmid DNA, while 50 nm particles did not. These exceptions underscore some of the difficulties associated with answering NIOSH's questions. When "the body" is replaced by "a single mammalian cell," the answers remain unknown, even for a well-studied nanomaterial like TiO_2. Additional studies are required before biotoxicity relationships can be understood in more complex human systems at the molecular level.

Acknowledgment

The authors acknowledge equipment support from the NSF MRSEC Grant number DMR00-80031. The authors also acknowledge support from DARPA/AFOSR Grant F49620-02-1-0307 and from the David and Lucille Packard Foundation Grant #2001-17715. We thank the National Science Foundation (CRAEMS) for support of the environmental chemistry aspects of this work.

References

1 DOWLING, A., Development of nanotechnologies. *Nanotoday.* **2004**, 30–35.

2 JOY, B., Why the future doesn't need us. *Wired.* **2000**, 8.04, 37.

3 ERHARDT, D., Materials conservation: Not-so-new technology. *Nat. Mater.* **2003**, 2, 509–510.

4 www.visionminerals.com.

5 www.cdc.gov.

6 www.fda.gov.

7 DONALDSON, K., STONE, V., Current hypothesis on the mechanisms of toxicity of ultrafine particles. *Ann. 1st Super Sanita.* **2003**, 39, 405–410.

8 DONALDSON, K., STONE, V., CLOUTER, A., RENWICK, L., MACNEE, W., Ultrafine particles. *Occup. Environ. Med.* **2001**, 58, 211–218.

9 HART, G., HESTERBERG, T., In vitro toxicity of respirable-size particles of diatomaceous earth and crystalline silica compared with asbestos and titanium dioxide. *J. Occup. Environ. Med.* **1998**, 40, 29–42.

10 U.S. Department of Labor, Bureau of Labor Statistics www.bls.gov.

11 ROCO, M., Broader societal issues of nanotechnology. *J. Nanopart. Res.* **2003**, 5, 181–189.

12 www.nano.gov.

13 *Federal Resister,* **2003**, 68, 42 068– 42 071.

14 SLIVKA, S., LANDEEN, L., ZEIGLER, F., ZIMBER, M., BARTEL, R., Characterization, barrier function, and drug metabolism of an in vitro skin model. *J. Invest. Dermatol.* **1993**, 100, 40–46.

15 BENNAT, C., MÜLLER-GOYMANN, C., Skin penetration and stabilization of formulations containing microfine

titanium dioxide as physical UV filter. *Inter. J. Cos. Sci.* **2000**, 22, 271–283.

16 PFLÜCKER, F., HOHENBERG, H., HÖLZLE, E., WILL, T., PFEIFFER, S., WEPF, R., DIEMBECK, W., WENCK, H., GERS-BARLAG, H., The outermost stratum corneum layer is an effective barrier against dermal uptake of topically applied micronized titanium dioxide. *Inter. J. Cos. Sci.* **1999**, 21, 399–411.

17 HALLMANS, G., LIDEN, S., Penetration of 65Zn through the skin of rats. *Acta Dermatol. Venereol.* **1979**, 59, 105–112.

18 KAPUR, S., BHUSSRY, B., RAO, S., HORMOUTH-HOENE, E. Percutaneous uptake on zinc in rabbit skin. *Proc. Soc. Exp. Biol. Med.* **1974**, 145, 932–937.

19 DUNFORD, R., SALINARO, A., CAI, L., SERPONE, N., HORIKOSHI, S., HIDAKA, H., KNOWLAND, J., Chemical oxidation and DNA damage catalysed by inorganic sunscreen ingredients. *FEBS Lett.* **1997**, 418, 87–90.

20 JACKSON, S., ELIAS, P., Skin as an organ of protection. In FITZPATRICK, T., EISEN, A., WOLFF, K., FREEDBERG, I., AUSTEN, K., eds. *Dermatol. Gen. Med.* **1993**, 1, 241–243.

21 TAN, M., COMMENS, C., BURNETT, L., SNITCH, P., A pilot study on the percutaneous absorption of microfine titanium dioxide from sunscreens. *Austral. J. Dermatol.* **1996**, 37, 185–187.

22 GHADIALLY, R., HALKIER-SORENSEN, L., ELIASS, P., Effects of petrolatum on stratum corneum structure and function. *J. Am. Acad. Dermatol.* **1992**, 26, 387–396.

23 NEMANIC, M., ELIAS, P., In situ precipitation: A novel cytochemical technique for visualization of permeability pathways in mammalian stratum corneum. *J. Histochem. Cytochem.* **1980**, 28, 573–578.

24 EDWARDS, S., DONNELLY, T., SAYRE, R., RHEINS, L. Quantitative in vitro assessment of phototoxicity using a human skin model, Skin2^{TM}. *Photodermatol. Photoimmunol. Photomed.* **1994**, 10, 111–117.

25 DONALDSON, K., LI, X., MACNEE, W., Ultrafine (nanometer) particle mediated lung injury. *J. Aersol. Sci.* **1998**, 29, 553–560.

26 SAITO, T., IWASE, T., HORIE, J., MORIOKA, T., Mode of photocatalytic bactericidal action of powdered semiconductor TiO_2 on mutans streptococci. *J. Photochem. Photobiol. B: Biol.* **1992**, 14, 369–379.

27 WHITESIDES, G., Nanoscience, nanotechnology, and chemistry. *Small.* **2005**, 1, 172–179.

28 SCHWERTMANN, U., CORNELL, R.M. *The Iron Oxides*, VCH Publishers, Weinheim, **1996**.

29 SCHWERTMANN, U., CORNELL, R.M. *Iron Oxides in the Laboratory: Preparation and Characterization*, VCH Publishers, Weinheim, **1991**.

30 LAWRENCE, R. Development and comparison of iron dextran products. *PDA J. Pharm. Sci. Technol.* **1998**, 52, 190–198.

31 BAILIE, G.R., JOHNSON, C.A., MASON, N.A. Parenteral iron use in the management of anemia in end-stage renal disease patients. *Am. J. Kidney Dis.* **2000**, 35, 1–12.

32 ANDRÄ, W., NOWAK, H. *Magnetism in Medicine: A Handbook*, Wiley-VCH, Berlin, **1998**.

33 HÄFELI, U., SCHÜTT, W., TELLER, J., ZBOROWSKI, M. *Scientific and Clinical Applications of Magnetic Microspheres.* Plenum Press, New York, **1997**.

34 SAFARIK, I., SAFARIKOVA, M. Magnetic nanoparticles in biosciences. *Mont. FurChem.* **2002**, 133, 737–759.

35 KOPPENOL, W.H. The centennial of the Fenton reaction. *Free Rad. Biol. Med.* **1993**, 15, 645–651.

36 www.polysciences.com.

37 GOETZE, T., GANSAU, C., BUSKE, N., ROERDER, M., GORNERT, P., BAHT, M. Biocompatible magnetic core/shell nanoparticles. *J. Magn. Magn. Mater.* **2002**, 252, 399–402.

38 LACAVA, Z.G.M., AZEVEDO, R.B., MARTINS, E.V., LACAVA, L.M., FREITAS, M.L.L., GARCIA, V.A.P., REBULA, C.A., LEMOS, A.P.C., SOUSA, M.H., TOURINHO, F.A., DA SILVA, M.F., MORAIS, P.C. *J. Magn. Magn. Mater.* **1999**, 201, 431–434.

39 MIKHAYLOVA, M., JO, Y.S., KIM, D.K., BOBRYSHEVA, N., ANDERSSON, Y., ERIKSSON, T., OSMOLOWSKY, M., SEMENOV, V., MUHAMMED, M. The effect of biocompatible coating layers on magnetic properties of superparamagnetic iron oxide nanoparticles. *Hyperfine Int.* **2004**, 156/157, 257–263.

40 GUPTA, A.K., GUPTA, M. Cytotoxicity suppression and cellular uptake enhancement of surface modified magnetic nanoparticles. *Biomaterials* **2005**, 26, 1565–1573.

41 PETRI-FINK, A., CHASTELLAIN, M., JUILLERAT-JENNERET, L., FERRARI, A., HOFMANN, H. Development of functionalized superparamagnetic iorn oxide nanoparticles for interaction with human cancer cells. *Biomaterials* **2005**, 26, 2685–2694.

42 STROH, A., ZIMMER, C., GUTZEIT, C., JAKSTADT, M., MARSCHINKE, F., JUNG, T., PILGRIMM, H., GRUNE, T. Iron oxide particles for molecular magnetic resonance imaging cause transient oxidative stress in rat macrophages. *Free Rad. Biol. Med.* **2004**, 36, 976–984.

43 FINKLEA, H.O. *Semiconductor Electrodes*, Chapter 2: TiO_2, Elsevier, New York **1988**.

44 LI, G., LI, L., BOERIO-GOATES, J., WOODFIELD, B.F. High purity anatase TiO_2 nanocrystals: near room temperature synthesis, grain growth kinetics, and surface hydration chemistry. *J. Am. Chem. Soc.* **2005**, 127, 8659–8666.

45 ZHANG, H., BANFIELD, J.F. Thermodynamic analysis of phase stability of nanocrystalline titania. *J. Mater. Chem.* **1998**, 8, 2073–2076.

46 ZHANG, H., BANFIELD, J.F. Understanding polymorphic phase transformation behavior during growth of nanocrystalline aggregates: Insights from TiO_2 *J. Phys. Chem. B* **2000**, 104, 3481–3487.

47 MATIJEVIC, E., BUDNIK, M., MEITES, L. Preparation and mechanism of formation of titanium dioxide hydrosols of narrow size distribution. *J. Colloid Interf. Sci.* **1977**, 61, 302–111.

48 WANG, C.-C., YING, J.Y. Sol-gel synthesis and hydrothermal processing of anatase and rutile titania nanocrystals. *Chem. Mater.* **1999**, 11, 3113–3120.

49 FU, G., VARY, P.S., LIN, C.-T. Anatase TiO_2 nancomposites for antimicrobial coatings. *J. Phys. Chem. B* **2000**, 104, 8889–8898.

50 NIRMAL, M., BRUS, L.E. Luminescence photophysics in semiconductor nanocrystals. *Acc. Chem. Res.* **1999**, 32, 407–414.

51 KAVA, L., GRATZEL, M., GILBERT, S.E., KLEMENZ, C., SCHEEL, H.J. Electrochemical and photoelectro-chemical investigations of single-crystal anatase. *J. Am. Chem. Soc.* **1996**, 118, 6716–6723.

52 FUJISHIMA, A., HONDA, K. Electro-chemical photolysis of water at a semiconductor electrode. *Nature* **1972**, 238, 37–38.

53 KHAN, A.U., WILSON, T. Reactive oxygen species as cellular messengers. *Curr. Biol.* **1995**, 2, 437–445.

54 MICIC, O., ZHANG, Y., CROMACK, K.R., TRIFUNAC, A., THURNAUER, D.M. Photoinduced hole transfer from titanium dioxide to methanol aqueous solution studied by electron paramagnetic resonance. *J. Phys. Chem.* **1993**, 97, 13 284–13 288.

55 DONALDSON, K., BESWICK, P., GILMOUR, P., Free radical activity associated with the surface of particles: a unifying factor in determining biological activity? *Toxicol. Lett.* **1996**, 88, 293–298.

56 HART, G., KATHMAN, L., HESTERBERG, T., In vitro cytotoxicity of asbestos and man-made vitreous fibers: roles of fiber length, diameter and composi-tion. *Carcinogen* **1994**, 15, 971–977.

57 HESTERBERG, T., BARRETT, J., Induction by asbestos fibers of anaphase abnormalities: Mechanism for aneuploidy induction and possible carcinogenesis. *Carcinogen* **1985**, 6, 473–475.

58 WOODWORTH, C., MOSSMAN, B., CRAIGHEAD, J., Induction of squamous metaplasia in organic cultures of hamster trachea by naturally occurring and synthetic fibers. *Cancer Res.* **1983**, 43, 4906–4912.

59 MAROUDAS, N., O'NEILL, C., STANTON, M., Fibroblast anchorage in carcinogenesis by fibres. *Lancet,* **1973**, I, 807–809.

60 PETERS, K., UNGER, R., KIRKPATRICK, J., GATTI, A., MONARI, E., Effects of nano-scaled particles on endothelial cell function in vitro: Studies on viability, proliferation and inflammation. *J. Mater. Sci.: Mater. Med.* **2004**, 15, 321–325.

61 www.promega.com.

62 SHANBHAG, A., JACOBS, J., BLACK, J., GALANTE, J., GLANT, T., Macrophage/ particle interactions: Effect of size, composition and surface area. *J. Biomed. Mater. Res.* **1994**, 28, 81–90.

63 STEARNS, R., PAULAUSKIS, J., GODLESKI, J., Endocytosis of ultrafine particles by A549 cells. *Am. J. Respir. Cell Mol. Biol.* **2001**, 24, 108–115.

64 WESTMORELAND, C., WALKER, T., MATTHEWS, J., MURDOCK, J., Preliminary investigations into the use of a human bronchial cell line (16HBE14o-) to screen for respiratory toxins in vitro. *Toxicol. in Vitro.* **1999**, 13, 761–764.

65 ABE, K., SAITO, H., Both oxidative stress-dependent and independent effects of amyloid β protein are detected by 3-(4,5-dimethylthiazol-2-yl)-2,5-diphenyltetrazolium bromide (MTT) reduction assay. *Brain Res.* **1999**, 830, 146–154.

66 KILGOUR, J., SIMPSON, S., ALEXANDER, D., REED, C., A rat nasal epithelial model for predicting upper respiratory tract toxicity: in vivo-in vitro correlations. *Toxicology.* **2000**, 145, 39–49.

67 MILLER, B., PUJADAS, E., GOCKE, E., Evaluation of the micronucleus test in vitro using Chinese hamster cells: Results of four chemicals weakly positive in the in vivo micronucleus test. *Environ. Mol. Muta.* **1995**, 26, 240–247.

68 Opinion concerning titanium dioxide, Colipa n° S75 adopted by the SCCNFP during the 14th plenary meeting of 24 October 2000. www.europa.eu.int/comm/health/ph_risk/committees/sccp.

69 UCHINO, T., TOKUNAGA, H., ANDO, M., UTSUMI, H., Quantitative determination of OH radical generation and its cytotoxicity induced by TiO_2-UVA treatment. *Toxicol. in Vitro.* **2002**, 16, 629–635.

70 ZHANG, A., SUN, Y., Photocatalytic killing effect of TiO_2 nanoparticles on Ls-174-t human colon carcinoma cells. *World J. Gastroenterol.* **2004**, 10, 3191–3193.

71 CAI, R., HASHIMOTO, K., ITOH, K., KUBOTA, Y., FUJISHIMA, A., Photokilling of malignant cells with ultrafine TiO_2 powder. *Bull. Chem. Soc. Jpn.* **1991**, 64, 1268–1273.

72 CAI, R., KUBOTA, Y., SHUIN, T., SAKAI, H., HASHIMOTO, K., FUJISHIMA, A., Induction of cytotoxicity by photoexcited TiO_2 particles. *Cancer Res.* **1992**, 52, 2346–2348.

73 SAKAI, H., ITO, E., CAI, R., YOSHIOKA, T., KUBOTA, Y., HASHIMOTO, K., FUJISHIMA, A., Intracellular Ca^{2+} concentration change of T24 cell under irradiation in the presence of TiO_2 ultrafine particles. *Biochim. Biophys. Acta.* **1994**, 1201, 259–265.

74 NAKAGAWA, Y., WAKURI, S., SAKAMOTO, K., TANAKA, N., The photogenotoxicity of titanium dioxide particles. *Mut. Res.* **1997**, 394, 125–132.

75 SINGH, N. P., MCCOY, M. T., TICE, R. R., SCHNEIDER, E. L., A simple technique for quantation of low levels of DNA damage in individual cells, *Exp. Cell Res.* **1988**, 175, 184–191.

76 HARTMANN, A., SPEIT, G., Comparative investigations of the genotoxic effects of metals in the single cell gel (SCG) assay and sister chromatid exchange (SCE) test, *Envir. Mol. Mutagen.* **1994**, 23, 299–305.

77 WARMER, W., YIN, J., WEI, R., Oxidative damage to nucleic acids photosensitized by TiO_2. *Free Rad. Biol. Med.* **1997**, 23, 851–858.

78 DIZDAROGLU, M., Formation of an 8-hydroxyguanine moiety in deoxyribonucleic acid on γ irradiation in aqueous solution. *Biochemistry* **1985**, 24, 4476–4481.

79 FLOYD, R., WEST, M., ENEFF, K., HOGSETT, W., TINGEY, D., Hydroxyl free radical mediated formation of 8-hydroxyguanine in isolated DNA. *Arch. Biochem. Biophys.* **1988**, 262, 266–272.

80 BLAKELY, W., FUCIARELLI, A., WEGHER, B., DIZDAROGLU, M., Hydrogen peroxide-induced base damage in deoxyribonucleic acid. *Radiat. Res.* **1990**, 121, 338–343.

81 KUBOTA, Y., SHUIN, T., KAWASAKI, C., HOSAKA, M., KITAMURA, H., CAI, R., SAKAI, H., HASHIMOTO, K., FUJISHIMA, A., Photokilling of T-24 human bladder cancer cells with TiO_2. *Br. J. Cancer* **1994**, 70, 1107–1111.

82 CAI, R., HASHIMOTO, K., KBOTA, Y., FUJISHIMA, A., Increment of photocatalytic killing of cancer cells using TiO_2 with the aid of superoxide dismutase. *Chem. Lett.* **1992**, 427–430.

83 CAI, R., HASHIMOTO, K., FUJISHIMA, A., KUBOTA, Y., Conversion of photogenerated superoxide anion into hydrogen peroxide in TiO_2 suspension system. *J. Electroanal. Chem.* **1992**, 326, 345–350.

84 JANG, H., KIM, S., Effect of particle size and phase composition of titanium dioxide nanoparticles on the photocatalytic properties. *J. Nanopart. Res.* **2001**, 3, 141–147.

85 CARAFOLI, E., Intracellular calcium homeostatis. *Annu. Rev. Biochem.* **1987**, 56, 395–433.

86 FARBER, J., The role of calcium in cell death. *Life Sci.* **1981**, 29, 1289–1295.

87 ORRENIUS, S., McCONKEY, D., BELLOMO, G., NICOTERA, P., Role of Ca^{2+} in toxic cell killing. *Trends Pharmacol. Sci.* **1989**, 10, 281–285.

88 KOMULAINEN, H., BONDY, S., Increased free intracellular Ca^{2+} by toxic agents: an index of potential neurotoxicity? *Trends Pharmacol. Sci.* **1988**, 8, 154–156.

89 NICOTERA, P., HARTZELL, P., BALDI, C., SVENSSON, S., BELLOMO, G., ORRENIUS, S., Cystamine induces toxicity in hepatocytes through the elevation of cytosolic Ca2+ and the stimulation of a nonlysosomal proteolytic system. *J. Biol. Chem.* **1986**, 261, 14 628–14 635.

90 KANE, A., STANTON, R., RAYMOND, E., DOBSON, E., KNAFELC, E., FARBER, J., Dissociation of intracellular lysosomal rupture from the cell death caused by silica. *J. Cell Biol.* **1980**, 87, 643–651.

91 LEMASTERS, J., DIGUISEPPI, J., NIEMINEN, A., HERMAN, B., Blebbing, free calcium and mitochondrial membrane preceding cell death in heptaocytes. *Nature.* **1987**, 325, 78–81.

92 LINNAINMAA, K., KIVIPENSAS, P., VAINIO, H., Toxicity and cytogenetic studies of ultrafine titanium dioxide in cultured rat liver epithelial cells. *Toxicol. in Vitro.* **1997**, 11, 329–335.

93 ULLAH KHAN, A., WILSON, T. Reactive oxygen species as cellular messengers. *Chem. Biol.* **1995**, 2, 437–445.

94 MANESS, P., SMOLINSKI, S., BLAKE, D., HUANG, Z., WOLFRUM, E., JACOBY, W., Bactericidal activity of photocatalytic TiO_2 reaction: toward an understanding of its killing mechanism. *Appl. Environ. Microbiol.* **1999**, 65, 4094–4098.

95 YAMAMOTO, A., HONMA, R., SUMITA, M., HANAWA, T., Cytotoxicity evaluation of ceramic particles of different sizes and shapes. *J. Biomed. Mater. Res.* **2004**, 68A, 244–256.

96 RAE, T., A study on the effects of particulate metals of orthopaedic interest on murine macrophages in vitro. *J. Bone Joint Surg.* **1975**, 57-B, 444–450.

97 RAE, T., The toxicity of metals used in orthopaedic prostheses. *J. Bone Joint Surg.* **1981**, 63-B, 435–440.

98 HANAWA, T., KAGA, M., ITOH, Y., ECHIZENYA, T., OGUCHI, H., OTA, M., Cytotoxicities of oxides, phosphates and sulphides of metals. *Biomaterials* **1992**, 13, 20–24.

99 LISON, D., LARDOT, C., HUAUX, F., ZANETTI, G., FUBINI, B., Influence of particle surface area on the toxicity of insoluble manganese dioxide dusts. *Arch. Toxicol.* **1997**, 71, 725–729.

100 LEHNERT, B., Defense mechanisms against inhaled particles and associated particle-cell interactions.,

In: GUTHRIE, G., MOSSMAN, B., *Review in Mineralogy, Health Effects in Mineral Dusts*, Mineralogical Society of America, Washington, DC, **1993**, p. 28.

101 KECELI, S., ALANYALI, H., A study on the evaluation of the cytotoxicity of Al_2O_3, Nb_2O_5, Ta_2O_5, TiO_2, and ZrO_2. *Turk. J. Eng. Environ. Sci.* **2004**, 28, 49–54.

102 CHIU, S., LEE, M., CHEN, H., CHOU, W., LIN, L., Germanium oxide inhibits the transition from G2 to M phase of CHO cells. *Chem. Biol. Interact.* **2002**, 141, 211–228.

103 http://www.biology.arizona.edu/cell_bio/tutorials/cell_cycle/cells2.html.

104 GAUDENZI, S., FURFARO, M., POZZI, D., SILVESTRI, I., CASTELLANO, A., Cell-metal interaction studied by cytotoxic and FT-IR spectroscopic methods.

Environ. Toxicol. Pharmacol. **2003**, 14, 51–59.

105 PIGOTT, G., PINTO, P., Effects of nonfibrous minerals in the V79-4 cytotoxicity test. *Environ. Health Perspect.* **1983**, 51, 173–179.

106 TIMBRELL, V., GILSON, J., WEBSTER, I., UICC standard reference samples of asbestos. *Int. J. Cancer.* **1968**, 3, 406–408.

107 DAVIES, R., The effect of ducts on enzyme release from macrophages. In: *The In Vitro Effects of Mineral Dusts.* Academic Press, London **1980**, 67–74.

108 PIGOTT, G., JUDGE, P., The effects of mineral dusts "in vitro": a comparison of the response of rat peritoneal macrophages and the P388D1 cell line. In: *The In Vitro Effects of Mineral Dusts.* Academic Press, London **1980**, 53–57.

2
Ecotoxicity of Engineered Nanomaterials

Eva Oberdörster, Patricia McClellan-Green, and Mary Haasch

2.1
Introduction

To date there has only been one thorough review of the nanotoxicology literature from a biological viewpoint [1], and that review through necessity was based on related work done primarily on the toxicology of ultrafine particulate matter in mammalian models. In this chapter we will not repeat the previous report, but will instead focus on new investigations of engineered nanomaterials in environmentally relevant species and models. Initial "eco-nano" considerations were focused on using nanomaterials in the environment for remediation, in the development of more accurate and sensitive biosensors, and for green energy production, for example (Table 2.1). With these initial efforts there was little concern for engineered nanomaterials functioning as toxicants themselves, and the focus was on technology development. Only since 2004 has the issues of nano-ecotoxicology of highly reactive, lipophilic engineered nanomaterials come to the front. The numerous benefits to society from the development of NP should not be minimized. Decreasing our dependence on highly toxic fossil fuels, remediating superfund sites, creating new and better drug delivery systems and green manufacturing are all processes whose benefits portend great promise. We should not approach these technologies wearing blinders, but rather be cognizant of the big picture. In other words: Be aware of the benefits and the costs.

One issue that immediately confronts any scientist in the area of nanoparticle toxicology is terminology. Standardized terminology is not yet in use, although efforts are underway by Rice University's CBEN to move forward on this front. In this chapter, we will use engineered nanoparticles (NP) to designate any man-made nanomaterial (one dimension < 100 nm) with specific chemical, size, and shape characteristics, including materials such as fullerenes (C_{60}, C_{70}), single-walled carbon nanotubes (SWNT), quantum dots, nano-wires-films, -textiles, and so forth. The more general term, nanosized particle (NSP), will include both the NP and naturally occurring particles that are less than 100 nm in one dimension, such as the ultrafine particles (UFP) in air pollution, and small bacteria and viruses. This chapter will focus on engineered nanoparticles (NP).

Nanotechnologies for the Life Sciences Vol. 5
Nanomaterials – Toxicity, Health and Environmental Issues. Edited by Challa S. S. R. Kumar
Copyright © 2006 WILEY-VCH Verlag GmbH & Co. KGaA, Weinheim
ISBN: 3-527-31385-0

Tab. 2.1. Some recent funding by the US EPA to develop applications of NP for use in the environment [38].

Type of NP used	Potential use	Lead PIs and Institutions
	Remediation	
NanoTiO$_2$	Photocatalysis of organic contaminants	D.D. Dionysiou Miami University-Oxford, OH; University of Cincinnatti, OH
Carbon nanostructures	Sorption of organics	M.B. Tomson Rice University
Nano-metal oxides	Control NO$_x$ production	S. Senkan UCLA
Nano-iron	Degradation of PAH-based contaminants	G.V. Lowry, S.A. Majetich, K. Matyjaszewski, R.D. Tilton Carnegie Mellon University
Nano-biopolymers	Control of heavy metals	W. Chen, M. Matsumoto, A. Mulchandani UC Riverside
Bi-metallic nano-Fe/Pd	Remediation of inorganics and organics	W.X. Zhang Lehigh University
Nano-crystalline zeolite	NO$_x$, photocatalytic oxidation of organics	S.C. Larsen, V.H. Grassian University of Iowa
Nano-magnetite	Groundwater contamination	M. Hull Luna Innovations, Inc.
	Filtration	
Ferromagnetic particles	Using nanocomposites to monitor and filter (smart particles)	W.M. Sigmund, D. Mazyck, C.Y. Wu University of Florida
Nano-crystalline catalysts	Disinfection by-product control in drinking water	S.J. Masten, M.J. Baumann Michigan State University
Nanostructured electrodes	Perchlorate from drinking water	S.M. Jaffe Material Methods LLC
	Sensors	
Carbon nanoparticle based microchip	Analytical chemistry of environmentally relevant endpoints	J. Wang New Mexico State University

Tab. 2.1 *(continued)*

Type of NP used	Potential use	Lead PIs and Institutions
Nanocrystalline metallic conductors	Gas sensor	V. Subramanian UC Berkeley
Colloidal-metal nanoparticles	Monitoring heavy metals	O. Sadik, J. Wang New Mexico State University
Polystyrene beads coated with peptides	Detection of aquatic toxins	R.E. Gawley University of Miami
Fullerene	Tracers for water pollution	J.B. Callegary University of Arizona
Green energy/manufacturing		
Nano-clay	Substitute petroleum-based products for nano-composites	L.T. Drzal, M. Misra, A.K. Mohanty Michigan Sate University
Nano-micelles	Replacing VOCs with nano-structured microemulsions	D.A. Sabatini, J.H. Harwell University of Oklahoma
Nano-plastic fibrils and crystals	Alternative to petroleum-based composites	W.T. Winter SUNY College of Environmental Science and Forestry
Nano-TiO$_2$	Photocatalyst for solar cells	G. Chumanov Clemson University
Semi-conducting nanoparticles	Catalyst fuel cells	N.Y. Dolney University of Michigan-Ann Arbor

When "nano" first became the hot new technology, immediate environmental applications were sought using these reactive materials. Numerous funding agencies encouraged development of NP use in the environment (applications, Table 2.1) and initially very little consideration was given to the unintended consequences or implications of nanomaterial production or use in the environment. The new technologies developed include a wide array of materials designed for remediation activities (some of which are in commercial use), the development of biosensors for chemicals or biological agents, the development of environmental filtration processes, and green manufacturing. Although one could argue that both applications and implications are important areas of research, the implica-

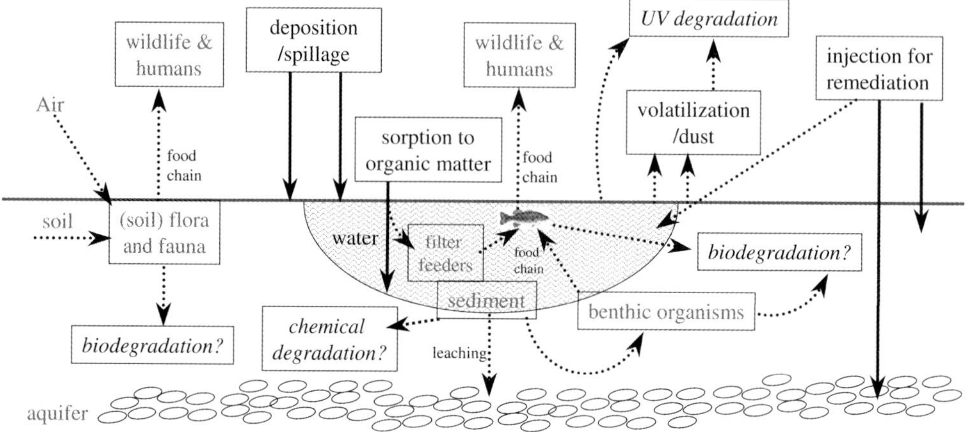

Fig. 2.1. Model of NP movement through the environment. Dotted lines indicate pathways not yet verified, while solid arrows indicate verified pathways. (From Oberdörster et al. [1].)

tions research of NPs has lagged behind. In this chapter, we focus on some of the recent environmental implications of NPs as toxicants, and will discuss issues that need to be addressed through future research.

The complexity of NP in ecotoxicology can be seen in Fig. 2.1. Particle movement through water, air and soils has been poorly studied, and biological uptake and food chain transport have not been considered. Biotransformation and chemical and UV breakdown have been given scant consideration and in-depth chemical analysis or other characterizations, including electron microscopic imaging of NP before and after environmental ageing, have not been conducted. In the following sections, we review these issues along with suggested areas of future research.

2.2
Water

Many NP are poorly soluble in water (e.g., SWNT, C_{60}). However, if coated with an appropriate molecule, such as peptides or proteins [2, 3], poly(ethylene glycol) (PEG), or other surfactants, even non-water-soluble NP can be rendered miscible with water, including water-containing humic acids and salts [4]. Another effective method to render lipophilic NP (such as C_{60}) water-soluble is to allow the particle powders (clumps of NP) to stir in water for a few days or up to several weeks (longer agitation times equal more dissolution). This slow procedure is a more environmentally relevant method of introducing NP into solution. Recent data shows that using organic solvents as an intermediary to render NP water-soluble leaves traces of the organic solvent in centers of fullerene clusters. For example, Andrievsky et al. [5], Brant et al. [6], and Fortner et al. [4] have demonstrated that using tetrahydrofuran (THF), a common organic solvent, to solubilize C_{60} in water resulted in formation of nC_{60} aggregates that contained residual amounts of THF. This residual THF is of biological concern and toxicologically relevant. In daphnia we have found that the 48-hour LC_{50} is orders of magnitude different between THF-solubilized nC_{60} (0.8 ppm) and stirred nC_{60} (>30 ppm, which was the highest concentration tested). (For a review of solubility levels of NP in various solvents please refer to Nakamura and Isobe [7].) Thus, when performing ecotoxicity tests, it is crucial to determine not only a valid range of concentrations for testing (will we really see ppm levels?), but also to use realistic methods of water-solubilization of NP.

There are several target areas of concern when discussing ecotoxicology of NP. Engineered NP will tend to agglomerate to each other or to larger particles in the environment, and will tend to sorb onto or associate with sediments. These sediments can then be ingested by benthos, creating a food-chain through which these NP can move (Fig. 2.1). In preliminary studies with the suspension-feeding worm *C. elegans*, we have shown that FITC-labeled SWNT (which can be easily tracked through the exposure dishes and inside the worms) move through the digestive tract (DG) and are not absorbed into the animal (unpublished data, laboratory of Eva Oberdörster in collaboration with Jim Waddell, Southern Methodist University and Ya-Ping Sun, Clemson University). This type of tracking has not been done for other NP, but is a crucial step in determining uptake into biota. Even if NP remain solely in the DG tract and do not bioaccumulate, they are still likely to move up the food chain as worms and other organisms are consumed by benthivores (Fig. 2.1). This scenario seems even more likely with a detritivore benthic organism like *Hyalella azteca* in which the nC_{60} LC_{50} is greater than 7 ppm. In fact, no toxic effects are observed in *Hyalella* even when the nC_{60} is mixed in the food [8].

In contrast to the *C. elegans* study, we have shown that filter-feeding crustaceans (*Daphnia magna*) can accumulate NP when exposed via the water column (Figs. 2.2 and 2.3). Nano-iron used in remediation is ingested by daphnia and can coat their carapace, including filtering apparatus and appendages (Fig. 2.3). Even though the daphnids were coated with nano-iron, they were able to survive in the laboratory test and were able to feed and reproduce. The toxicity of nano-iron was the same as that for bulk iron, approximately 55 ppm (Fig. 2.3). The daphnids containing nano-iron in the gut and on the carapace are much darker in coloration than daphnids without the nano-iron. Since many daphnid predators (fish) are visual feeders, it would be interesting to determine whether the darker daphnids are more likely to be preyed upon than lighter daphnids, similar to what has been shown with melanized (darker) vs. lighter-colored daphnids [9].

Daphnids are generalist filter feeders specializing in larger-sized phytoplankton [9]. However, numerous species are specialized filter feeders, including many rotifers that specialize in nanosized prey, such as Archaea and other small bacteria. The differential impact of NP on filter feeders that are generalists vs. specialists still needs to be determined. A study by Conova [10] has shown that some filter-

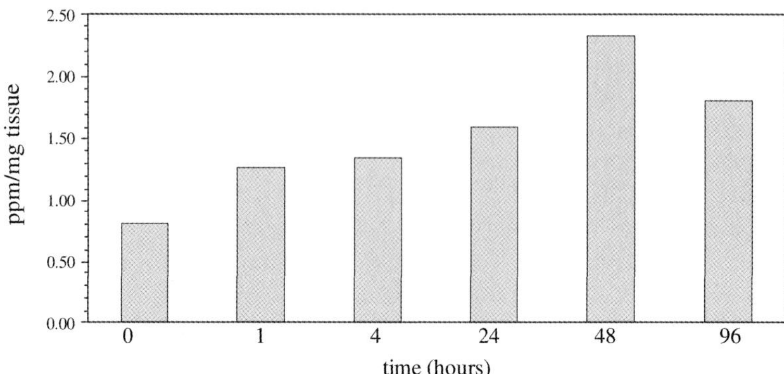

Fig. 2.2. Uptake of nC$_{60}$ into *Daphnia magna.* Approximately 15 daphnids were collected at each time point, and were rinsed 5× with reconstituted hard water during 1 h. Water was removed by blotting, and a wet-weight was taken. The nC$_{60}$ was oxidized, using 1 mL bleach, and extracted overnight into toluene. Absorbance was read at 332 nm and compared to a standard curve.

feeders select their prey by surface chemistry, not necessarily size. Therefore, coating NP to make them more "biocompatible" may make them easier for certain species to selectively filter. In addition, many aquatic and marine organisms, especially crustaceans, carry out their feeding and reproductive behaviors through the use of chemosensory organs. Adherence of NP to the surface of the chemosensory structures such as seen in Fig. 2.3 could disrupt their growth, development or reproduction by interfering both physically or physiologically with these structures. These are not minor considerations given that zooplankton is the basis of aquatic food chains. Specific impacts on zooplankton can significantly alter predator/prey balance and lead to shifts in ecosystem health.

Vertebrates would be exposed not only via the food chain (either by ingesting sediments directly or by ingesting NP-contaminated prey) but also through gill and skin. Observations by Tjälve [11, 12] and Oberdörster [13] have shown that translocation of toxicants, including NP, via the olfactory neuron into the brain is likely in several species of fish. Although most NP will tend to sorb to sediments or onto phytoplankton, NP will likely move up the food chain due to benthos and filter-feeding invertebrates. These types of studies – systematic bioaccumulation/ biodistribution – have not been done to date. Considering the current restrictions on fish consumption for humans due to PCBs and methyl mercury [14], it is not only an ecosystem health issue, but also a human-health issue.

Although movement of NP through the food chain is likely, recent studies by Lecoanet [15] have shown that NP are of very low mobility in aquifers. Even though NP are currently injected into aquifers and ground-water for remediation, it has been hypothesized that they will not move far from the injection point. However, the rate of movement of NP in real-life applications has not been tested. The size,

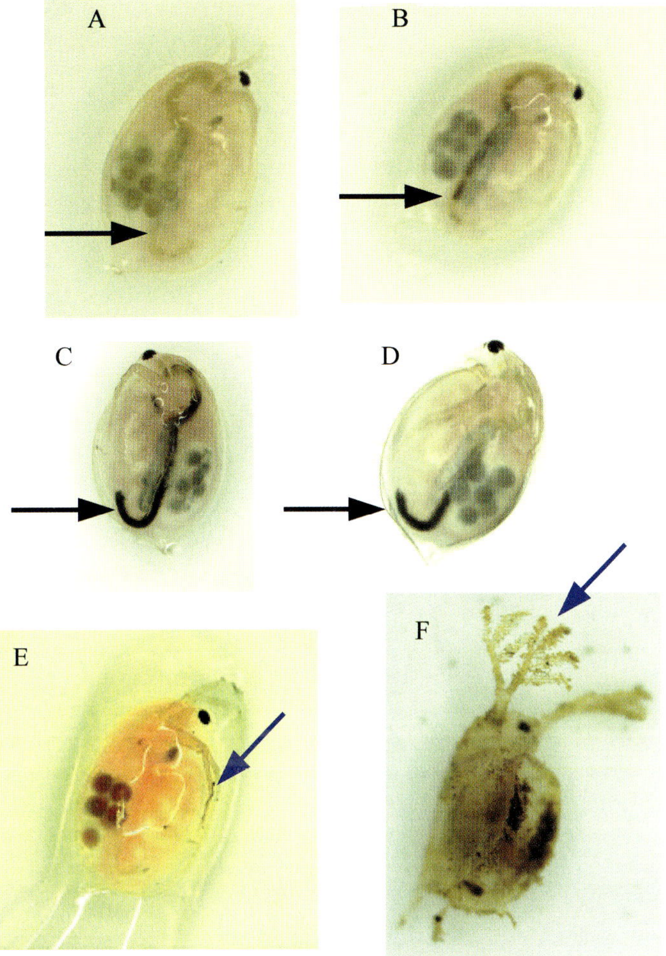

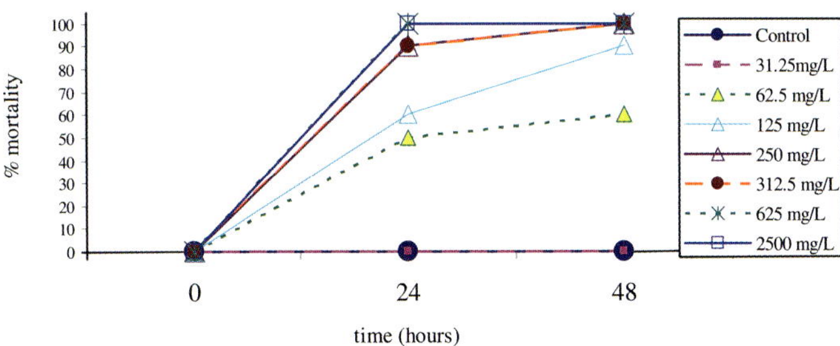

Fig. 2.3. *Daphnia magna* exposed to various concentrations of nano-iron used in remediation. A = control; B = 3; C = 7.5; D = 15; E = 30; F = 125 mg L^{-1} (dead daphnid). All daphnids shown are 21-days-old and eggs are visible in their brood pouches (green circles). Note the darkening of the digestive tract from A (normal greenish color) to D with increased ingestion of nano-iron particles (black arrows). Antennae become clogged with nano-iron in E and F (blue arrows). The 24 and 48 h mortality curves are also shown.

shape and surface chemistry of NPs that make them so attractive in various technologies will also influence their behavior in the environment. Pharmaceutical NPs that have been coated to make them more soluble or to improve the solubility of attached drugs or compounds will surely enter the waste-water stream, similar to what has been observed with other drugs and medicines [16]. Once there, NP will be transported, broken down or accumulated based on their physicochemical properties. For example, polyhydroxylated C_{60} (fullernols) or surfactant modified nanotubes are specifically engineered to increase their solubility in water, and could therefore remain in the aqueous phase. But these modifications also increase NP movement through porous media [17]. Movement of specific NPs in environmental media has not been thoroughly investigated, although researchers at Rice University are beginning to address this issue.

2.3
Air

Most research on airborne NSP has been on ultrafine particles (UFP) of various chemical and size compositions, and of NP in the workplace [1]. NSP can either agglomerate to each other or to other particles in the air, depending on particle number density and time. Particle sizes of less than 5 nm in diameter will behave more like gases, while larger sized NSP behave more like bulk particles; for a review see Ref. [1]. NSP are produced from combustion by-products, and can deposit as dust particles, and can also be re-suspended by wind (Fig. 2.1). Global movement of NSP through atmospheric deposition and re-suspension is likely, similar to what has been found in other gaseous and particle pollutants, such as CFCs. Biswas and Wu have recently reviewed the NP/NSP literature [18]; please refer to that reference and to Chapters 4 and 7 of this text for more details on airborne NSP effects, especially as they relate to workplace exposure.

2.4
Soils

Soils are a complex mixture of organic and inorganic compounds, and microbial and other living organisms. Soils are so complex that standard "soils" are used in ecotoxicology testing. To date, no studies have been performed using these standard soil protocols, but one recent study shows that at relatively high doses (ppm) microbial growth (*E. coli* and *B. subtilis*) was reduced with nano-TiO_2 and nC_{60} [4]. Since these are relatively high levels of nanomaterials, modeling exercises are needed to determine relevant doses before being able to decide whether there are risks involved in release of NP into soils.

Another source of NP in the environment (especially soils and sediments) are NP-containing matrices that function as slow-release agents for various biocides [19]. These matrices can be applied to various surfaces to inhibit biological growth (e.g., of fungi), but as the matrices weather or wear off, NP will be released into the environment. What happens to these matrices (how fast do they break down? are NP released from them as they break down?) is unknown.

Toxic NSP can be inadvertently created in soils, as bacteria absorb toxicants, creating "biological" toxic NSP. Pollmann et al. [20] have demonstrated that the outer layer of some bacterial species function as selective matrices for the binding of toxic metals. They demonstrated the S-layer of *Bacillus sphaericus* JG-A12, through its hydrophobic construction and the presence of phosphorylated proteins, possesses an extremely high and reversible binding capacity for toxic metals such as uranium and palladium. This binding creates NSP that can be removed from the environment. But what happens to the inadvertently created NSP? Where do they go? And how does sorption of toxicants, including NP, affect microbial communities?

In addition to effects on microbial communities, it is likely that plants can take-up and bioaccumulate NP. Some interesting solar-cell applications have been developed using synthetic chlorophyll and fullerene [21], but it has not yet been determined whether fullerene can interfere with or enhance natural photosynthesis by bypassing the usual electron transport chain. Given that solar-power research is focused on using nanomaterials, it is critical to determine whether natural solar-power (i.e., photosynthesis) can be disrupted or enhanced by NP. Preliminary studies in our laboratory indicate that nC_{60} may influence (enhance) the growth rates of blue-green algae (*Anabaena* sp.) (unpublished observation, B. Craig and P. McClellan-Green). The mechanism behind the change in growth is unknown. The nC_{60} might act as a nutrient source, facilitate uptake of media nutrients, or possibly interact with photosynthesis to accelerate the process. Although it is difficult to predict the types of challenges that could arise due to NP in soils, another area of concern is that these NP could interfere with cell signaling, such as with root nodulation of nitrogen-fixing bacteria. Such interference has been shown with pesticides [22], and could lead to unintended agricultural consequences.

Movement of NP through soil food chains is likely. Owing to the tendency to sorb to particles, NP will likely be ingested or absorbed by soil organisms (bacteria, worms, insects, plants, fungi, etc.) and could move up the food chain (Fig. 2.1). No studies have been performed on soil-food chain transport, but this will likely be an important future area of research given the human manipulation of environmental NP and NSP, and cycling between air deposition and re-suspension in air and water by dust and debris (Fig. 2.1).

2.5
Weathering

Currently, studies on NP and ecologically-relevant endpoints have been carried out under laboratory conditions without allowing for action by abiotic factors, such as UV, other chemicals, and dissolved oxygen levels/anoxia. Few studies have been carried out on UV interactions with NP that are either coated or covalently linked to molecules that render the NP less toxic. These studies have shown that even a

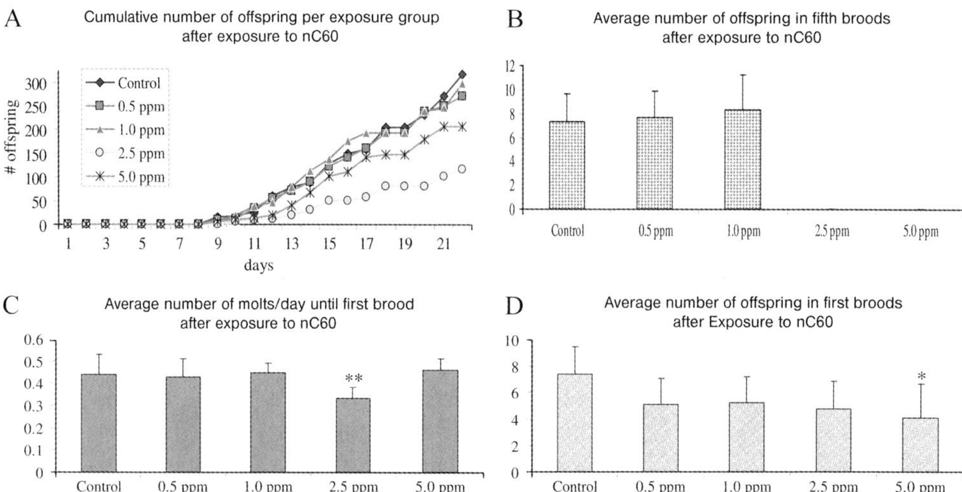

Fig. 2.5. Exposure of *Daphnia magna* to sublethal levels of nC_{60} delays reproduction (A), results in one fewer brood (no fifth brood) over 21 days (B), slows molting (C), and reduces the number of offspring in the first brood (D); $* p < 0.05$, $** p < 0.01$. Therefore, sub-lethal endpoints need to be investigated when studying NP effects in environmentally relevant species. nC_{60} used in this study was prepared by stirring. No organic solvents were used to solubilize the nC_{60}.

sights into mechanisms of action. For example, in daphnia we have found that sub-LC_{50} doses of nC_{60} can inhibit reproduction by delaying brood production and brood size (Fig. 2.5). Gene chips can also give insights into sub-lethal effects and mechanisms of action. Clearly, a combination of population-level data (such as reproductive output) and protein and gene expression can give a clearer picture of environmental risks posed by NP. To date this information is not available, although several researchers are addressing these issues.

2.7
Conclusions

NP are already in the environment, used either for remediation, or from normal use and wear of products containing NP (tires, clothing, sporting equipment, cosmetics, etc.). As there is almost no data on the toxicity of NP to environmentally relevant species, it is difficult to predict ecosystem risks. Although steps are being taken to remedy this lack of knowledge, several key research areas will need to be addressed. These include, but are not limited to:

1. How far can NP move through air, water and soil? How do size, shape, surface chemistry and agglomeration state affect this movement?

2. What are the most likely routes of exposure for environmentally relevant species (ingestion, dermal, inhalation, root uptake)?
3. Can NP interfere with photosynthesis, with microbial communities, or with inter-species communication (such as root nodulation)?
4. Can NP bioaccumulate?
5. Can NP be metabolized to more or less toxic forms?
6. What biomarkers are relevant for measuring NP exposure levels?
7. What end-points are significant for determining risk of NP?
8. What are the mechanisms of toxicity of NP in environmentally relevant systems?
9. Does the presence of NP in the environment affect the toxicity of other compounds and vice versa?

Many of these questions are currently being addressed by scientists around the world. We may well be able to come to a better consensus on eco-risks of NP once these basic questions are answered. Until then, the numerous benefits of NP should not be underestimated! Decreasing our dependence on highly toxic fossil fuels, remediating superfund sites, creating new and better drug delivery systems and green manufacturing are all processes whose benefits need to be considered alongside any toxic effects of NP.

References

1 OBERDÖRSTER, G., E. OBERDÖRSTER, J. OBERDÖRSTER, Nanotoxicology: An emerging discipline evolving from studies of ultrafine particles. *Environ. Health Persp.*, **2005**. 113(7), 823–839.

2 OBERDÖRSTER, E., A. ORTIZ-ACEVEDO, H. XIE, P. PANTANO, R.H. BAUGHMAN, G.R. DIECKMANN, I.H. MUSSELMAN, R.K. DRAPER, Exposure of fathead minnow to fullerene and single-walled carbon nanotubes. *The toxicologist CD*, An official Journal of the Society of Toxicology, **2005**. 84(S1), 325.

3 DIECKMANN, G., A. DALTON, P. JOHNSON, J. RAZAL, J. CHEN, G. GIORDANO, E. MUNOZ, I. MUSSELMAN, R. BAUGHMAN, R. DRAPER, Controlled assembly of carbon nanotubes by designed amphiphilic peptide helices. *J. Am. Chem. Soc.*, **2003**. 125(7), 1770–1777.

4 FORTNER, J.L., LYON, D.Y., C.M. SAYES, A.M. BOYD, J. FALKNER, E. HOTZE, L. ALEMANY, Y. TAO, K. AUSMAN, V. COLVIN, J. HUGHES, C_{60} in water: Nanocrystal formation and microbial response. *Environ. Sci. Technol.*, **2005**. 39(11), 4307–4316.

5 ANDRIEVSKY, G., V. KLOCHKOV, A. BORDYUH, G. DOVBESHKO, Comparative analysis of two aqueous-colloidal solutions of C_{60} fullerene with help of FTIR reflectance and UV-vis spectroscopy. *Chem. Phys. Lett.*, **2002**. 364, 8–17.

6 BRANT, J., H. LECOANET, M. HOTZE, M. WIESNER, Comparison of electrokinetic properties of colloidal fullerenes (nC_{60}) formed using two procedures. *Environ. Sci. Technol.*, **2005**. 39(17): 6343–6351.

7 NAKAMURA, E., H. ISOBE, Functionalized fullerenes in water. The first 10 years of their chemistry, biology, and nanoscience. *Acc. Chem. Res.*, **2003**. 36(11), 807–815.

8 OBERDÖRSTER, E., S. ZHU, T.M. BLICKLEY, P. MCCLELLAN-GREEN, M.L. HAASCH, Ecotoxicology of carbon-

based engineered nanoparticles: Effects of fullerne (C_{60}) on aquatic organisms. *Carbon*, **2005**. in press, available on-line December 22, 2005 (but no issue or page numbers) *doi:10.1016/j.carbon.2005.11.008.*

9 Dodson, S., *Introduction to Limnololgy.* 1st edn, **2004**. McGraw-Hill, New York, NY, USA.

10 Conova, S., Role of particle wettability in capture by suspension-feeding crab (*Emerita talpoida*). *Marine Biol.*, **1999**. 133, 419–428.

11 Tjälve, H., J. Henriksson, Uptake of metals in the brain via olfactory pathways. *Neurotoxicology*, **1999**. 20(2–3), 181–196.

12 Tjälve, H., C. Mejare, K. Borg-Neczak, Uptake and transport of manganese in primary and secondary olfactory neurones in pike. *Pharmacol. Toxicol.*, **1995**. 77(1), 23–31.

13 Oberdörster, E., Manufactured nanomaterials (fullerenes, C_{60}) induce oxidative stress in brain of juvenile largemouth bass. *Environ. Health Persp.*, **2004**. 112(10), 1058–1062.

14 US EPA, *Fish Advisories*, **2005**. http://www.epa.gov/ost/fish/, accessed July 24, 2005.

15 Lecoanet, H., J. Bottero, M. Wiesner, Laboratory assessment of the mobility of nanomaterials in porous media. *Environ. Sci. Technol.*, **2004**. 38, 5164–5169.

16 Kolpin, D., E. Furlong, M. Meyer, E. Thurman, S. Zaugg, L. Barber, H. Buxton, Pharmaceuticals, hormones, and other organic wastewater contaminants in US streams, 1999–2000: A national reconnaissance. *Environ. Sci. Technol.*, **2002**. 36, 1202–1211.

17 Lecoanet, H., M. Wiesner, Velocity effects on fullerene and oxide nanoparticle deposition in porous media. *Environ. Sci. Technol.*, **2004**. 38(16), 4377–4382.

18 Biswas, P., C.-Y. Wu, Nanoparticles and the environment. *J. Air Waste Manage. Assoc.*, **2005**. 55, 708–746.

19 Cioffi, N., L. Torsi, N. Ditaranto, L. Sabbatini, P.G. Zambonin, G. Tantillo, L. Ghibelli, M. D'Alessio, T. Bleve-Zacheo, E. Traversa, Antifungal activity of polymer-based copper nanocomposite coatings. *Appl. Phys. Lett.*, **2004**. 85(12), 2417–2419.

20 Pollmann, K., J. Raff, M. Merroun, K. Fahmy, S. Selenska-Pobell, Metal binding by bacteria from uranium mining waste piles and its technological applications. *Biotechnol Adv.*, **2006**. 24(1) 58–68.

21 Kureishi, Y., H. Tamiaki, H. Shiraishi, K. Maruyama, Photoinduced electron transfer from synthetic chlorophyll analogue to fullerene C_{60} on carbon paste electrode. Preparation of a novel solar cell. *Bioelectrochem. Bioenerg.*, **1999**. 48(1), 95–100.

22 Fox, J., M. Starcevic, P. Jones, M. Burrow, J. McLachlan, Phytoestrogen signaling and symbiotic gene activation are disrupted by endocrine-disrupting chemicals. *Environ. Health Persp.*, **2004**. 112(6), 672–677.

23 Kamat, J., T. Devasagayam, K. Priyadarsini, H. Mohan, J. Mittal, Oxidative damage induced by the fullerene C_{60} on photosensitization in rat liver microsomes. *Chem. Biol. Interact.*, **1998**. 114(3), 145–159.

24 Rancan, F., S. Rosan, F. Boehm, A. Cantrell, M. Brellreich, H. Schoenberger, A. Hirsch, F. Moussa, Cytotoxicity and photocytotoxicity of a dendritic (C_{60}) mono-adduct and a malonic acid (C_{60}) tris-adduct on Jurkat cells. *J. Photochem. Photobiol. B*, **2002**. 67(3), 157–162.

25 Hamano, T., T. Mashino, M. Hirobe, Oxidation of [C_{60}] fullerene by cytochrome P450 chemical models. *Chem. Commun.*, **1995**. 1537–1538.

26 Santos, L. The effects of fullerenes compounds on the microsomal cytochrome P450-monooxygenase system present in human liver microsomes. In *223rd Amercian Chemical Society National Meeting*. **2005**. Orlando, FL, American Chemical Society, Washington, DC, 389.

27 Ueng, T., J. Kang, H. Wang, Y. Cheng, L. Chiang, Suppression of microsomal cytochrome P450-dependent monooxygenases and mitochondrial oxidative phosphorylation by fullerenol, a polyhydroxylated fullerene C_{60}. *Toxicol. Lett.*, **1997**. 93(1), 29–37.

28 Wang, H., L. Chiang, Inhibition of drug-metabolizing enzymes in mouse liver by a water soluble fullerene C_{60}. *Fullerene Sci. Technol.*, **1999**. 7(4), 681–694.

29 Sayes, C., J. Fortner, W. Guo, D. Lyon, A. Boyd, K. Ausman, Y. Tao, B. Sitharaman, L. Wilson, J. Hughes, J. West, V. Colvin, The differential cytotoxicity of water-soluble fullerenes. *Nano Lett.*, **2004**. 4(10), 1881–1887.

30 Fischer, A., R. Hoch, D. Moy, M. Lu, M. Martin, C. Niu, N. Ogata, H. Tennent, F. Jameison, P. Liang, D. Simpson, et al. SWNT can be functionalized by P450s. US. Patent Office, Hyperion Catalysis International, Inc: USA. Kramer Levin Naftalis & Frankel, LLP, Attorneys & Agents. **2004**. International Patent #D01F 009/12; C07C 063/333.

31 Zhu, S., E. Oberdörster, M. Haasch, Toxicity of an engineered nanoparticle (fullerene, C_{60}) in two aquatic species, *Daphnia* and fathead minnow. *Marine Environ. Res.*, **2006**. in press.

32 Haasch, M. L., P. McClellan-Green, E. Oberdörster, Consideration of the toxicity of manufactured nanoparticles. In XIX International Winterschool/Euroconference on Electronic Properties of Novel Materials. American Institute of Physics (AIP), Kirchberg, Tirol, Austria. H. Kuzmany, J. Fink, M. Mehring, S. Roth, eds. **2005**. 786, 586–589.

33 Oberdörster, E., Informal information on the toxicity of engineered nanomaterials. **2005**. www.nanotox.info. accessed 9/25/05.

34 Kamat, J., T. Devasagayam, K. Priyadarsini, H. Mohan, Reactive oxygen species mediated membrane damage induced by fullerene derivatives and its possible biological implications. *Toxicology*, **2000**. 155(1–3), 55–61.

35 Nakajima, N., C. Nishi, F. Li, Y. Kada, Photo-induced cytotoxicity of water-soluble fullerene. *Fullerene Sci. Technol.*, **1996**. 4, 1–19.

36 Yamakoshi, Y., N. Umezawa, A. Ryu, K. Arakane, N. Miyata, Y. Goda, T. Masumizu, T. Nagano, Active oxygen species generated from photoexcited fullerene (C_{60}) as potential medicines: O2-* versus 1O2. *J. Am. Chem. Soc.*, **2003**. 125(42), 12 803–12 809.

37 Zhang, T., P. Lu, F. Wang, G. Wang, Reaction of [60] fullerene with free radicals generated from active methylene compounds by manganese(III) acetate dihydrate. *Org. Biomol. Chem.*, **2003**. 1(24), 4403–4407.

38 US EPA, NCER STAR grants related to nanotechnology. **2005**. http://cfpub.epa.gov/ncer_abstracts/index.cfm/fuseaction/searchControlled.main?RequestTimeout=180&records_per_page=ALL&abstyperesearch=on&abstypefellowship=on&abstypegrants=on&abstypesmallBiz=on&identifier=on&institute=on&annual=on&pubcount=on&principal=on&EPARep=on&grantamt=on&proposedstart=on&addRptOption=on&hiliteOption=on&refreshPage=True&txtSearch=nanotechnology. Accessed 9/26/05.

II
Health

3
Possible Health Impact of Nanomaterials

Peter H. M. Hoet, Irene Brüske-Hohlfeld, and Oleg V. Salata

3.1
Introduction

Nanotechnology is often portrayed as a force that will help to materialize ultimate solutions to today's technological problems. Nanomaterials are the first nanotechnological products hitting the markets. Widespread use of nanomaterials in the consumer and industrial products is also causing some health concerns [1, 2]. Proponents of nanotechnology [3] as well as its opponents find it hard to argue their case due to the limited information available.

How much do we know? To try to answer this question, we start this chapter by looking at the scale and current sources of nanomaterials engineered by men. Next, we use the relative wealth of research data available from the epidemiological studies of the technogenic nanoparticles to highlight the apparent health effects associated with the inhalation of ultrafine particulate matter. The inhalation of ultrafine particles is a well established entry route; hence we discuss the potential entry points of nanoparticles into the human body via airways, and also alternative paths through the skin and gastrointestinal tract. Then, we explore their likely pathways inside the body, the effects associated with nanoparticle interactions on the cellular level, and analyze the origins of the bioactivity of nanomaterials. Nanofibers, a special case of nanomaterials that are known to be hazardous in the micron domain because of their shape, are given some thought. In addition, recent observations on nanoparticle penetration through the blood–brain barrier are evaluated. Finally, the implications of our findings for the field of nanotechnology are discussed. This chapter is one of the very first attempts to overview a rapidly developing field of nanotoxicology, and to sum up and reflect upon recent experimental findings in this field.

3.2
Sources of Nanoparticles

Nanoparticles can be classified into three groups: natural, anthropogenic and man-made (or artificial). The natural kind is produced, for example, during forest fires

Nanotechnologies for the Life Sciences Vol. 5
Nanomaterials – Toxicity, Health and Environmental Issues. Edited by Challa S. S. R. Kumar
Copyright © 2006 WILEY-VCH Verlag GmbH & Co. KGaA, Weinheim
ISBN: 3-527-31385-0

or volcanic eruptions; anthropogenic particles are quite often a by-product of industrial activities like welding or polishing. Diesel exhaust particles are also placed in this group. The last group includes engineered nanomaterials deliberately produced because of their technologically beneficial properties caused by the reduction in particle size. These novel properties of common materials observable only at nano-scale dimensions already have commercial applications [4]. For example, nanomaterials can be found in sunscreens, toothpastes, sanitary ware coatings and even food. The production volumes of man-made nanoparticles range from the multi-ton for carbon black and fumed silica used in plastic fillers and car tires to the microgram quantities of fluorescent quantum dots used as markers in biological imaging.

Following massive investments [5, 6], efforts to exploit the unique properties of everyday materials at the sub-micrometer scale are truly world-wide [7, 8] and consumer products relying on nanotechnology will experience a steady growth [9].

3.3
Epidemiological Evidence

Man-made nanoparticles are a relatively recent phenomena. As no data are available yet to evaluate the long-term risks of engineered nanoparticles, the epidemiological evidence on adverse health effects of ultrafine particles will be overviewed as a surrogate source of information.

Environmental air pollution consists of a complex mixture of compounds in gaseous, liquid and solid phases, the latter usually referred to as particulate matter (PM). In general, ambient levels of particulate matter are characterized as total suspended matter (TSP), and particulate matter with an effective aerodynamic diameter of less than 10 μm (PM$_{10}$) or 2.5 μm (PM$_{2.5}$). Particles in the sub-micrometer ranges, particularly in the range < 100 nm, are labeled as ultrafine particles in epidemiological studies. Ultrafine particles in ambient air vary in chemical composition and size, as do technically produced nanoparticles. The number concentration of these small particles exceeds by far that of larger ones in urban area, but their contribution to the total mass concentration is relatively low. Therefore, it is standard to measure PM$_{10}$ and PM$_{2.5}$ in mass concentration (μg m^{-3}). For ultrafine particles, the number concentration (cm^{-3}) or surface area concentration (μ^2 m^{-3}) or particle length concentration (mm cm^{-3}) is more relevant. Particles in ambient air are generated by numerous sources: motor vehicles, power plants, wind blown dust, photochemical processes, cigarette smoking, nearby quarry operation, etc. Some particles are introduced from the source into the air in solid or liquid form, while others are formed in the air by gas into particle conversion.

In the United States, the Environmental Protection Agency set National Ambient Air Quality Standards for particulate matter. According to the 1987 standard of PM$_{10}$, the maximal allowable 24-hour concentration was set at 150 μg m^{-3} and the maximal annual mean was set at 50 μg m^{-3}. From 1988 to 1993, the averages of the annual mean PM$_{10}$ concentrations at 799 sites monitored by the US EPA declined by 20%. Despite these improvements in air quality, Samet and coworkers [10] reported associations between particle concentrations and the number of deaths per day in 20 of the largest cities and metropolitan areas in the United States from 1987 to 1994 with mean 24-hour PM_{10} concentrations well below the standard. Analysis of the daily number of deaths occurring within an urban region has shown that 10 $\mu g\ m^{-3}$ PM_{10} were associated with an increase of 0.2%. The result is based on a recent reevaluation of the National Mortality Morbidity Air Pollution Study (NMMAPS) that included 90 urban areas of United States in these analyses [11]. In 29 European cities, an increase of 0.6% in daily mortality was observed in association with an increase of 10 $\mu g\ m^{-3}$ in the study by the Air Pollution and Health Effect Association (APHEA) [12]. Studies on particles mass concentration indicate that there is a linear relationship between PM_{10} and $PM_{2.5}$ and various health indicators (like cough, symptom exacerbation, bronchodilator use, hospital admissions and mortality [13]) for concentration levels between 0 and 200 $\mu g\ m^{-3}$, and no threshold in particle concentrations below which health would not be jeopardized.

Within most established monitoring networks, ambient particulate matter is measured as either PM_{10} or $PM_{2.5}$. The epidemiological research has therefore focused on the links between these mass characteristics of ambient particles and adverse health effects. However, with reductions in particulate emissions from industry and power stations, the relevance of the number concentrations of ultrafine particles increased (mainly from traffic emissions). Not much was known about their impact on health. Panel morbidity studies with asthmatic subjects indicated that both fine and ultrafine particles were negatively associated with the respiratory health of the exposed population [14]. A decrease of respiratory functions, e.g., peak expiratory flow [15], and an increase in symptoms and medication use [16], was associated with elevated particle concentrations of ultrafine particles, independently from fine particles. Inflammatory events in the lungs took several days to develop. It was considered as likely that a lag time existed between exposure to ultrafine particles and the acute respiratory health effects of the exposed population. Cumulative effects over 5 days seemed to be stronger than same-day effects. There was an indication that the acute effects of the number of ultrafine particles on respiratory health were stronger than those of the mass of the fine particles [17, 18].

To improve our knowledge on human exposure to particulate matter of different sizes and of different chemical composition in Europe, and to develop standards for air quality in Europe, the ULTRA project was initiated. Specifically, the project aimed to improve exposure assessment to fine particles by assessing the size distributions, including ultrafine particles, and elemental compositions of fine particles in ambient air in three European cities with different sources of particulate air pollution. Three panel studies were carried out, in Amsterdam, the Netherlands, Erfurt, Germany, and Helsinki, Finland, during winter and spring 1998–1999 [19–21]. In all three cities, about 50 elderly persons with coronary heart disease were followed up for six months with bi-weekly intensive examinations, which included measurements of the function of the heart and lungs, blood pressure and of biomarkers for lung damage from urine. The subjects also kept daily symptom diaries. These studies were limited to the investigation of the acute health effects of short-term exposure by evaluating the impact of day-to-day variation in ambient pollution on health through correlating mortality and morbidity with daily pollution levels. There is an association between exposure to ultrafine particles and cardiovascular morbidity in the population with chronic heart diseases. In Helsinki [22] independent associations between both fine and ultrafine particles and the risk of ST-segment depression in their ECG were observed among subjects with coronary heart disease. ST-segment depression is regarded as an indicator of myocardial ischemia. The study reported increased odds ratios for 45 subjects, ranging from 1.03 to 3.29, with 95% confidence intervals ranging from 0.54 to 6.32. Several plausible mechanistic pathways have been described, including enhanced coagulation/thrombosis, a propensity for arrhythmias, acute arterial vasoconstriction, systemic inflammatory responses, and the chronic promotion of atherosclerosis [23].

A study conducted in Erfurt, Germany, on daily mortality showed comparable and independent increases in mortality in association with fine and ultrafine particles [24]. All particles had a strong seasonal dependency, with maximal concentrations in winter. The concentrations of ultrafine particles showed a pronounced day of the week effect with concentrations during the weekend 40% lower than during the week. This and a clear increase of the ultrafine particles concentrations during the rush hours suggest that the main source for ultrafine particles was automobile traffic. Associations between health effects and particle number and particle mass concentrations have been observed in different size classes, and both immediate effects (lags 0 or 1 days) and delayed effects (lags 4 or 5 days) were found. The effects could be found for total mortality and also for respiratory and cardiovascular causes. There was a tendency for more immediate effects on respiratory causes and more delayed effects for cardiovascular causes. Mortality increased in association with ambient particles after adjustment for season, influenza epidemics, day of week and meteorology, and sensitivity analyses showed the results to be stable.

In summary, both fine and ultrafine particles are associated with respiratory and cardiovascular morbidity and mortality and appear to be so independently of each other. There is also epidemiological evidence of similar responses to fine and ultrafine particles, although the size of the effects is often larger for ultrafine than for fine particles (at least on a per mass basis). One can expect that similar effects can be induced due to the presence of man-made nanoparticles.

3.4
Entry Routes into the Human Body

The above-mentioned health effects result from the inhalation of ultrafine particles. In general, compounds or materials can enter the body via three "natural" portals: skin, intestinal tract and respiratory tract (nose, airways and alveoli), or via intentional delivery through injection, intravenous (i.v.), intraperitoneal (i.p.) or intramuscular (i.m.). Although our knowledge in this field is partly built on studies concerning drug delivery (pharmaceutical research) and toxicology (xenobiotics) of an intentional dose, in this chapter we will mainly concentrate on the health effects of nanomaterials entering the body via one of the natural portals.

The skin acts as a strict barrier between the body and the environment; no essential elements are taken up through the skin (except solar radiation necessary to build up vitamin D).

The respiratory tract and the intestinal tract allow transport (passive and/or active) of various substances like water, nutrients and gasses. The lungs exchange oxygen and carbon dioxide with the environment, and some water escapes with the warm exhaled air. The intestinal tract is in close contact with all the materials taken up orally; here all nutrients (except gasses) are exchanged between the body and the environment. The anatomy and histology of the three organs in contact with the environment differ significantly.

The skin of an adult human is roughly 1.5 m^2 in area, and is at most places covered with a relatively thick first barrier (10 μm) built of strongly keratinized dead cells. This first barrier is difficult to pass for ionic compounds as well as water-soluble molecules.

The respiratory tract consists of three different parts: nose, airways (transporting the air in and out the lungs) and alveoli (gas exchange areas). The nose and the airways are a relatively robust barrier, built of an active epithelium protected with a viscous layer of mucus. In the gas exchange area, the barrier between the alveolar wall and the capillaries is very thin. The air in the lumen of the alveoli is only 0.5 μm (500 nm) away from the blood flow. The large surface area of the alveoli, 140 m^2 in adults, and the intense air–blood contact in this region make the alveoli less well protected than the airways against environmental damage.

The intestinal tract is a more complex barrier – exchange side, it is the portal for macromolecules to enter the body. From the stomach, only small molecules can diffuse through the epithelium. The epithelium of the small and large intestines, in close contact with ingested material, allows and controls the uptake of nutrients such as disaccharides, peptides, fatty acids, and monoglycerides generated by digestion. The overall surface available to exchange nutrients is about 200 m^2 in adults.

In the following sub-sections, interactions of the three portals with nanomaterials are briefly discussed. Two critical aspects, from the health effect point of view, will be discussed for each. First, how can nanomaterials have a local effect in each of these organs and, second, can nanomaterials move from the portal into the body.

3.4.1
Lung

3.4.1.1 Inhalation, Deposition and Pulmonary Clearing of Insoluble Solids

Inhalation and Deposition The deposition of solid material in the respiratory tract depends on the physical characteristics of the material, such as particle size and shape, relative weight, and on the anatomy of the respiratory tract, such as diameter of the airways, air speed, branching angle etc. [25–27].

Spherical solid material can be inhaled when its aerodynamic diameter is less than 10 μm. The smaller the particulates the deeper they can travel into the lung – particles < 2.5 μm will even reach the alveoli. Ultrafine particles (nanoparticles with an aerodynamic diameter of less than 100 nm) are deposited mainly in the alveolar region, largely by diffusion (Brownian movement). Models have shown that the deposition efficiency at the three pulmonary regions is not linear with size: particles of between 5 and 50 nm are deposited mainly in the alveoli, smaller and larger ones are more efficiently deposited in the higher regions [25, 28].

Fibers are defined, in pulmonary sciences, as solid materials with a length-to-diameter ratio of at least 3:1. Their aerodynamic diameter can be used to judge their penetration into the lungs. Fibers with a small diameter will penetrate deeper into the lungs, while very long fibers (≫20 μm) are easily stuck in the higher airways, although some long fibers can enter the alveolar space [29–34].

Clearance The removal of solid material from the lungs is carried out by two distinct mechanisms. The mucociliary escalator dominates the clearance from the airways and the nose; in the alveolar region the clearance takes predominantly place by macrophage phagocytosis.

The mucociliary escalator, driven by the cilia of airway epithelium, is an efficient transport system, pushing the mucus, which covers the airways, together with the trapped solid materials towards the mouth.

The phagocytosis of particles and fibers results in activation of macrophages and induces the release of chemokines, cytokines, reactive oxygen species, and other mediators; this can lead to sustained inflammation and eventually fibrotic changes [35, 36]. The phagocytosis efficiency can be affected by the (physical-chemical) characteristics of the solid material [37] (see below); moreover, fibers too long to be phagocytized (fibers longer than the diameter of the alveolar macrophage, depending on the species studied) will not (or very slowly) be cleared [32, 38–40].

Laboratory exposure studies have shown that if the inhaled concentrations are low, such that the deposition rate of the inhaled particles is less than the clearance rate, then the retention half-time is about 70 days. For fine and nanoparticulates, the alveolar macrophage-mediated clearance is the limiting factor. If the deposition rate of the inhaled particles exceeds this clearance rate, the retention half-time is significantly increased, reflecting an impaired or prolonged alveolar macrophage-mediated clearance function with continued accumulation of lung burden (overload) [41–43].

Clearance from the lung depends not only on the total mass of particles inhaled but also on the particle size and, by implication, on particle surface, as shown in the following studies. A sub-chronic 3 months inhalation exposure of rats to ultrafine (∼20 nm) and fine (∼200 nm) titanium dioxide (TiO_2) particles demonstrated that the ultrafine particles cleared significantly slower and showed more translocation to interstitial sites and to regional lymph nodes than the fine TiO_2 particles [25].

To summarize, most nanosized spherical solid materials are likely to enter the lungs and reach the alveoli. These particles can be cleared from the lungs, as long as the clearance mechanisms are not affected by the particles themselves or by any other cause. Nanosized particles are more likely to hamper the clearance, resulting in a higher burden [44], possibly amplifying any related chronic effects caused by these particles. Notably, specific particle surface area is probably a better indication for maximum tolerated exposure level than total mass [28, 45], suggesting that the biological effects are linked to surface reactivity.

3.4.1.2 Biopersistence of Inhaled Solid Material

The main determinants of biopersistence are species-specific physiological clearance and material specific bio-durability (physical-chemical processes).

In the alveoli, the rate at which fibers are cleared depends on the ability of alveolar macrophages to phagocytose them. Macrophages containing fibers longer than their own diameter (in humans longer than 20 μm) may not be mobile and will be unable to clear the fibers from the lung [39].

The bio-durability of a fiber depends on its dissolution and leaching as well as mechanical breaking and splitting [46, 47]. Biopersistent fibers such as amosite asbestos (brown asbestos) [39, 48], where breakage occurs longitudinally, result in more fibers of the same length but smaller diameter. Other types of fibers (e.g., amorphous) break perpendicular to their long axis, resulting in fibers that can be engulfed by the macrophages [49].

Self-evidently, the slower the fibers are cleared (high biopersistence), the higher is the tissue burden and the longer the fibers reside in a tissue the higher is the probability of an adverse response [29]. Despite the crucial role played by the length of the fibers (Stanton hypothesis) [50], it does not strictly indicate that all fibers longer than the lower threshold are equally active or that shorter fibers are not. Although fibers less than 5 μm long did not appear to contribute to lung cancer risk in exposed rats [39], fibers more than 40 μm long impose the highest risk (recent review by Schins [38]).

Inhaled fibers, which are persistent in the alveoli, can further interact with the pulmonary epithelial cells or even penetrate the alveolar wall and enter the lung tissue. These fibers are often described as being in the "interstitial" because they may lie between or within the cells making up the alveolar walls. Biopersistent solid materials, certainly those containing mutagenic potency and which remain for years in the lungs, increase the risk of developing cancer [32, 33].

Not much is known on the long-term health effects of fibrous purpose-made nanomaterials. There are no indications that the bio-durability of fibers with a diameter < 100 nm will differ from larger inhalable fibers. Therefore, great caution must be taken in the case of contact with nanofibers; bio-durability tests must be performed before releasing any products containing them.

Technologically, carbon nanotubes are an important group of nanofibers. Recently, they have been reported to show signs of toxicity in the lung of laboratory animals [51]. This is confirmed in two independent publications, by Warheit et al. [52] and Lam et al. [53], which demonstrated the pulmonary effects of single-walled carbon nanotubes *in vivo* after intratracheal instillation, in both rats and mice. Both groups reported granuloma formation, and some interstitial inflammation. Warheit et al. [52] concluded that these findings (multifocal granulomas) may not have physiological relevance, and may be related to the instillation of a bolus of agglomerated nanotubes. The other group [53] suggested that if carbon nanotubes reach the lungs they are much more toxic than carbon black and can be more toxic than quartz. These studies have to be read with some caution because a study by the National Institute for Occupational Safety and Health (NIOSH) showed that none or only a small fraction of the nanotubes present in the air can be inhaled [54] (see short review of Donaldson and Tran [55]).

As noted above, at similar lung burdens, spherical TiO_2 ultrafine particles cleared significantly more slowly from the alveoli and showed more translocation to interstitial sites and to regional lymph nodes than did fine particles. Thus, besides the greater biological effects (see below) of ultrafine particles, the difference in toxicokinetics in the lung results in a higher burden [44].

3.4.1.3 Systemic Translocation of Inhaled Particles

The impact of inhaled particles on other organs has been reported in several epidemiological studies. Most research has concentrated on the possible consequences of particle related malfunction of the cardiovascular system, such as arrhythmia, coagulation [56] etc. However, the autonomic nervous system [57, 58] as well as the olfactory nerves may be a target for inhaled particulates [28].

Until recently, the possible passage of xenobiotic particles has not attracted much attention, although the concept is now gaining acceptance in pharmacology for the administration of macromolecular drugs by inhalation [59].

In evaluating the health effects of inhaled nanoparticles, translocation to the systemic circulation is an important issue. Several para- and trans-cellular mechanisms have been described in the pulmonary epithelium, but it is unclear which one allows the translocation of nanoparticulates. Conhaim and coworkers [60] found that the lung epithelial barrier was best fitted by a three-pore-sized model, including a small number (2%) of large-sized pores (pore radius 400 nm), an intermediate number (30%) of medium-sized pores (40-nm pore radius), and a very large number (68%) of small-sized pores (1.3-nm pore radius). The exact anatomical location of this structure, however, remains to be established (see the review by Hermans and Bernard [61]). Possible endocytic pathways have been reviewed by Rejman et al. [62]. Of all endocytic pathways caveolae seem (the most) important portals for large molecules to enter cells or to cross the epithelial border [63–65]. Caveolae allow internalization of particles as large as 500 nm in diameter, though it depends on the surface coating.

In humans, translocation of inhaled ultrafine technetium (^{99m}Tc) labeled carbon particles into the blood circulation has been studied independently by Nemmar et al. [66] and Kawakami et al. [67]. However, the translocation mechanism is still unclear. Nemmar et al. demonstrated that technetium (^{99m}Tc) labeled carbon particles, which are very similar to the ultrafine fraction of actual pollutant particles, diffused rapidly – within 5 min – into the systemic circulation [66]. The authors concluded, therefore, that it was unlikely that phagocytosis by macrophages and/or endocytosis by epithelial and endothelial cells are solely responsible for particle translocation to the blood, but that a paracellular mechanism probably also plays a role. More recently, Kato et al. [68] showed, morphologically, that inhaled polystyrene particles are transported into the pulmonary capillary space, presumably by transcytosis.

Aerosolized insulin gives a rapid therapeutic effect [69], although the pathways for this translocation are still unclear [70]. In addition to human studies, extra-pulmonary translocation of ultrafine particles after intratracheal instillation or inhalation has been reported in experimental animal studies [66, 71–73]. However, the amount of ultrafine particles that translocate into blood and extra-pulmonary organs was different. Following intranasal delivery, polystyrene microparticles (1.1 μm) can translocate to tissues in the systemic compartment [74].

Oberdörster et al. have explored another alley of translocation from the respiratory tract towards other organs [31]. In inhalation experiments with rats, using ^{13}C-labeled particles, they found that nanosized particles (25 nm) were present in several organs 24 hours after exposure. The most extraordinary finding was the discovery of particles in the central nervous system (CNS). The authors examined this phenomenon further and found that particles, after being taken up by the nerve cells, can be transported via nerves (in this experiment via the olfactory nerves) at 2.5 mm h^{-1} [72].

Passage of solid material from the pulmonary epithelium to the circulation seems not to be restricted to nanoparticles, as shown by Kato et al. [68], and depends on the surface characteristics of the material. The issue of particle translocation still needs to be clarified: both the trans-epithelial transport in the alveoli and the transport via nerve cells. Thus, the role of factors governing particle translocation, such as the way of exposure, dose, size, surface chemistry and time course, should be investigated. For instance, it would be very important to know how and to what extent the extra-pulmonary translocation of particles is modulated by the lung inflammation.

3.4.2
Intestinal Tract

3.4.2.1 Deposition and Translocation
Already in 1926, Kumagai recognized that particles could translocate from the lumen of the intestinal tract via aggregations of intestinal lymphatic tissue (PP) containing M-cells (specialized phagocytic enterocytes). Particulate uptake happens not only via the M-cells in the PP and the isolated follicles of the gut-associated lymphoid tissue but also via the normal intestinal enterocytes. There have been several excellent reviews on the intestinal uptake of particles [75, 76]. Uptake of inert particles occurs trans-cellulary through normal enterocytes and in PP via M-cells, and, to a lesser extent, across paracellular pathways [77]. Initially it was assumed that the PP did not discriminate strongly in the type and size of the absorbed particles. Subsequently, it has been shown that modifying characteristics, such as particle

size [78] the surface charge of particles [79, 80], attachment of ligands [81, 82] or coating with surfactants [83], offers possibilities of site-specific targeting to different regions of the gastrointestinal tract (GIT), including the PP [84].

The kinetics of particle translocation in the intestine depends on diffusion and accessibility through mucus, initial contact with enterocyte or M-cell, cellular trafficking, and post-translocation events. Charged particles, such as carboxylated polystyrene nanoparticles [80] or those composed of positively charged polymers, exhibit poor oral bioavailability [85].

Specific studies on nanomaterials are rather scarce; in general, they show that most of them simply pass through the GIT and are rapidly eliminated. In one study [79], the body distribution after translocation of polystyrene particles was examined in some detail. Polystyrene spheres (ranging from 50 nm to 3 μm) were fed by gavage to female Sprague–Dawley rats daily for 10 days at a dose of 1.25 mg kg^{-1}. As much as 34% and 26% of the 50 and 100 nm particles, respectively, were absorbed. Those larger than 300 nm were absent from blood. No particles were detected in heart or lung tissue. In another study, the oral uptake of radiolabeled functionalized C_{60} fullerenes (water solubilized using albumin and PEG) in rats resulted in a 98% clearance (faeces) within 48 h, and the rest was eliminated via the urine, which is an indication for systemic uptake [86]. Kreyling et al. [71], using ultrafine ^{192}Ir, did not find any significant nanoparticle uptake in the GI tract.

3.4.2.2 Intestinal Translocation and Disease

Crohn's disease is characterized by transmural inflammation of the gastrointestinal tract. It is of unknown aetiology, but it is suggested that a combination of genetic predisposition and environmental factors play a role. Particles (0.1–1.0 μm) are associated with the disease [87] and indicated as potent adjuvants in model antigen-mediated immune responses. A double-blind randomized study showed that a diet low in calcium and exogenous microparticles alleviates the symptoms of Crohn's disease [88].

Other studies found that material uptake (endocytosis) capacity of M cells is induced under various immunological conditions, e.g., a greater uptake of particles (0.1, 1 and 10 μm diameter) has been demonstrated in the inflamed colonic mucosa of rats compared to non-ulcerated tissue [89, 90] and inflamed esophagus [91].

Clearly, from the literature cited above, engineered nanoparticles can be taken up via the intestinal tract. In general, the intestinal uptake of particles is better understood and studied in more detail than pulmonary and skin uptake. Because of this advantage, it may be possible, with caution, to predict the behavior of some particles in the intestines.

3.4.3
Skin

3.4.3.1 Deposition and Penetration through the Skin

Skin is an important barrier, protecting against insult from the environment. The skin is structured in three layers: the epidermis, the dermis and the subcutaneous layer. The outer layer of the epidermis, the stratum corneum (SC), covers the entire outside of the body. In the SC we find only dead cells, which are strongly keratinized. For most chemicals, the SC is the rate-limiting barrier to percutaneous absorption (penetration). The skin of most mammalian species is covered with hair on most parts of the body.

At the sites where hair follicles grow, the barrier capacity of the skin differs slightly from the "normal" stratified squamous epidermis [92]. Most studies concerning penetration of materials into the skin have focused on whether drugs penetrate through the skin using different formulations containing chemicals and/or particulate materials as a vehicle [93]. The main types of particulate materials commonly used are liposomes, solid poorly soluble materials such as TiO_2, polymer particulates, and submicron emulsion particles, such as solid lipid nanoparticles. The penetration of these particulate carriers has not been studied in detail.

TiO_2 particles are often used in sunscreens to absorb UV light and therefore to protect skin against sunburn or genetic damage. Lademann et al. have reported [94] that micrometer-sized particles of TiO_2 get through the human stratum corneum and even into some hair follicles, including their deeper parts.

Tinkle et al. have demonstrated that 0.5 and 1.0 µm particles, in conjunction with motion, penetrate the stratum corneum of human skin and reach the epidermis and, occasionally, the dermis [95]. It has been hypothesized that the lipid layers within the cells of the stratum corneum form a pathway by which the particles can move [96] into the skin and be phagocytized by the Langerhan's cells. In this study, the penetration of particles was limited to a particle diameter of 1 µm or less. Nevertheless, other studies reported penetration through the skin to the dermis using particles with diameters of 3–8 µm [92, 94, 97] but only limited penetration was found, often clustered at the hair follicle (see above). This can lead to an interaction with the immune system [93].

Penetration of non-metallic solid materials such as biodegradable poly(D,L-lactic-*co*-glycolic acid) (PLGA) microparticles, 1 to 10 µm with a mean diameter of 4.61 ± 0.8 µm, has been studied after application on porcine skin. The number of microparticles in the skin decreased with depth (measured from the airside towards the subcutaneous layer). At 120 µm depth (where viable dermis is present) a relatively high number of particles was found, at 400 µm (dermis) some microparticles were still seen. At a depth of 500 µm no microparticles were found [98]. In the skin of individuals who had an impaired lymphatic drainage of the lower legs, soil microparticles, frequently 0.4–0.5 µm, were found, and particles as large as 25 µm in diameter were seen in the dermis of the foot of a patient with endemic elephantiasis. The particles are seen to be in the phagosomes of macrophages or in the cytoplasm of other cells. The failure to conduct lymph to the node produces a permanent deposit of silica in the dermal tissues (a parallel is drawn with similar deposits in the lung in pneumoconiosis). This indicates that soil particles penetrate through (damaged) skin, most probably in every individual, and normally are removed via the lymphatic system [99, 100].

According to Hostynek [101] the uptake of metals through the skin is complex. Both exogenous (e.g., dose, vehicle, protein reactivity, and valence) and endogenous factors (e.g., age of skin, anatomical site, and homeostatic control) are involved. Attempts to define rules governing skin penetration to give predictive quantitative structure–diffusion relationships for metallic elements for risk assessment purposes have been unsuccessful, and penetration of the skin still needs to be determined separately for each metal species, either by *in vitro* or *in vivo* assays.

From the limited literature on nanoparticles penetrating the skin some conclusions can be drawn. Firstly, penetration of the skin barrier is size dependent – nanosized particles are more likely to enter more deeply into the skin than larger ones. Secondly, different types of particles are found in the deeper layers of the skin and, at present, it is impossible to predict the behavior of a particle in the skin. Thirdly, materials that can dissolve or leach from a particle (e.g., metals) can possibly enter the systemic circulation.

3.4.3.2 **Irritation of Skin**

Glass fibers and Rockwool fibers are widely used man-made mineral fibers, mainly as thermal insulation materials, which have become important as a replacement for asbestos fibers. While in contact with the skin, these fibers can induce dermatitis simply resulting from mechanical irritation. Why these fibers are such a strong irritant has not been examined in detail. In occlusion irritant patch tests in humans, Rockwool fibers with a diameter of 4.20 ± 1.96 μm were found to be more irritating than those with a mean diameter of 3.20 ± 1.50 μm [102].

Some recent experimental work that exposed human epidermal keratinocytes (HEK) to carbon nanotubes (CNT) indicates that caution should be taken in handling these materials [103, 104]. Single-wall (SW) CNT induce apoptosis and decrease cellular adhesion ability, in a dose and time dependent manner [104]. Multi-wall (MW) CNT are taken up by the HEK, in vacuoles present within the cytoplasm, but no CNT were found in the nuclei of the cells [103]. In the same study the release, into the culture medium, of IL-8, a marker of irritation in human skin, was found to be dose dependent.

3.5
What Makes Nanoparticles Dangerous?

Several mechanisms have been proposed to explain the adverse health effects of nanomaterials. In "nanotoxicology", probably two distinct characteristics would play a role: on the one hand, the material-specific and intrinsic toxicity and, on the other hand, more general but specific nanoparticle-induced responses [1, 45].

Material-specific responses can often be understood and/or explained by material-specific toxic responses, local stimulation of irritant receptors, covalent modification of key enzyme receptors, etc. More general nanoparticle dependent responses, certainly in lung and liver, can often be categorized as inflammatory responses concurrent with cytokine and chemokine release, production of white blood cells, free-radical production, etc. [28, 40]. Certainly, it would be incorrect to separate these two responses too much as, most often, after exposure to a nanomaterial multiple responses can be observed that can influence each other.

Another aspect, which has not been studied in any detail in respect to nanomaterials, is the deposition of nanomaterials at any specific sink in the body [71, 72]. In kidney toxicity, the precipitation of chemicals and formation of crystals, certainly in chronic exposure, can lead to tissue damage [105].

The next section discusses some material characteristics and toxic mechanisms important in the adverse health effects of nanomaterials.

3.5.1
Particle Size – Surface and Body Distribution

Reports on the surface properties of nanoparticles, both physical and chemical, stress that nanoparticles differ from bulk materials. The biological effects do not just depend on the intrinsic toxicity of the material itself but on the size and surface area the nanoparticles made out of this material. Nanoparticles are not merely small crystals but an intermediate state of matter placed between bulk and molecular material. Independently of the particle size, two other parameters play dominant roles: the charges carried by the particle in contact with the cell membranes and the chemical reactivity of the particle [28, 45, 106–108].

3.5.1.1 Effect of Size

Two samples of carbon black, which can be considered as a relatively inert material, of similar size and composition but with significantly different specific surface areas (300 versus 37 $m^2 g^{-1}$) showed biological effects (inflammation, genotoxicity, and histology) that depend on the specific surface area and not on particle mass. Similar findings were reported in earlier studies on tumorigenic effects of inhaled particles. In the lung, tumor incidence of chronically inhaled TiO_2 of nanosized particles (20 nm diameter) at low exposure (10 mg m^{-3}) was significantly higher than for high exposure (250 mg m^{-3}) of 300 nm particles [109]. Tumor incidence correlates better with specific surface area than with particle mass [25, 110]. *In vivo* and *in vitro*, nanosized particles inhibit phagocytosis when compared to fine particles [111] and can change the chemotactic behavior of macrophages significantly [112].

Size is also a critical parameter in the distribution of particles in the body.

Oral uptake (gavage) of polystyrene spheres of different sizes (50 nm to 3 μm) in female Sprague–Dawley rats (for 10 days at a dose of 1.25 mg kg^{-1} day^{-1}) resulted in systemic distribution of the nanoparticles. About 7% (50 nm) and 4% (100 nm) was found in the liver, spleen, blood and bone marrow. Particles larger than 100 nm did not reach the bone marrow and those larger than 300 nm were absent from blood. No particles were detected in heart or lung tissue [80].

3.5.1.2 Effect of Surface Charges

Beside particle size, surface characteristics play a dominant role in the distribution of material in the body.

Coating poly(methyl methacrylate) nanoparticles with different types and concentrations of surfactants significantly changes their body distribution [113]. Coating these nanoparticles with $\geq$0.1% poloxamine 908, a non-ionic surfactant, reduces their liver concentration significantly (from 75% to 13% of total amount of particles administrated) 30 min after i.v. injection. Another surfactant, polysorbate 80, was effective above 0.5%. A different report showed that modification of the nanoparticle surface with a cationic compound, didodecyldimethylammonium bromide (DMAB), facilitates the arterial uptake 7–10-fold [114]. The authors noted that the DMAB surface-modified nanoparticles had a mean zeta potential of +22.1 mV, which is significantly different from the original -27.8 ± 0.5 mV (mean $\pm$ sem, $n = 5$). The mechanism for the altered biological behavior is unclear, but surface modifications have possible applications for intra-arterial drug delivery.

Polycationic macromolecules show a strong interaction with cell membranes *in vitro*. The Acramin F textile paint system is a good example. Three polycationic paint components exhibited considerable cytotoxicity (LC_{50} generally below 100 mg mL^{-1} for an incubation of 20–24 h) in diverse cell cultures, such as primary cultures of rat and human type II pneumocytes, and alveolar macrophages and human erythrocytes. The multiple positive charges play, speculatively, an important role in the toxic mechanism [115, 116].

A study of the biocompatibility (cytotoxicity) of polycationic materials [117] as a function of molecular weight found that with increasing molecular weight some macromolecules, such as DEAE-dextran and poly-l-lysine (PLL) [118, 119], dendrimers [120] and polyethylenimine (PEI) [121], become more toxic. The toxic mechanism is not fully understood but membrane integrity plays a role.

Dekie et al. [122] concluded that a primary amine group on poly(L-glutamic acid) derivatives has a significant toxic effect on red blood cells, causing them to agglutinate. Not only the type of amino function but also the charge density resulting from the number and special arrangement of the cationic residues is important for cytotoxicity. Ryser [123] has suggested that a three-point attachment is necessary to elicit a biological response on cell membranes, and speculated that the activity of a polymer will decrease when the space between reactive amine groups is increased. The arrangement of cationic charges depends on the three-dimensional structure and flexibility of the macromolecules and determines the accessibility of their charges to the cell surface.

Branched molecules are more efficient in neutralizing the cell surface charge than polymers with linear or globular structure; the latter are more rigid and so have more difficulty attaching to the membranes [124]. Therefore, high cationic charge densities and highly flexible polymers should cause higher cytotoxic effects than those with low cationic charge densities. Globular polycationic polymer structures [catonised Human Serum Albumin (cHSA), ethylenediamine-core poly(amidoamine) dendrimers (PAMAM)] exhibit good biocompatibility (low cytotoxicity) whereas polymers with a more linear or branched and flexible structure [poly(diallyldimethylammonium chloride) (DADMAC), PLL, PEI] showed higher cell-damaging effects.

The serum half-life and body distribution of CdSe quantum dots with different surface characteristics, coatings with short-chain or long-chain PEG, have been studied by Ballou et al. [125]. The mPEG-750 coated quantum dots were, 24 h after dosage, found in lymph nodes and the spleen. The long-chain (PEG-5000) coated quantum dots were less apparent in lymph nodes but more in the liver, spleen, and bone marrow. This type of coating allowed a slow clearance from the body and the particles were still observed after 133 days.

Regardless of uptake route, the body distribution of particles is most dependent on the surface characteristics and size of the particles. This is important in drug design in order to help to deliver medication to the right target.

3.5.2
Nanoparticles, Thrombosis and Lung Inflammation

3.5.2.1 Prothrombotic Effect

Epidemiological studies have reported a close association between particulate air pollution and cardiovascular adverse effects [126] such as myocardial infarction [127]. The latter results from rupture of an atherosclerotic plaque in the coronary artery, followed by rapid thrombus growth caused by exposure of highly reactive sub-endothelial structures to circulating blood, thus leading to additional or complete obstruction of the blood vessel [127].

Nemmar et al. have studied the possible effects of particles on haemostasis, focusing on thrombus formation as a relevant endpoint [128–130]. Polystyrene particles 60 nm in diameter (surface modifications: neutral, negatively or positively charged) had a direct effect on haemostasis after intravenous injection. Positively charged amine-particles led to a marked increase in prothrombotic tendency, resulting from platelet activation. These observations have been confirmed recently by Silva et al. [131] in a comparable model.

A similar effect could be obtained after the intratracheal administration of these positively charged polystyrene particles, which also caused lung inflammation [132]. Importantly, the pulmonary instillation of larger (400 nm) positive particles caused a definite pulmonary inflammation (of similar intensity to 60 nm particles), but they did not lead to a peripheral thrombosis within the first hour of exposure. This lack of effect of the larger particles on thrombosis, despite their marked effect on pulmonary inflammation, suggests that pulmonary inflammation by itself was insufficient to influence peripheral thrombosis. Consequently, the effect found with the smaller, ultrafine particles is most probably due, at least in part, to their systemic translocation from the lung into the blood.

Using pollutant particles, namely diesel exhaust particles (DEP), it was shown that, within an hour after their deposition in the lungs, DEP cause a marked pulmonary inflammation. Moreover, intratracheal instillation of DEP promotes femoral venous and arterial thrombosis in a dose-dependent manner, already starting at a dose of 5 µg per hamster (ca. 50 µg kg^{-1}). Subsequent experiments showed that prothrombotic effects persisted at 6 and 24 h after instillation (50 µg per animal) and confirmed that peripheral thrombosis and pulmonary inflammation are not always associated [129, 133].

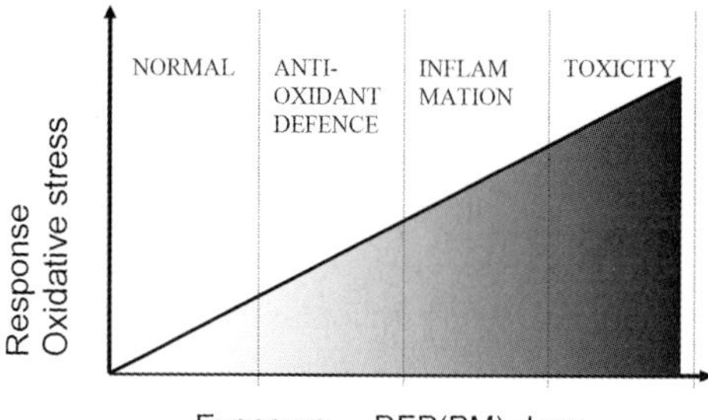

Fig. 3.1. Hierarchic oxidative stress model (Bernstein et al. [137] and Li et al. [134]). At low concentrations of PM a small change in oxidative stress can be observed without the induction of changes at cellular or tissue level. At a higher level of exposure, anti-oxidant defense mechanisms are triggered. With increasing concentrations, an inflammatory response is induced. Finally, at the highest concentration, toxicity and cellular and/or tissue damage can be observed.

3.5.2.2 Oxidative Stress, Inflammation and Endotoxins

Particle-induced pulmonary inflammation can induce protective and adverse cellular responses in a dose-dependent manner. Li et al. have proposed a "hierarchical oxidative stress model" in response to DEP exposure (Fig. 3.1) [134]. This model suggests that at a lower oxidative stress level (tier 1) PM induces cytoprotective responses, e.g., through the activation of antioxidant response elements, inducing the expression of several antioxidant and phase II drug metabolizing enzymes (e.g., heme oxygenase 1 and glutathione-S-transferase). If this level of protection fails, the oxidative stress (tier 2) will lead to mitogen-activated protein kinase/ nuclear factor kB activation and pro-inflammatory effects. Further escalation (tier 3), will trigger disturbance of the mitochondrial function, resulting in cellular apoptosis or necrosis. A weakened antioxidant defense can increase the susceptibility toward PM-induced airway inflammation, to infection, and maybe to asthma; it can explain the existence of susceptible human subsets. Xioa et al. showed that the hierarchical oxidative stress model can be applied in a macrophage cell line [135]. The authors demonstrated that in the dose range 10–100 g mL^{-1} organic DEP extracts induce a progressive decline in the cellular GSH/GSSG ratio. In parallel, it causes a linear increase in newly expressed proteins, including antioxidant enzymes (e.g., heme oxygenase-1 and catalase), pro-inflammatory components (e.g., 38 MAPK and Rel A), and products of intermediary metabolism that are regulated by oxidative stress.

In vivo in rats, Arimoto et al. showed that exposure to DEP and Lipopolysaccarides (LPS) (intratracheal co-instillation) resulted in synergistic enhancement of free radical generation in the lungs, paralleled by a synergistic increase in total protein and by infiltration of neutrophils in the bronchoalveolar lavage fluid of the lungs [136]. The free radicals result from activated macrophages; more specifically, because of enhanced xanthine xanthine-oxidase activity.

3.5.3
Nanoparticles and Cellular Uptake

Reviewing the literature, there are several reports on cellular uptake of micro- and nanosized particles and CNT. Reports on particle uptake by endothelial cells [137, 138], pulmonary epithelium [68, 139, 140], intestinal epithelium [75, 91] alveolar macrophages [41, 73, 111, 141–143], other macrophages [88, 99, 144, 145], nerve cells [146, 147] and other cells [71, 148] are available. This is an expected phenomenon for phagocytic cells (macrophages) and cells that function as a barrier and/or transport for (large) compounds. Except for macrophages, the health effects of cellular uptake of nanoparticles have not been studied in depth.

In designing quantum dots specifically to enter cells, endocytosis is highly size-dependent, and an optimal size of around 50 nm has been suggested [149]. Besides the size, the surface plays a role: quantum dots with amine-modified coating were more efficiently internalized into the various human cells examined [150].

3.5.4
Nanoparticles and the Blood–Brain Barrier

One of the promising avenues of nanotechnology is organ- or cell-specific drug delivery mediated by nanoparticles [151–153]. Transport of nanoparticles across the blood–brain barrier (BBB) is expected to be possible by either passive diffusion or by carrier-mediated endocytosis. Coating of particles with polysorbates (e.g., polysorbate-80) results in anchoring of apolipoprotein E (apo E) or other blood components. Surface-modified particles seem to mimic Low Density Lipoproteins (LDL) particles and can interact with the LDL receptor, leading to uptake by endothelial cells. Hereafter, the drug (which was loaded in the particle) may be released in these cells and diffuse into the brain interior or the particles may be trans-cytosed.

Also, other processes such as tight junction modulation or P-glycoprotein (Pgp) inhibition may occur [154]. The translocation of inhaled nanoparticles via the olfactory nerves to the brain has been reported by Oberdörster et al. [72] and Fechter et al. [155].

3.6
Summary and Discussion

In general, in the search for potential adverse effects of a new product, toxicologists initially look into basic mechanisms such as acute toxicity (cytotoxicity), uptake and distribution and excretion of the material in organisms (pharmaco- or toxicokinetics). In a somewhat later phase, the mode of action at the target organs, tissues, and cells is studied.

Reviewing the knowledge collected concerning health effects of nanomaterials we have to conclude that it is still premature to draw final conclusions, simply because too little has been investigated.

A first important remark is that current knowledge is mainly based on epidemiological and experimental work concerning environmental particle pollution, most often referred to as coarse (10–2.5 μm), fine (2.5–0.1 μm) and ultrafine particulates (UFP) (<0.1 μm). We can certainly learn from this research, but it has to be taken into account that exposure to man-made nanomaterials differs from exposure to environmental particles in several ways. UFP, often arising from combustion, have a complex composition, have no uniform size and some compounds are soluble in biological systems, while man-made nanoparticles often have a uniform (crystalline) structure and size, and are often not soluble. Beside these differences it has to be remembered that some effects of particulate matter will be more generic, not discriminating between the nature (chemical composition) of the materials, e.g., activation (or inhibition) of phagocytosis, cellular uptake or other cellular interactions. Also, with decreasing particle diameter their surface area increases significantly, resulting in increased surface activity; health effects can often be better correlated with the total surface area of the material in the exposure rather than with its total mass.

In conclusion, some important observations are summarized below:

- Of the three exposure routes, inhalation of nanomaterials is easily the most troublesome for two reasons: (1) the lung itself is a target organ because the inhaled particulates can easily induce inflammation and oxidative stress locally, and the particulates are not always efficiently cleared from the alveoli and (2) the lung is, as far as investigated, a portal to enter the systemic circulation.
- Exposure to skin will not, as far as we know, lead to systemic uptake but penetration into the dermis can induce immunological effects.
- Oral exposure can result in systemic penetration – a feature that can be used in medicine, but unintended penetration from the intestinal tract, given the current state of knowledge, is not a worrying issue.
- Penetration, independent of portal, and the subsequent distribution are related to the size and surface properties of the particles.

From previous studies, using environmental particles, we know that oxidative mechanisms, local inflammation, thrombotic effects, and, only recently reported, uptake into sensory nerves are the important issues. Some of the reported effects will be specific for the complex DEP or Urban Particulate Matter (UPM), but new previously unobserved effects can be expected for man-made nanomaterials, therefore research focused on novel effects due to the specific nanomaterials will certainly be required.

Besides the expected and predictable risks, nanomaterials can induce a biologic response in a way we are not familiar with from previous studies of known compounds. For example, it has recently been observed that green-light-emitting quantum dots are more toxic *in vitro* than red-light-emitting dots, simply because of the difference in DNA damage by the emitted light [156].

Moreover, nanomaterials are often defined as materials with a dimension smaller than 100 nm. The definition proposed by the European Academy at Bad Neuenahr,

Fig. 3.2. "From exposure to decease" flowchart.

Germany is probably more appropriate: "Nanotechnology is dealing with functional systems based on the use of subunits with specific size-dependent properties of the individual sub-units or of a system of those" [157]. This definition takes into account size-dependent activity and/or effect rather than a certain size; following this it is expected that for many materials the border between "bulk" and "nano" will be mainly situated at a size range smaller than 20 nm, thus defining the size window more strictly.

Finally, although nanotoxicology will discover some new or specific toxicological mechanisms, the general concepts describing the process from exposure to disease, shown in Fig. 3.2, will not change. This paradigm can be used for any compound and it stresses the most important steps in exposure related diseases.

Without exposure, no health effects can develop, even not from very harmful compounds. From the portal of exposure, the compound must have the capability to enter the body, and then be distributed to the target tissue(s). In contact with the target tissue the compound can induce malfunction, most often subtle at first, but resulting in irreversible changes after chronic exposure.

This scheme can easily be adopted for nanomaterials; the challenge will be to find those (few) nanomaterials out of the large pool of newly produced materials, with specific sizes, composition and coating, that would induce significant health effects.

3.7
What Can be Done?

Although few direct reports are available on the health implications of exposure to man-made nanoparticles, the indirect evidence assembled here from epidemiological sources, drug delivery studies, as well as some *in vivo* and *in vitro* results, suggests that potential health risks can not be neglected. With growing volumes of production and their incorporation into more and more products, the chances of exposure of the general public to the engineered nanoparticles are likely to grow. Seemingly safe bulk materials, when reduced to nanoparticles, can drastically change their chemical, biological and catalytic activities, and become toxic. Moreover, their minute dimensions often help to overcome the existing biological barriers and body defense mechanisms. The whole issue is complicated by the difficulties associated with the detection and monitoring of nanoparticles.

On the bright side, most nanomaterials are currently used as additives or property enhancers that are highly diluted in a matrix material. The nanomaterials are bound to the matrix and unlikely to be released in serious quantities. In addition, most material manufacturers prefer to use liquid dispersions of nanomaterials (as opposed to dry powders) as these are much easier to handle. Unfortunately, some nanomaterials can only be produced in either a dry atmosphere or under vacuum. This raises a series of questions on their safe collection, handling, dispersion, cleaning, disposal, recycling, and environmental protection. At the time of writing, the issue of potential health risks that might be associated with nanomaterials is getting both an adequate press coverage and reasonable attention from the regulatory bodies and governments.

University laboratories are often at the forefront of nanomaterial research. Their laboratory procedures and practices should perhaps be critically re-evaluated in light of the material presented in this chapter. Both small and big manufacturers who state on the MSDS forms that nanomaterials are "safe because the bulk material is considered safe" should, perhaps, reconsider and try to get some proof. Finally, the public should be learning about the benefits and risks of nanotechnology not from scandal-driven tabloids but from the scientific community directly engaged in nanotechnology.

References

1 HOET, P.H., BRUSKE-HOHLFELD, I., SALATA, O.V. Nanoparticles – known and unknown health risks. *J. Nanobiotechnol.* **2004**, 2, 12.

2 UK Royal Society and Royal Academy of Engineering, Nanoscience and nanotechnologies: Opportunities and uncertainties. Final Report **2004** [http://www.nanotec.org.uk/finalReport.htm].

3 Anonymous. Nanotech is not so scary. *Nature* **2003**, 421, 299.

4 SALATA, O. Applications of nanoparticles in biology and medicine. *J. Nanobiotechnol.* **2004**, 2, 3.

5 MAZZOLA, L. Commercializing nanotechnology. *Nat. Biotechnol.* **2003**, 21, 1137–1143.

6 PAULL, R., WOLFE, J., HEBERT, P., SINKULA, M. Investing in nanotechnology. *Nat. Biotechnol.* **2003**, 21, 1144–1147.

7 FEYNMAN, R.P. There's Plenty of Room at the Bottom, *Science* **1991**, 254, 1300–1301.

8 BALL, P. Roll-up for the revolution. *Nature* **2001**, 414, 142–144.

9 NanoBusiness Alliance. **2003**. www.nanobusiness.org.

10 SAMET, J.M., DOMINICI, F., CURRIERO, F.C., COURSAC, I., ZEGER, S.L. Fine particulate air pollution and mortality in 20 U.S. cities, 1987–1994. *N. Engl. J. Med.* **2000**, 343, 1742–1749.

11 DOMINICI, F., MCDERMOTT, A., DANIELS, M., ZEGER, S.L., AND SAMET, J.M. Revised Analysis of the National Morbidity, Mortality, and Air Pollution Study (NMMAPS), Part II: Mortality Among Residents of 90 Cities. In: Revised Analyses of Time-series Studies of Air Pollution and Health, Anon, Boston: Health Effects Institute, **2003**, p. 9–24.

12 KATSOUYANNI, K., TOULOUMI, G., SAMOLI, E., GRYPARIS, A., LE TERTRE, A., MONOPOLIS, Y., ROSSI, G., ZMIROU, D., BALLESTER, F., BOUMGHAR, A., ANDERSON, H.R., WOJTYNIAK, B., PALDY, A., BRAUNSTEIN, R., PEKKANEN, J., SCHINDLER, C., SCHWARTZ, J. Confounding and effect modification in the short-term effects of ambient particles on total mortality: Results from 29 European cities within the APHEA2 project. *Epidemiology* **2001**, 12, 521–531.

13 *WHO Air Quality Guidelines for Europe.* 2nd edn **2000**. WHO, Regional Office for Europe, Copenhagen.

14 PETERS, A., WICHMANN, H.E., TUCH, T., HEINRICH, J., HEYDER, J. Respiratory effects are associated with the number of ultrafine particles. *Am. J. Respir. Crit. Care Med.* **1997**, 155, 1376–1383.

15 PEKKANEN, J., TIMONEN, K.L., RUUSKANEN, J., REPONEN, A., MIRME, A. Effects of ultrafine and fine particles in urban air on peak expiratory flow among children with asthmatic symptoms. *Environ. Res.* **1997**, 74, 24–33.

16 VON KLOT, S., WOLKE, G., TUCH, T., HEINRICH, J., DOCKERY, D.W., SCHWARTZ, J., KREYLING, W.G., WICHMANN, H.E., PETERS, A. Increased asthma medication use in association with ambient fine and ultrafine particles. *Eur. Respir. J.* **2002**, 20, 691–702.

17 PENTTINEN, P., TIMONEN, K.L., TIITTANEN, P., MIRME, A., RUUSKANEN, J., PEKKANEN, J. Number concentration and size of particles in urban air: Effects on spirometric lung function in adult asthmatic subjects. *Environ. Health Perspect.* **2001**, 109, 319–323.

18 PENTTINEN, P., TIMONEN, K.L., TIITTANEN, P., MIRME, A., RUUSKANEN, J., PEKKANEN, J. Ultrafine particles in urban air and respiratory health among adult asthmatics. *Eur. Respir. J.* **2001**, 17, 428–435.

19 IBALD-MULLI, A., TIMONEN, K.L., PETERS, A., HEINRICH, J., WOLKE, G., LANKI, T., BUZORIUS, G., KREYLING, W.G., DE HARTOG, J., HOEK, G., TEN BRINK, H.M., PEKKANEN, J. Effects of particulate air pollution on blood pressure and heart rate in subjects with cardiovascular disease: A multicenter approach. *Environ. Health Perspect.* **2004**, 112, 369–377.

20 DE HARTOG, J.J., HOEK, G., PETERS, A., TIMONEN, K.L., IBALD-MULLI, A., BRUNEKREEF, B., HEINRICH, J., TIITTANEN, P., VAN WIJNEN, J.H., KREYLING, W., KULMALA, M., PEKKANEN, J. Effects of fine and ultrafine particles on cardiorespiratory symptoms in elderly subjects with coronary heart disease: The ULTRA study. *Am. J. Epidemiol.* **2003**, 157, 613–623.

21 RUUSKANEN, J., TUCH, T., TEN BRINK, H.M., PETERS, A., KHLYSTOV, A., MIRME, A., KOS, G.P., BRUNEKREEF, B., WICHMANN, H.E., BUZORIUS, G., VALLIUS, M., PEKKANEN, J. Concentrations of ultrafine, fine and PM2.5 particles in three European cities. *Atmos. Environ.* **2001**, 35, 3729–3738.

22 PEKKANEN, J., PETERS, A., HOEK, G., TIITTANEN, P., BRUNEKREEF, B., DE HARTOG, J., HEINRICH, J., IBALD-MULLI, A., KREYLING, W.G., LANKI, T., TIMONEN, K.L., VANNINEN, E. Particulate air pollution and risk of ST-segment depression during repeated submaximal exercise tests among subjects with coronary heart disease: The Exposure and Risk Assessment for Fine and Ultrafine Particles in Ambient Air (ULTRA) study. *Circulation* **2002**, 106, 933–938.

23 BROOK, R.D., FRANKLIN, B., CASCIO, W., HONG, Y., HOWARD, G., LIPSETT, M., LUEPKER, R., MITTLEMAN, M., SAMET, J., SMITH, S.C., JR., TAGER, I. Air pollution and cardiovascular disease: A statement for healthcare professionals from the Expert Panel on Population and Prevention Science of the American Heart Association. *Circulation* **2004**, 109, 2655–2671.

24 WICHMANN, H.E., SPIX, C., TUCH, T., WOLKE, G., PETERS, A., HEINRICH, J., KREYLING, W.G., HEYDER, J. Daily mortality and fine and ultrafine particles in Erfurt, Germany part I: Role of particle number and particle mass. *Res. Rep. Health Eff. Inst.* **2000**, 5–86.

25 OBERDÖRSTER, G., FERIN, J., LEHNERT, B.E. Correlation between particle size, in vivo particle persistence, and lung injury. *Environ. Health Perspect.* **1994**, 102(Suppl 5), 173–179.

26 BORM, P.J. Particle toxicology: From coal mining to nanotechnology. *Inhal. Toxicol.* **2002**, 14, 311–324.

27 SANFELD, A., STEINCHEN, A. Does the size of small objects influence chemical reactivity in living systems? *C. R. Biol.* **2003**, 326, 141–147.

28 OBERDÖRSTER, G., OBERDÖRSTER, E., OBERDÖRSTER, J. Nanotoxicology: An emerging discipline evolving from studies of ultrafine particles. *Environ. Health Perspect.* **2005**, 113, 823–839.

29 LIPPMANN, M. Effects of fiber characteristics on lung deposition, retention, and disease. *Environ. Health Perspect.* **1990**, 88, 311–317.

30 MOORE, M.A., BROWN, R.C., PIGOTT, G. Material properties of MMVFs and their time-dependent failure in lung environments. *Inhal. Toxicol.* **2001**, 13, 1117–1149.

31 OBERDÖRSTER, G. Pulmonary effects of inhaled ultrafine particles. *Int. Arch. Occup. Environ. Health* **2001**, 74, 1–8.

32 OBERDÖRSTER, G. Toxicokinetics and effects of fibrous and nonfibrous particles. *Inhal. Toxicol.* **2002**, 14, 29–56.

33 OBERDÖRSTER, G. Determinants of the pathogenicity of man-made vitreous fibers (MMVF). *Int. Arch. Occup. Environ. Health* **2000**, 73 Suppl, S60–S68.

34 WARHEIT, D.B., HART, G.A., HESTERBERG, T.W., COLLINS, J.J., DYER, W.M., SWAEN, G.M., CASTRANOVA, V., SOIEFER, A.I., KENNEDY, G.L., JR. Potential pulmonary effects of man-made organic fiber (MMOF) dusts. *Crit. Rev. Toxicol.* **2001**, 31, 697–736.

35 DONALDSON, K., TRAN, C.L. Inflammation caused by particles and fibers. *Inhal. Toxicol.* **2002**, 14, 5–27.

36 SEATON, A., MACNEE, W., DONALDSON, K., GODDEN, D. Particulate air pollution and acute health effects. *Lancet* **1995**, 345, 176–178.

37 BROWN, D.M., WILSON, M.R., MACNEE, W., STONE, V., DONALDSON, K. Size-dependent proinflammatory effects of ultrafine polystyrene particles: A role for surface area and oxidative stress in the enhanced activity of ultrafines. *Toxicol. Appl. Pharmacol.* **2001**, 175, 191–199.

38 SCHINS, R.P. Mechanisms of genotoxicity of particles and fibers. *Inhal. Toxicol.* **2002**, 14, 57–78.

39 BERMAN, D.W., CRUMP, K.S., CHATFIELD, E.J., DAVIS, J.M., JONES, A.D. The sizes, shapes, and mineralogy of asbestos structures that induce lung tumors or mesothelioma in AF/HAN rats following inhalation. *Risk Anal.* **1995**, 15, 181–195.

40 KREYLING, W.G., SEMMLER, M., MOLLER, W. Dosimetry and toxicology of ultrafine particles. *J. Aerosol Med.* **2004**, 17, 140–152.

41 OBERDÖRSTER, G. Lung particle overload: Implications for occupational exposures to particles. *Regul. Toxicol. Pharmacol.* **1995**, 21, 123–135.

42 MUHLE, H., MANGELSDORF, I. Inhalation toxicity of mineral particles: Critical appraisal of endpoints and study design. *Toxicol. Lett.* **2003**, 140–141, 223–228.

43 MOSSMAN, B.T. Mechanisms of action of poorly soluble particulates in overload-related lung pathology. *Inhal. Toxicol.* **2000**, 12, 141–148.

44 MOOLGAVKAR, S.H., BROWN, R.C., TURIM, J. Biopersistence, fiber length, and cancer risk assessment for inhaled fibers. *Inhal. Toxicol.* **2001**, 13, 755–772.

45 DONALDSON, K., STONE, V. Current hypotheses on the mechanisms of toxicity of ultrafine particles. *Ann. Ist. Super. Sanita* **2003**, 39, 405–410.

46 HESTERBERG, T.W., CHASE, G., AXTEN, C., MILLER, W.C., MUSSELMAN, R.P., KAMSTRUP, O., HADLEY, J., MORSCHEIDT, C., BERNSTEIN, D.M., THEVENAZ, P. Biopersistence of synthetic vitreous fibers and amosite asbestos in the rat lung following inhalation. *Toxicol. Appl. Pharmacol.* **1998**, 151, 262–275.

47 WARHEIT, D.B., REED, K.L., WEBB, T.R. Man-made respirable-sized organic fibers: What do we know about their toxicological profiles? *Ind. Health* **2001**, 39, 119–125.

48 WYLIE, A.G., SKINNER, H.C., MARSH, J., SNYDER, H., GARZIONE, C., HODKINSON, D., WINTERS, R., MOSSMAN, B.T. Mineralogical features associated with cytotoxic and

proliferative effects of fibrous talc and asbestos on rodent tracheal epithelial and pleural mesothelial cells. *Toxicol. Appl. Pharmacol.* **1997**, 147, 143–150.

49 SEARL, A. A comparative study of the clearance of respirable para-aramid, chrysotile and glass fibres from rat lungs. *Ann. Occup. Hyg.* **1997**, 41, 217–233.

50 STANTON, M.F., WRENCH, C. Mechanisms of mesothelioma induction with asbestos and fibrous glass. *J. Natl. Cancer Inst.* **1972**, 48, 797–821.

51 SERVICE, R.F. American Chemical Society meeting. Nanomaterials show signs of toxicity. *Science* **2003**, 300, 243.

52 WARHEIT, D.B., LAURENCE, B.R., REED, K.L., ROACH, D.H., REYNOLDS, G.A., WEBB, T.R. Comparative pulmonary toxicity assessment of single wall carbon nanotubes in rats. *Toxicol. Sci.* **2003**, 77, 117–125.

53 LAM, C.W., JAMES, J.T., McCLUSKEY, R., HUNTER, R.L. Pulmonary toxicity of single-wall carbon nanotubes in mice 7 and 90 days after intratracheal instillation. *Toxicol. Sci.* **2003**, 77, 126–134.

54 MAYNARD, A.D., BARON, P.A., FOLEY, M., SHVEDOVA, A.A., KISIN, E.R., CASTRANOVA, V. Exposure to carbon nanotube material: Aerosol release during the handling of unrefined single-walled carbon nanotube material. *J. Toxicol. Environ. Health A* **2004**, 67, 87–107.

55 DONALDSON, K., TRAN, C.L. An introduction to the short-term toxicology of respirable industrial fibres. *Mutat. Res.* **2004**, 553, 5–9.

56 YEATES, D.B., MAUDERLY, J.L. Inhaled environmental/occupational irritants and allergens: Mechanisms of cardiovascular and systemic responses. Introduction. *Environ. Health Perspect.* **2001**, 109(Suppl 4), 479–481.

57 GOLD, D.R., LITONJUA, A., SCHWARTZ, J., LOVETT, E., LARSON, A., NEARING, B., ALLEN, G., VERRIER, M., CHERRY, R., VERRIER, R. Ambient pollution and heart rate variability. *Circulation* **2000**, 101, 1267–1273.

58 LIAO, D., CREASON, J., SHY, C., WILLIAMS, R., WATTS, R., ZWEIDINGER, R. Daily variation of particulate air pollution and poor cardiac autonomic control in the elderly. *Environ. Health Perspect.* **1999**, 107, 521–525.

59 BROWN, J.S., ZEMAN, K.L., BENNETT, W.D. Ultrafine particle deposition and clearance in the healthy and obstructed lung. *Am. J. Respir. Crit. Care Med.* **2002**, 166, 1240–1247.

60 CONHAIM, R.L., EATON, A., STAUB, N.C., HEATH, T.D. Equivalent pore estimate for the alveolar-airway barrier in isolated dog lung. *J. Appl. Physiol.* **1988**, 64, 1134–1142.

61 HERMANS, C., KNOOPS, B., WIEDIG, M., ARSALANE, K., TOUBEAU, G., FALMAGNE, P., BERNARD, A. Clara cell protein as a marker of Clara cell damage and bronchoalveolar blood barrier permeability. *Eur. Respir. J.* **1999**, 13, 1014–1021.

62 REJMAN, J., OBERLE, V., ZUHORN, I.S., HOEKSTRA, D. Size-dependent internalization of particles via the pathways of clathrin- and caveolae-mediated endocytosis. *Biochem. J.* **2004**, 377, 159–169.

63 ELJAMAL, M., NAGARAJAN, S., PATTON, J.S. In situ and in vivo methods for pulmonary delivery. *Pharm. Biotechnol.* **1996**, 8, 361–374.

64 PATTON, J.S. Mechanisms of macro-molecule absorption by the lungs. *Adv. Drug Deliv. Rev.* **1996**, 19, 3–36.

65 SMITH, A.E., HELENIUS, A. How viruses enter animal cells. *Science* **2004**, 304, 237–242.

66 NEMMAR, A., VANBILLOEN, H., HOYLAERTS, M.F., HOET, P.H., VERBRUGGEN, A., NEMERY, B. Passage of intratracheally instilled ultrafine particles from the lung into the systemic circulation in hamster. *Am. J. Respir. Crit. Care Med.* **2001**, 164, 1665–1668.

67 KAWAKAMI, K., IWAMURA, A., GOTO, E., MORI, Y., ABE, T., HIRASAW, Y., ISHIDA, H., SHIMADA, T., TOMINAGA, G. Kinetics and clinical application of 99mTc-technegas. *Kaku Igaku* **1990**, 27, 725–733.

68 Kato, T., Yashiro, T., Murata, Y., Herbert, D.C., Oshikawa, K., Bando, M., Ohno, S., Sugiyama, Y. Evidence that exogenous substances can be phagocytized by alveolar epithelial cells and transported into blood capillaries. *Acc. Chem. Res.* **2003**, 311, 47–51.

69 Steiner, S., Pfutzner, A., Wilson, B.R., Harzer, O., Heinemann, L., Rave, K. Technosphere/Insulin – proof of concept study with a new insulin formulation for pulmonary delivery. *Exp. Clin. Endocrinol. Diabetes* **2002**, 110, 17–21.

70 Patton, J.S., Bukar, J.G., Eldon, M.A. Clinical pharmacokinetics and pharmacodynamics of inhaled insulin. *Clin. Pharmacokinet.* **2004**, 43, 781–801.

71 Kreyling, W., Semmler, M., Erbe, F., Mayer, P., Schulz, H., Oberdörster, G., Ziesenis, A. Translocation of ultrafine insoluble iridium particles from lung epithelium to extra-pulmonary organs is size dependent but very low. *J. Toxicol. Environ. Health A* **2002**, 65, 1513–1530.

72 Oberdörster, G., Sharp, Z., Atudorei, V., Elder, A., Gelein, R., Lunts, A., Kreyling, W., Cox, C. Extrapulmonary translocation of ultrafine carbon particle following whole-body inhalation exposure of rats. *J. Toxicol. Environ. Health A* **2002**, 65, 1531–1543.

73 Takenaka, S., Karg, E., Roth, C., Schulz, H., Ziesenis, A., Heinzmann, U., Schramel, P., Heyder, J. Pulmonary and systemic distribution of inhaled ultrafine silver particles in rats. *Environ. Health Perspect.* **2001**, 109(Suppl 4), 547–551.

74 Eyles, J.E., Bramwell, V.W., Williamsson, E.D., Alpar, H.O. Microsphere translocation and immunopotentiation in systemic tissues following intranasal administration. *Vaccine* **2001**, 19, 4732–4742.

75 Florence, A.T., Hussain, N. Transcytosis of nanoparticle and dendrimer delivery systems: Evolving vistas. *Adv. Drug Deliv. Rev.* **2001**, 50(Suppl 1), S69–S89.

76 Hussain, N., Jaitley, V., Florence, A.T. Recent advances in the understanding of uptake of microparticulates across the gastrointestinal lymphatics. *Adv. Drug Deliv. Rev.* **2001**, 50, 107–142.

77 Aprahamian, M., Michel, C., Humbert, W., Devissaguet, J.P., Damge, C. Transmucosal passage of polyalkylcyanoacrylate nanocapsules as a new drug carrier in the small intestine. *Biol. Cell* **1987**, 61, 69–76.

78 Hillyer, J.F., Albrecht, R.M. Gastrointestinal persorption and tissue distribution of differently sized colloidal gold nanoparticles. *J. Pharm. Sci.* **2001**, 90, 1927–1936.

79 Jani, P., Halbert, G.W., Langridge, J., Florence, A.T. Nanoparticle uptake by the rat gastrointestinal mucosa: Quantitation and particle size dependency. *J. Pharm. Pharmacol.* **1990**, 42, 821–826.

80 Jani, P., Halbert, G.W., Langridge, J., Florence, A.T. The uptake and translocation of latex nanospheres and microspheres after oral administration to rats. *J. Pharm. Pharmacol.* **1989**, 41, 809–812.

81 Hussain, N., Florence, A.T. Utilizing bacterial mechanisms of epithelial cell entry: Invasin-induced oral uptake of latex nanoparticles. *Pharm. Res.* **1998**, 15, 153–156.

82 Hussain, N., Jani, P.U., Florence, A.T. Enhanced oral uptake of tomato lectin-conjugated nanoparticles in the rat. *Pharm. Res.* **1997**, 14, 613–618.

83 Hillery, A.M., Jani, P.U., Florence, A.T. Comparative, quantitative study of lymphoid and non-lymphoid uptake of 60 nm polystyrene particles. *J. Drug Target* **1994**, 2, 151–156.

84 Woodley, J.F. Lectins for gastrointestinal targeting – 15 years on. *J. Drug Target* **2000**, 7, 325–333.

85 Szentkuti, L. Light microscopical observations on luminally administered dyes, dextrans, nanospheres and microspheres in the pre-epithelial mucus gel layer of the rat distal colon. *J. Control. Release* **1997**, 46, 233–242.

86 Yamago, S., Tokuyama, H.,

NAKAMURA, E., KIKUCHI, K., KANANISHI, S., SUEKI, K., NAKAHARA, H., ENOMOTO, S., AMBE, F. In vivo biological behavior of a water-miscible fullerene: 14C labeling, absorption, distribution, excretion and acute toxicity. *Chem. Biol.* **1995**, 2, 385–389.

87 GATTI, A.M. Biocompatibility of micro- and nano-particles in the colon. Part II. *Biomaterials* **2004**, 25, 385–392.

88 LOMER, M.C., THOMPSON, R.P., POWELL, J.J. Fine and ultrafine particles of the diet: Influence on the mucosal immune response and association with Crohn's disease. *Proc. Nutr. Soc.* **2002**, 61, 123–130.

89 POWELL, J.J., HARVEY, R.S., ASHWOOD, P., WOLSTENCROFT, R., GERSHWIN, M.E., THOMPSON, R.P. Immune potentiation of ultrafine dietary particles in normal subjects and patients with inflammatory bowel disease. *J. Autoimmun.* **2000**, 14, 99–105.

90 KUCHARZIK, T., LUGERING, A., LUGERING, N., RAUTENBERG, K., LINNEPE, M., CICHON, C., REICHELT, R., STOLL, R., SCHMIDT, M.A., DOMSCHKE, W. Characterization of M cell development during indomethacin-induced ileitis in rats. *Aliment. Pharmacol. Ther.* **2000**, 14, 247–256.

91 HOPWOOD, D., SPIERS, E.M., ROSS, P.E., ANDERSON, J.T., McCULLOUGH, J.B., MURRAY, F.E. Endocytosis of fluorescent microspheres by human oesophageal epithelial cells: Comparison between normal and inflamed tissue. *Gut* **1995**, 37, 598–602.

92 LADEMANN, J., OTBERG, N., RICHTER, H., WEIGMANN, H.J., LINDEMANN, U., SCHAEFER, H., STERRY, W. Investigation of follicular penetration of topically applied substances. *Skin Pharmacol. Appl. Skin Physiol.* **2001**, 14, 17–22.

93 KREILGAARD, M. Influence of microemulsions on cutaneous drug delivery. *Adv. Drug Deliv. Rev.* **2002**, 54(Suppl 1), S77–S98.

94 LADEMANN, J., WEIGMANN, H., RICKMEYER, C., BARTHELMES, H., SCHAEFER, H., MUELLER, G., STERRY, W. Penetration of titanium dioxide microparticles in a sunscreen formulation into the horny layer and the follicular orifice. *Skin Pharmacol. Appl. Skin Physiol.* **1999**, 12, 247–256.

95 TINKLE, S.S., ANTONINI, J.M., RICH, B.A., ROBERTS, J.R., SALMEN, R., DePREE, K., ADKINS, E.J. Skin as a route of exposure and sensitization in chronic beryllium disease. *Environ. Health Perspect.* **2003**, 111, 1202–1208.

96 MENON, G.K., ELIAS, P.M. Morphologic basis for a pore-pathway in mammalian stratum corneum. *Skin Pharmacol.* **1997**, 10, 235–246.

97 ANDERSSON, K.G., FOGH, C.L., BYRNE, M.A., ROED, J., GODDARD, A.J., HOTCHKISS, S.A. Radiation dose implications of airborne contaminant deposition to humans. *Health Phys.* **2002**, 82, 226–232.

98 DE JALON, E.G., BLANCO-PRIETO, M.J., YGARTUA, P., SANTOYO, S. PLGA microparticles: Possible vehicles for topical drug delivery. *Int. J. Pharm.* **2001**, 226, 181–184.

99 BLUNDELL, G., HENDERSON, W.J., PRICE, E.W. Soil particles in the tissues of the foot in endemic elephantiasis of the lower legs. *Ann. Trop. Med. Parasitol.* **1989**, 83, 381–385.

100 CORACHAN, M. Endemic non-filarial elephantiasis of lower-limbs – Podoconiosis. *Med. Clin.* **1988**, 91, 97–100.

101 HOSTYNEK, J.J. Factors determining percutaneous metal absorption. *Food Chem. Toxicol.* **2003**, 41, 327–345.

102 EUN, H.C., LEE, H.G., PAIK, N.W. Patch test responses to rockwool of different diameters evaluated by cutaneous blood flow measurement. *Contact Dermatitis* **1991**, 24, 270–273.

103 MONTEIRO-RIVIERE, N.A., NEMANICH, R.J., INMAN, A.O., WANG, Y.Y., RIVIERE, J.E. Multi-walled carbon nanotube interactions with human epidermal keratinocytes. *Toxicol. Lett.* **2005**, 155, 377–384.

104 SHVEDOVA, A.A., CASTRANOVA, V., KISIN, E.R., SCHWEGLER-BERRY, D., MURRAY, A.R., GANDELSMAN, V.Z.,

MAYNARD, A., BARON, P. Exposure to carbon nanotube material: Assessment of nanotube cytotoxicity using human keratinocyte cells. *J. Toxicol. Environ. Health* **2003**, 66, 1909–1926.

105 MARKOWITZ, G.S., PERAZELLA, M.A. Drug-induced renal failure: A focus on tubulointerstitial disease. *Clin. Chim. Acta* **2005**, 351, 31–47.

106 SCHINS, R.P., DUFFIN, R., HOHR, D., KNAAPEN, A.M., SHI, T., WEISHAUPT, C., STONE, V., DONALDSON, K., BORM, P.J. Surface modification of quartz inhibits toxicity, particle uptake, and oxidative DNA damage in human lung epithelial cells. *Chem. Res. Toxicol.* **2002**, 15, 1166–1173.

107 HOET, P.H., NEMMAR, A., NEMERY, B. Health impact of nanomaterials? *Nat. Biotechnol.* **2004**, 22, 19.

108 BECKETT, W.S., CHALUPA, D.F., PAULY-BROWN, A., SPEERS, D.M., STEWART, J.C., FRAMPTON, M.W., UTELL, M.J., HUANG, L.S., COX, C., ZAREBA, W., OBERDÖRSTER, G. Comparing inhaled ultrafine versus fine zinc oxide particles in healthy adults: A human inhalation study. *Am. J. Respir. Crit. Care Med.* **2005**, 171, 1129–1135.

109 LEE, K.P., KELLY, D.P., SCHNEIDER, P.W., TROCHIMOWICZ, H.J. Inhalation toxicity study on rats exposed to titanium tetrachloride atmospheric hydrolysis products for two years. *Toxicol. Appl. Pharmacol.* **1986**, 83, 30–45.

110 DRISCOLL, K.E., DEYO, L.C., CARTER, J.M., HOWARD, B.W., HASSENBEIN, D.G., BERTRAM, T.A. Effects of particle exposure and particle-elicited inflammatory cells on mutation in rat alveolar epithelial cells. *Carcinogenesis* **1997**, 18, 423–430.

111 RENWICK, L.C., DONALDSON, K., CLOUTER, A. Impairment of alveolar macrophage phagocytosis by ultrafine particles. *Toxicol. Appl. Pharmacol.* **2001**, 172, 119–127.

112 RENWICK, L.C., BROWN, D., CLOUTER, A., DONALDSON, K. Increased inflammation and altered macrophage chemotactic responses caused by two ultrafine particle types. *Occup. Environ. Med.* **2004**, 61, 442–447.

113 ARAUJO, L., LOBENBERG, R., KREUTER, J. Influence of the surfactant concentration on the body distribution of nanoparticles. *J. Drug Target* **1999**, 6, 373–385.

114 LABHASETWAR, V., SONG, C., HUMPHREY, W., SHEBUSKI, R., LEVY, R.J. Arterial uptake of biodegradable nanoparticles: Effect of surface modifications. *J. Pharm. Sci.* **1998**, 87, 1229–1234.

115 HOET, P.H., GILISSEN, L., NEMERY, B. Polyanions protect against the in vitro pulmonary toxicity of polycationic paint components associated with the Ardystil syndrome. *Toxicol. Appl. Pharmacol.* **2001**, 175, 184–190.

116 HOET, P.H., GILISSEN, L.P., LEYVA, M., NEMERY, B. In vitro cytotoxicity of textile paint components linked to the "Ardystil syndrome". *Toxicol. Sci.* **1999**, 52, 209–216.

117 FISCHER, D., LI, Y., AHLEMEYER, B., KRIEGLSTEIN, J., KISSEL, T. In vitro cytotoxicity testing of polycations: Influence of polymer structure on cell viability and hemolysis. *Biomaterials* **2003**, 24, 1121–1131.

118 MORGAN, D.M., LARVIN, V.L., PEARSON, J.D. Biochemical characterisation of polycation-induced cytotoxicity to human vascular endothelial cells. *J. Cell Sci.* **1989**, 94(Pt 3), 553–559.

119 MORGAN, D.M., CLOVER, J., PEARSON, J.D. Effects of synthetic polycations on leucine incorporation, lactate dehydrogenase release, and morphology of human umbilical vein endothelial cells. *J. Cell Sci.* **1988**, 91(Pt 2), 231–238.

120 HAENSLER, J., SZOKA, F.C., JR. Polyamidoamine cascade polymers mediate efficient transfection of cells in culture. *Bioconj. Chem.* **1993**, 4, 372–379.

121 FISCHER, D., BIEBER, T., LI, Y., ELSASSER, H.P., KISSEL, T. A novel non-viral vector for DNA delivery based on low molecular weight, branched polyethylenimine: Effect of molecular weight on transfection efficiency and cytotoxicity. *Pharm. Res.* **1999**, 16, 1273–1279.

122 DEKIE, L., TONCHEVA, V., DUBRUEL, P., SCHACHT, E.H., BARRETT, L., SEYMOUR, L.W. Poly-L-glutamic acid derivatives as vectors for gene therapy. *J. Control. Release* **2000**, 65, 187–202.

123 RYSER, H.J. A membrane effect of basic polymers dependent on molecular size. *Nature* **1967**, 215, 934–936.

124 SINGH, A.K., KASINATH, B.S., LEWIS, E.J. Interaction of polycations with cell-surface negative charges of epithelial cells. *Biochim. Biophys. Acta* **1992**, 1120, 337–342.

125 BALLOU, B., LAGERHOLM, B.C., ERNST, L.A., BRUCHEZ, M.P., WAGGONER, A.S. Noninvasive imaging of quantum dots in mice. *Bioconjug. Chem.* **2004**, 15, 79–86.

126 SCHULZ, H., HARDER, V., IBALD-MULLI, A., KHANDOGA, A., KOENIG, W., KROMBACH, F., RADYKEWICZ, R., STAMPFL, A., THORAND, B., PETERS, A. Cardiovascular effects of fine and ultrafine particles. *J. Aerosol Med.* **2005**, 18, 1–22.

127 PETERS, A., DOCKERY, D.W., MULLER, J.E., MITTLEMAN, M.A. Increased particulate air pollution and the triggering of myocardial infarction. *Circulation* **2001**, 103, 2810–2815.

128 NEMMAR, A., HOYLAERTS, M.F., HOET, P.H., NEMERY, B. Possible mechanisms of the cardiovascular effects of inhaled particles: Systemic translocation and prothrombotic effects. *Toxicol. Lett.* **2004**, 149, 243–253.

129 NEMMAR, A., HOET, P.H., DINSDALE, D., VERMYLEN, J., HOYLAERTS, M.F., NEMERY, B. Diesel exhaust particles in lung acutely enhance experimental peripheral thrombosis. *Circulation* **2003**, 107, 1202–1208.

130 NEMMAR, A., HOYLAERTS, M.F., HOET, P.H., DINSDALE, D., SMITH, T., XU, H., VERMYLEN, J., NEMERY, B. Ultrafine particles affect experimental thrombosis in an in vivo hamster model. *Am. J. Respir. Crit. Care Med.* **2002**, 166, 998–1004.

131 SILVA, V.M., CORSON, N., ELDER, A., OBERDÖRSTER, G. The rat ear vein model for investigating in vivo thrombogenicity of ultrafine particles (UFP). *Toxicol. Sci.* **2005**, 85, 983–989.

132 NEMMAR, A., HOYLAERTS, M.F., HOET, P.H., VERMYLEN, J., NEMERY, B. Size effect of intratracheally instilled particles on pulmonary inflammation and vascular thrombosis. *Toxicol. Appl. Pharmacol.* **2003**, 186, 38–45.

133 NEMMAR, A., HOET, P.H., VERMYLEN, J., NEMERY, B., HOYLAERTS, M.F. Pharmacological stabilization of mast cells abrogates late thrombotic events induced by diesel exhaust particles in hamsters. *Circulation* **2004**, 110, 1670–1677.

134 LI, N., HAO, M., PHALEN, R.F., HINDS, W.C., NEL, A.E. Particulate air pollutants and asthma. A paradigm for the role of oxidative stress in PM-induced adverse health effects. *Clin. Immunol.* **2003**, 109, 250–265.

135 XIAO, G.G., WANG, M., LI, N., LOO, J.A., NEL, A.E. Use of proteomics to demonstrate a hierarchical oxidative stress response to diesel exhaust particle chemicals in a macrophage cell line. *J. Biol. Chem.* **2003**, 278, 50 781–50 790.

136 ARIMOTO, T., KADIISKA, M.B., SATO, K., CORBETT, J., MASON, R.P. Synergistic production of lung free radicals by diesel exhaust particles and endotoxin. *Am. J. Respir. Crit. Care Med.* **2005**, 171, 379–387.

137 BERNSTEIN, J.A., ALEXIS, N., BARNES, C., BERNSTEIN, I.L., BERNSTEIN, J.A., NEL, A., PEDEN, D., DIAZ-SANCHEZ, D., TARLO, S.M., WILLIAMS, P.B. Health effects of air pollution. *J. Allergy Clin. Immunol.* **2004**, 114, 1116–1123.

138 AKERMAN, M.E., CHAN, W.C., LAAKKONEN, P., BHATIA, S.N., RUOSLAHTI, E. Nanocrystal targeting in vivo. *Proc. Natl. Acad. Sci. U.S.A.* **2002**, 99, 12 617–12 621.

139 JUVIN, P., FOURNIER, T., BOLAND, S., SOLER, P., MARANO, F., DESMONTS, J.M., AUBIER, M. Diesel particles are taken up by alveolar type II tumor cells and alter cytokines secretion. *Arch. Environ. Health* **2002**, 57, 53–60.

140 BOLAND, S., BAEZA-SQUIBAN, A., FOURNIER, T., HOUCINE, O., GENDRON, M.C., CHEVRIER, M.,

JOUVENOT, G., COSTE, A., AUBIER, M., MARANO, F. Diesel exhaust particles are taken up by human airway epithelial cells in vitro and alter cytokine production. *Am. J. Physiol.* **1999**, 276, L604–L613.

141 HOET, P.H., NEMERY, B. Stimulation of phagocytosis by ultrafine particles. *Toxicol. Appl. Pharmacol.* **2001**, 176, 203.

142 LUNDBORG, M., JOHARD, U., LASTBOM, L., GERDE, P., CAMNER, P. Human alveolar macrophage phagocytic function is impaired by aggregates of ultrafine carbon particles. *Environ. Res.* **2001**, 86, 244–253.

143 MOSSMAN, B.T., SESKO, A.M. In vitro assays to predict the pathogenicity of mineral fibers. *Toxicology* **1990**, 60, 53–61.

144 POWELL, J.J., AINLEY, C.C., HARVEY, R.S., MASON, I.M., KENDALL, M.D., SANKEY, E.A., DHILLON, A.P., THOMPSON, R.P. Characterisation of inorganic microparticles in pigment cells of human gut associated lymphoid tissue. *Gut* **1996**, 38, 390–395.

145 FERNANDEZ-URRUSUNO, R., FATTAL, E., FEGER, J., COUVREUR, P., THEROND, P. Evaluation of hepatic antioxidant systems after intravenous administration of polymeric nanoparticles. *Biomaterials* **1997**, 18, 511–517.

146 OBERDÖRSTER, G., SHARP, Z., ATUDOREI, V., ELDER, A., GELEIN, R., KREYLING, W., COX, C. Translocation of inhaled ultrafine particles to the brain. *Inhal. Toxicol.* **2004**, 16, 437–445.

147 OBERDÖRSTER, G., SHARP, Z., ATUDOREI, V., ELDER, A., GELEIN, R., LUNTS, A., KREYLING, W., COX, C. Extrapulmonary translocation of ultrafine carbon particles following whole-body inhalation exposure of rats. *J. Toxicol. Environ. Health A* **2002**, 65, 1531–1543.

148 PRATTEN, M.K., LLOYD, J.B. Uptake of microparticles by rat visceral yolk sac. *Placenta* **1997**, 18, 547–552.

149 OSAKI, F., KANAMORI, T., SANDO, S., SERA, T., AOYAMA, Y. A quantum dot conjugated sugar ball and its cellular uptake. On the size effects of endocytosis in the subviral region. *J. Am. Chem. Soc.* **2004**, 126, 6520–6521.

150 HASEGAWA, U., NOMURA, S.M., KAUL, S.C., HIRANO, T., AKIYOSHI, K. Nanogel-quantum dot hybrid nanoparticles for live cell imaging. *Biochem. Biophys. Res. Commun.* **2005**, 331, 917–921.

151 ALYAUDTIN, R.N., REICHEL, A., LOBENBERG, R., RAMGE, P., KREUTER, J., BEGLEY, D.J. Interaction of poly(butylcyanoacrylate) nanoparticles with the blood-brain barrier in vivo and in vitro. *J. Drug Target* **2001**, 9, 209–221.

152 PULFER, S.K., CICCOTTO, S.L., GALLO, J.M. Distribution of small magnetic particles in brain tumor-bearing rats. *J. Neurooncol.* **1999**, 41, 99–105.

153 SCHROEDER, U., SOMMERFELD, P., ULRICH, S., SABEL, B.A. Nanoparticle technology for delivery of drugs across the blood-brain barrier. *J. Pharm. Sci.* **1998**, 87, 1305–1307.

154 KREUTER, J. Nanoparticulate systems for brain delivery of drugs. *Adv. Drug Deliv. Rev.* **2001**, 47, 65–81.

155 FECHTER, L.D., JOHNSON, D.L., LYNCH, R.A. The relationship of particle size to olfactory nerve uptake of a non-soluble form of manganese into brain. *Neurotoxicology* **2002**, 23, 177–183.

156 LOVRIC, J., BAZZI, H.S., CUIE, Y., FORTIN, G.R., WINNIK, F.M., MAYSINGER, D. Differences in subcellular distribution and toxicity of green and red emitting CdTe quantum dots. *J. Mol. Med.* **2005**, 83, 377–385.

157 SCHMID, G., DECKER, M., ERNST, H., FUCHS, H., GRÜNWALD, W., GRUNWALD, A., HOFMANN, H., MAYOR, M., RATHGEBER, W., SIMON, U., WYRWA, D. Small dimensions and material properties. A definition of nanotechnology. *Gaue Reihe* **2003**, 35, 1–134.

4
Dosimetry, Epidemiology and Toxicology of Nanoparticles[1]

Wolfgang G. Kreyling, Manuela Semmler-Behnke, and Winfried Möller

4.1
Introduction

4.1.1
Overview

Nanoparticles are increasingly used in a wide range of applications in science, technology and medicine. Since they are produced for specific purposes that cannot be met by larger particles and bulk material they are likely to be highly reactive, in particular with biological systems. However, a large body of know-how in environmental sciences is available from toxicological effects of ultrafine particles after inhalation. Since nanoparticles feature similar reactivity to ultrafine particles a sustainable development of new emerging nanoparticles is required. This chapter briefly reviews the dosimetry of nanoparticles, including deposition in the various regions of the respiratory tract and systemic translocation and uptake in secondary target organs, epidemiologic associations with health effects and toxicology of inhaled nanoparticles. General principles and current paradigms to explain the specific behavior of nanoparticles in toxicology are discussed. Since the evidence for health risks of ultrafine and nanoparticles after inhalation has been increasing over the last decade, this chapter attempts to extrapolate these findings and principles observed in particle inhalation toxicology into recommendations for an integrated concept of risk assessment of nanoparticles for a broad range of use in science, technology and medicine.

4.1.2
General Background

Definition: Since the term "nanoparticle" is used heterogeneously in current discussion we want to define that they are shorter than 100 nm at least in one dimen-

---

1 This chapter is based on a recently published article: *Kreyling, W.G., Semmler-Behnke, M., Moeller, W.*, Health implications of nanoparticles. J. Nanoparticle Res. DOI 10.1007/s11051-005-9068-z.

Nanotechnologies for the Life Sciences Vol. 5
Nanomaterials – Toxicity, Health and Environmental Issues. Edited by Challa S. S. R. Kumar
Copyright © 2006 WILEY-VCH Verlag GmbH & Co. KGaA, Weinheim
ISBN: 3-527-31385-0

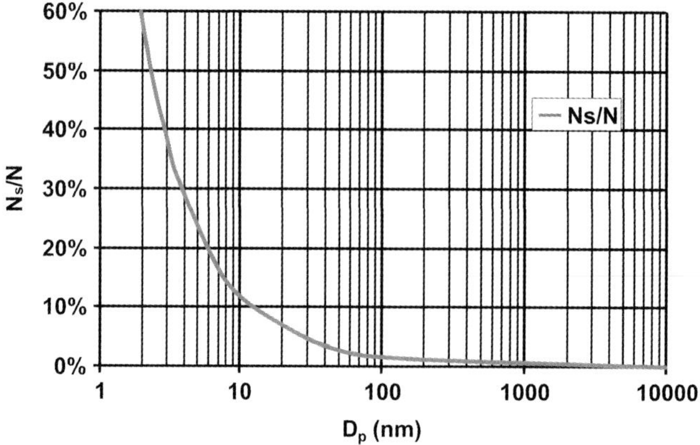

Fig. 4.1. Dependence of the percentage of surface molecules, relative to total number of molecules, of spherical nanoparticles on the diameter of the particles. Note: below 100 nm this ratio increases steeply. (From Ref. [7].)

sion, according to a recent suggested definition within the European Union [1]. In addition, nanoparticles are so-called "intended" particles, intentionally produced for specific use in science, technology, medicine, industries and many day-to-day applications. Therefore, they have a well-defined composition and are structured to fit the anticipated properties. As a result they differ from ultrafine particles (UP), a term frequently used in environmental sciences for ambient or occupational particles that are less than 100 nm in size in each dimension and which are unintended particles, originating often from combustion processes, of diverse sources and gas–particle interactions in their environment. Therefore, ultrafine ambient particles are often composed of a multitude of compounds, which may be structured in a highly complex manner. This is an additional basic difference between well-defined intended nanoparticles and unintended ultrafine particles.

Particles in the nanometer size range have two particular properties: (a) anything smaller than about 50 nm is no longer subject to the laws of classical physics but of quantum physics. This means that nanoparticles can exhibit optical, magnetic or electrical capabilities that distinguish them clearly from larger particles or bulk material; (b) with decreasing size the ratio between mass and surface area increases rapidly, i.e., the ratio of atoms or molecules at the surface to the total number of atoms or molecules rises steeply with decreasing particle size (Fig. 4.1). According to their very large specific surface area (surface area per mass), nanoparticles exert a stronger effect on their environment, i.e., with any adjacent materials. In other words, nanoparticles can catalyze chemical reactions at the surface; a given mass of material in nanoparticulate form will be much more reactive than the same mass of material made up of larger particles. For instance, crystalline nanoparticles have abundant atoms on their surface that are less strongly bonded than those in the interior of the particle. Given their unstable situation in the curvature of the surface, the atoms will try to change their binding: they are reactive towards their environment. This may well be desirable and is usually the purpose of their generation; however, exposure to such particles, for instance by inhalation or ingestion, may have harmful consequences [2–9].

Although the economic and societal health benefits of the introduction of nanomaterials have been welcomed, concerns have been expressed that properties that are being exploited by researchers and industry might have negative health effects and environmental impacts and, particularly, those might result in greater toxicity. Hence, the rapid development of a multitude of nanoparticle applications needs to be complemented by assessing possible implications, assuring a safe and sustainable handling of those nanoparticles. The challenge of an integrated application development and implication assessment is pro-active collaboration at the earliest stage to optimize functionality of the nanoparticle and to minimize its side effects without losses in terms of costs and time because of one-sided mismanagement or unfocussed or delayed initiation of risk assessment.

To understand the potential risks to humans from nanoparticles, it is necessary to understand the dosimetry and to consider the body's defenses against particles in general and the properties that particles require to overcome these defenses, as discussed in Chapter 4.2. Throughout much of their evolutionary history, humans have been exposed to small particles, often in very high concentrations, and the mechanisms evolved for defense against microorganisms are also used to defend the body against such particles. Generally, access to the human body can occur through the lungs, the skin or the gastrointestinal tract. Each organ presents a barrier to penetration by microorganisms or particles.

Chapter 4.3 discusses toxicological data that show a specific toxicological response of ultrafine particles, *in vitro* and *in vivo*, that is not found using fine particles of the same composition.

The general approach to assessing and controlling risk, as discussed in Chapter 4.4, involves identification of hazards (the potential of a nanoparticle or parts of it to cause harm) and then a structured approach to determining the probability of exposure to the hazard and the associated consequences. As in any new technology, foresight of possible risks depends on a consideration of the entire life cycle of a new nanoparticle being produced. This involves understanding the processes and materials used in manufacture, the likely interactions between the product and individuals or the environment during its manufacture and useful life, and the methods used in its eventual disposal.

4.1.3
Epidemiological Evidence for Health Effect Associations with Ambient Particulate Matter

Since safety issues are not yet very well developed in the field of science and technology of nanoparticles, this young rapidly developing interdisciplinary field may make use of existing knowledge. In fact, for over a decade, environmental risk assessment has continuously investigated potential health effects that may be associated with exposure to ultrafine particles in the environment [10, 11]. The need for those toxicological studies came from epidemiological investigations that had shown consistent associations between exposure to particulate air pollution in urban areas and acute increases in morbidity and mortality rates, especially for persons with obstructive lung and cardiovascular diseases [12]. The relative risk is surprisingly similar in many of these studies although the studies have been performed at very different global locations – predominantly in North America and Western Europe. These data have recently been collected and summarized by the US-EPA in a review of the Air Quality Criteria Document on Particulate Matter in 2004 [13].

Interestingly, epidemiological studies on health effects of ultrafine particles provided the first evidence that this particle fraction may induce adverse health effects independent of those of larger fine particles (<2.5 μm aerodynamic diameter) and other air toxics, as reviewed by Wichmann and Peters [10]. Furthermore, none of these single compounds are present at sufficiently high concentrations in the environmental aerosol that they may be considered toxicologically relevant based on occupational hygiene. Therefore, it appeared reasonable to investigate the interactions of complex mixtures of compounds with biological systems. Although the risks from these mixtures of compounds may be low to an individual, the large number of persons at risk can make these compounds an important public health threat. Frequently, health effects are only manifested in specific risk groups, i.e., persons predisposed by genetic susceptibility, age, and/or disease. Translating these insights to the use of nanoparticles, the more widespread the use by the general population the fewer implications are allowed. In particular, particulate nanomedicines specifically generated for treatment of patients – i.e., subjects predisposed by their disease – require extremely high safety standards, taking the specific physicochemical properties of all components on the scale of nanoparticles into account.

4.1.4
Toxicological Evidence for Ambient Particulate Matter Induced Adverse Health Effects

The adverse health effects shown in epidemiological studies have inspired scientists to use various techniques to study the toxicological mechanisms that form the biological background for adverse health effects associated with gaseous and particulate pollutants. At the beginning of the last decade, classical attempts like lung function tests, etc. were chosen in toxicological studies. This has changed substantially in recent years. An early key study demonstrated that ultrafine TiO_2 nanoparticles caused more inflammation in rat lungs than exposure to the same airborne mass concentration of larger, so-called "fine" TiO_2 [14]. Before this study, TiO_2 had been considered as a non-toxic dust and indeed had served as an innoxious control dust in many studies on the toxicology of particles. Therefore, this report was highly influential in highlighting that a material with low toxicity in the form of fine particles could be toxic in the form of ultrafine particles.

Although many questions are not yet fully answered, it is reasonable to apply this acquired knowledge to nanoparticles. With ambitious expectations of the widespread introduction, use and application of nanoparticles into nearly *everything*, it cannot be emphasized enough that accompanying safety measures must be devised early enough in their development. A major accident or a development similar to that of, for example, asbestos fibers related to health effects may turn public perception negative with the disastrous result that the promising potential of nanosciences and nanotechnology may be jeopardized in part, or even largely, by emotional arguments of the public that are not based on rational cost–benefit calculations.

This chapter summarizes key issues of risk assessment based on studies on the interaction of inhaled ultrafine particles found in urban atmospheres as well as on some nanoparticles that had been considered innoxious since their larger counterparts had not shown any toxic effect. We outline current understanding of modes of actions and underlying mechanisms involved in the pathogenesis of adverse health effects that eventually lead to disease. Therefore, we only consider the use of nanoparticles as long as they have access to the environment. There is no reason to consider firmly fixed nanoparticles in macroscopic entities that are unlikely to cause any harm. Inhalative exposure needs to distinguish between that of healthy adult workers during their work shift at the workplace in science, technology and industry and rather uncontrolled, eventually continuous, exposure of the entire public, including susceptible individuals such as infants, children, and the elderly, as well as diseased and genetically predisposed subjects.

4.2
Inhaled Nanoparticle Dosimetry

4.2.1
Particle Measures

In the past, particle mass concentration was by far the most common metric used. Daily averages range nowadays from 20 to 50 μg m^{-3} in most cities of industrialized countries. Taking the size range of particulate matter (PM) over more than four decades (1 nm–30 μm) into account, mass concentration overestimates large PM in the coarse fraction and basically neglects ultrafine particles < 100 nm in size. The limitation of this metric is illustrated by the fact that the water solubility of ambient PM may vary from 20 to 80% of PM mass and yet the toxicity of soluble compounds is unlikely to be similar to that of the insoluble fraction. With clearer insights into particle–lung interactions, other measures such as the particle number concentration and/or surface area need to be taken into account, depending on whether ultrafine or larger particles are to be considered. However, exposure measures may be inadequate, since it may be the number of deposited particles per unit surface area of airways and bifurcations and of alveoli, or dose to a specific cell such as macrophages or epithelial cells, that determines the response of specific regions. Therefore, the use of a metric depends on specific questions posed, requiring specifically defined measures.

4.2.2
Deposition of Ultrafine Particles in the Respiratory System

Particle deposition of ultrafine particles in the respiratory tract is determined predominantly by diffusional motion (thermal motion of air molecules) distorting particles from their stream lines of the inhaled and exhaled air towards the airway walls, where they deposit once they have touched the walls. The diffusion mechanism affects particle deposition through three important components of aerosol properties and respiratory tract physiology during breathing: (a) particle dynamics, including size and shape, and possible dynamic change during breathing; (b) geometry of the branching airways and the alveolar structures; and (c) breathing pattern, which determines the airflow velocity and the residence time in the respiratory tract, and includes nose versus mouth breathing [15].

Regarding regional particle deposition in the respiratory tract, the tract can be considered as a series of filters (Fig. 4.2), starting with the nose or mouth via the various diameters of airways to the alveoli [16, 17]. Figure 4.2 displays the deposition probability of particles of different sizes in the larger and smaller airways, as well as the alveolar region. This means that the toxicity of particles of different sizes can have different effects in different parts of the lungs. This may be particularly important in children with developing lungs or with asthma, which mainly affects the larger airways, or with COPD, which affects both large and small airways and alveoli. These diseases may also cause an up to several-fold increase in deposition of PM in diseased parts of the lungs, which may deteriorate their functions [18].

The ultrafine particle density per airway surface area may, notably, often exceed that of the gas exchange region because the alveolar surface area (adult lungs ∼

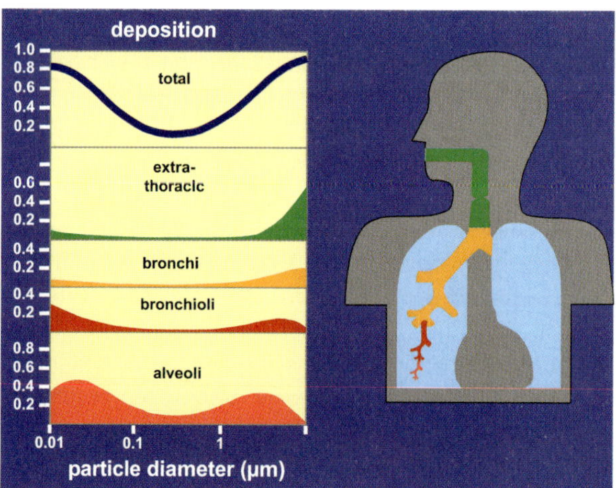

Fig. 4.2. Regional deposition of particles in the human respiratory tract during mouth-breathing at rest. (According to Ref. [16].)

140 m^2) is 100-fold larger than that of airways. In general, ultrafine particles are very capable of reaching the fragile structures of alveoli; their deposition in the alveoli increases with decreasing diameter until 20 nm. Particles < 20 nm deposit less in alveoli since their high diffusivity leads to their deposition in the airways.

4.2.3
Fate of Particles in the Lungs

On the walls of the respiratory tract (epithelium) particles contact first the mucous or serous lining fluid and its surfactant layer on top. Therefore, the fate of particle compounds soluble in this lining fluid need to be distinguished from slowly dissolving or even insoluble compounds.

4.2.3.1 Soluble Particle Compounds
Soluble particle compounds may either be lipid soluble or water soluble. They will be dissolved and rapidly diluted while spreading in the mucous layer or in the serous lining fluid or in the cellular sol. According to their chemical properties solutes and soluble components can undergo absorption and diffusion or binding to proteins and membranous or subcellular structures. Solutes and their metabolic products will eventually be transferred to the blood and lymphatic circulation, undergoing further metabolization, with a potential to reach any organ and to produce toxic effects, far from their site of entry in the lungs [19, 20].

4.2.3.2 Slowly Dissolving and Insoluble Particles Deposited on the Airway Wall
Slowly dissolving and insoluble particles deposited on the airway wall will be mostly moved by mucociliary transport or by cough within 1–2 days to the throat (larynx), where they are swallowed and taken up by the gastrointestinal tract for further metabolization or excretion. While this clearance mechanism removes basically all particles larger than 5 μm, the fraction of long-term retained particles in the airways increases with decreasing particle size such that the uncleared fraction of ultrafine particles is ∼80% of those deposited in the airways [21].

4.2.3.3 Slowly Dissolving and Insoluble Particles Deposited in the Alveolar Region
Slowly dissolving and insoluble particles deposited in the alveolar region will be taken up and digested by specialized defense cells in the alveoli called macrophages within a few hours after deposition – at least under physiological conditions in healthy lungs. Therefore, alveolar macrophages will determine the fate of these particles. While alveolar macrophages are well suited to recognize and phagocytize particles > 200 nm within several hours after deposition, mechanisms of recognition are increasingly less effective for ultrafine particles with decreasing size [21].

Note, cells and solutes of body fluids, like proteins interacting with an insoluble particle, will not recognize what is inside the particle but only react with the molecules according to their structure at the particle surface. In other words, the vast amount of a reactive molecular species attached to the surface of insoluble particles and of an insoluble particle core (remaining after dissolution of the soluble components) may be the ultimate metric that determines adverse responses, although this species may add only a small fraction to PM mass.

4.2.3.4 Macrophage-mediated Particle Transport

Macrophage-mediated particle transport: we list below the major pathways and actions of macrophages from the human alveolar epithelium and indicate fractions of deposited insoluble particles undergoing these pathways, as reviewed earlier [21]. Particularly, the fractions differ consistently from those observed in rodents as the most common experimental animal models. Macrophage-mediated particle transport is directed:

1. Towards ciliated airways on the epithelium for further removal by ciliary action, passage through the gut and excretion (about a third of deposited insoluble particles).
2. To storage on the epithelium or uptake into the lining cells [together with (3) more than half of deposited insoluble particles].
3. Across the epithelial lining cells towards the spaces between underlying cells.
4. Across the epithelial lining cells towards the lymphatic drainage system (1–10% of deposited insoluble particles).
5. Into and across the epithelial lining cells and eventually into the blood vessels towards secondary target organs. There is evidence for the uptake of ultrafine particles into the blood circulation, depending on the physical structure and chemical composition of their surface. Fractions and rates of uptake are currently under debate. Identified secondary target organs are liver, spleen, kidneys, vasculature and heart, the immune system and the central nervous system.

Ultrafine particles are less effectively taken up by macrophages but interact to a greater extent with epithelial lining cells than large particles. Due to their vast numbers, they provide a very large surface area, which is the interface by which they interact with biological systems. Generally, depending on their molecular surface composition, nanoparticles may have a greater capacity to induce or mediate more adverse effects than larger particles.

4.2.4
Translocation of Ultrafine Particles into Systemic Circulation

While a growing number of reports confirms that there is translocation of ultrafine particles into blood circulation and subsequent uptake in secondary target organs, the size of the translocated fraction, the transport mechanisms and the rate-determining parameters are under debate.

4.2.4.1 Studies of Systemic Particle Translocation in Humans

Conflicting translocation data are reported in human studies. Nemmar et al. [22] have demonstrated a rapid 3–5% uptake of radiolabeled carbonaceous ultrafine particles into the bloodstream within minutes after exposure and subsequent uptake in the liver. In this study, leaching of the radiolabel from the particles was not considered. In contrast, also using radiolabeled carbonaceous ultrafine particles, Brown et al. [18] could not find any detectable particulates (<2% of inhaled ultrafine particles, limit of detection) beyond the lungs when the data were corrected for leaching of the radiolabel off the particles. In another study on healthy subjects as well as mild asthmatics and smokers we found no significant clearance from the lungs of 100 nm carbonaceous particles, for which we carefully controlled radiolabel leaching off the particles to be <2% within 24 h [23]. Due to the limit of detection used in this study systemic translocation and subsequent uptake in organs like the liver was below 1% of the deposited particles.

4.2.4.2 Studies of Systemic Particle Translocation in Animals

Oberdörster and coworkers have observed rapid translocation towards the liver of more than 50% of ^{13}C labeled ultrafine carbonaceous particles (26 nm size) within 24 h in a rat model [24]. Takenaka et al. [25] showed about 5–10% ultrafine silver particle translocation to the liver. Kreyling et al. [26], however, observed only minute (<1%) translocation of iridium ultrafine particles (15–20 nm and 80 nm in size) into the blood of rats. However, these test particles did not only accumulate in the liver, but also in spleen, kidneys, brain and heart to similar fractions. Uptake of the 15–20 nm particles in secondary target organs was about a factor of 2–3 higher than for 80 nm particles. In this study we challenged the question of particle dissolution and were able to show that these ultrafine particles dissolved to a very small extent, for which the particle data were corrected. Interestingly, in a long-term retention study Semmler et al. [27] have shown that ultrafine iridium particle contents in each of these secondary target organs did not increase with increasing retention time after a single 1-h exposure but peaked after one week at about 0.5% in each secondary organ. Thereafter, fractions declined again and remained detectable but below 0.1% of the initial deposit throughout the six-months-period of observation, indicating clearance mechanisms in these organs.

However, even though the mass fractions of iridium particles were rather low in secondary target organs the number of particles is impressively high. Data at one week and six months after the single inhalation are shown for lungs and all secondary target organs – liver, spleen, heart, brain and kidneys – in Table 4.1 (data taken from Refs. [26, 27]). More than one billion particles were found in each of the secondary target organs one week after a single 1-h exposure; and more than 100 million particles were still determined six months after the inhalation.

Importantly, particle uptake in the brain was not via the neural pathway from the olfactorial epithelium in the nose, since extrathoracic airways of the rats were bypassed when ventilated through an endotracheal tube during exposure. Therefore, two principal pathways from the lung epithelium towards the brain are possible: (1) along neural axons and synapses [28–30], as reviewed by Oberdörster et al. [7] or (2) via systemic circulation. While the second route, via blood circulation, intuitively seems likely by analogy to uptake in other secondary organs the neural route should be kept in mind when designing the next generation of investigations. For

Tab. 4.1. Retained mass fractions as well as the corresponding numbers of insoluble iridium particles in the lungs and in secondary target organs one week and six months after single 1 h inhalation of 15 nm-sized iridium particles by WKY rats [26, 27]. The corresponding surface area of the retained particles is also calculated, based on the BET surface area of 123 m^2 g^{-1} or 1500 m^2 cm^{-3} (per mass or per volume of particles, respectively).

Organs	Retained mass fraction		Retained particles number		Particle surface area (cm^2)	
	One week	Six months	One week	Six months	One week	Six months
Lungs	0.6	0.06	7.00×10^{11}	7.00×10^{10}	1.26×10^2	12.6
Liver	0.006	0.0005	7.00×10^9	5.83×10^8	1.26	0.105
Spleen	0.004	0.0003	4.66×10^9	3.50×10^8	0.842	6.32×10^{-2}
Heart	0.004	0.0005	4.66×10^9	5.83×10^8	0.842	0.105
Brain	0.003	0.0005	3.50×10^9	5.83×10^8	0.632	0.105
Kidney	0.006	0.0001	7.00×10^9	1.17×10^8	1.26	2.11×10^{-2}

instance, Hunter and Undem [31] have demonstrated the transport of fluorescent 40 nm polystyrene nanoparticles from the nerve endings in the tracheal epithelium along their neurons to their cell body in the ganglion nodosum and jugular ganglia in the neck of guinea pigs along neurons innervating the trachea. Furthermore, translocation along olfactory nerves from the olfactory epithelium to the olfactory bulb was first reported by Howe and Bodian [32] for 0.03-μm polio virus in monkeys, and was later described for nasally deposited colloidal 0.05-μm gold particles, moving into the olfactory bulb of squirrel monkeys [29]. Carbonaceous ultrafine particles were reported to translocate also along the same pathway to the central nervous system (CNS), based on their presence in the olfactory bulb of rats after inhalation [30, 33].

In addition, the surface area of the retained iridium particles was calculated based on the BET specific surface area of 123 m^2 g^{-1} or 1500 m^{-2} cm^{-3} per mass or volume of particles [34]; (BET surface area was determined by nitrogen absorption measurements). Since there is now evidence that the iridium particles are covered with iridium oxide [35], the density of iridium oxide (11.7 g cm^{-3}) and not that of iridium (22.4 g cm^{-3}) was used for calculations. One week after inhalation the total surface area of the retained iridium particles is close to 1 cm^2 and after six months data are still a tenth of the one-week data. Compared to 1-μm-sized particles this retained surface area of the 15 nm particles in secondary target organs is five orders of magnitude larger because of their very large specific surface area. It remains to be investigated what impact that may have, and whether the large number of retained particles and the accordingly large particle surface area may have any adverse effects on the surrounding biological microstructures, like proteins, extracellular fluids, cells and their multiple functional and structural compartments as well as whole tissue of organs that usually are not considered to be exposed to such foreign bodies.

Recent studies of Nemmar and coworkers have shed light on possible prothrombotic effects in the systemic circulation as a result of activation of platelets [36, 37]. In their hamster model of experimentally induced thrombus formation they observed thrombotic clots in peripheral veins and arteries after intravenous and intratracheal administration of positively charged 60 nm polystyrene nanoparticles. They emphasized the importance of size and surface charge of test particles. In fact, the induced thrombus formation was not detectable after administration of negatively charged or neutral 60 nm as well as 400 nm sized positively charged polystyrene test particles; astonishingly, they observed similar thrombus formation after application of diesel exhaust particles via both routes [36]. These observations were confirmed by another group using a slightly modified approach [38]. From the fact that they observed peripheral thrombus formation even after intratracheal nanoparticle instillation into the lungs they concluded particle translocation from the lung epithelium to circulation. Whether particle translocation really occurred or whether thrombus formation was indirectly triggered by mediators released in the presence of the particles in the lungs needs to be proven in the next generation of studies. Nevertheless, the presence of these ultrafine particles with specific properties obviously was able to trigger biological responses that may initiate adverse health effects.

The fact that surface properties and possibly the particle matrix may play an important role in systemic translocation was demonstrated in a previous study [39, 40] in which ultrafine titanium dioxide (TiO_2) particles were produced basically by the same method as the iridium particles, yielding primary particles of about 5 nm, very similar to those of iridium. Figure 4.3 shows the aggregation of both particle agglomerates of about 20 nm size. TiO_2 agglomerates were inhaled by endotracheally intubated and ventilated WKY rats and the lungs were morphometrically studied in great detail immediately after the 1-h inhalation and 24 h later to determine the location of the retained TiO_2 particles within the various lung compartments. Interestingly, a substantial fraction was already found at each time point in the vascular compartment, including the endothelium and the vascular lumen, indicating a rapid systemic translocation, which was obviously not the case after the inhalation of iridium particles. There are hints that these TiO_2 show a positive zeta potential similar to commercially available TiO_2 nanoparticles and as opposed to the negative zeta potential of iridium particles.

In studies using radioactively labeled ultrafine particles, the observed translocated fractions always appeared to be higher when less care was taken to minimize particle dissolution and/or the leaching of the radiolabel from the particles. Likewise, analysis of the stable ^{13}C isotope-label is hampered by the fact that natural ^{13}C occurs in all biological tissues at the level of 1% of the other stable isotope ^{12}C, and ultrafine particle deposition in the lungs of the exposed rats made up only about a tenth of the natural ^{13}C in the lungs; therefore, slight variability of the

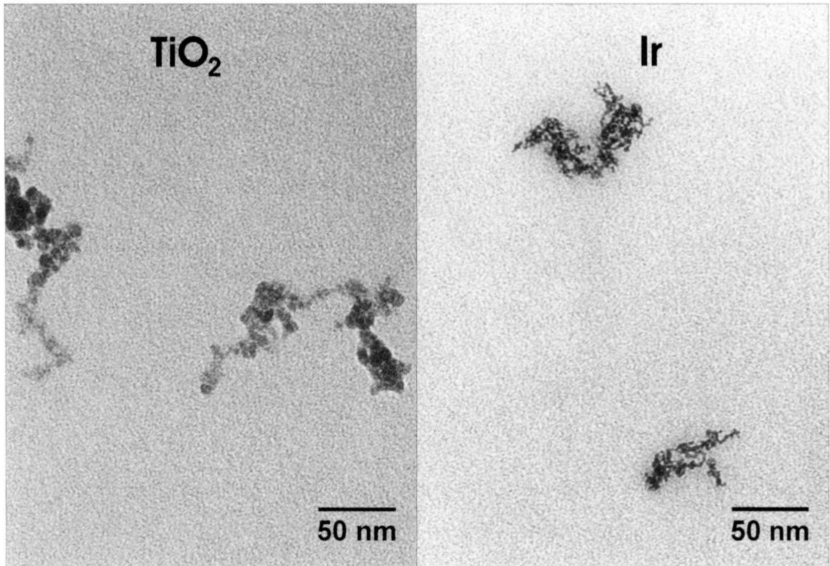

Fig. 4.3. Agglomerates of primary titanium dioxide particles (left) and iridium particles (right); primary particles are in the size range of 5 nm.

abundant natural ^{13}C or uncertainties of the estimate of total carbon content in a given organ or tissue may lead to erroneous estimates.

Although the first data on susceptible humans – asthmatics and adults – did not alter the translocation pathway, a recent mechanistic study on excised blocks of heart and lungs of rats indicated increased translocation rates when either or both was compromised experimentally by induced epithelial inflammation or histamine activated endothelium [41].

Recently, more attention has been paid to surface characteristics and charges that could influence this process, as is well known in drug delivery [42]. Studies of drug delivery across the blood–brain barrier further confirmed the importance of surface properties, showing that particle surface components may bind to the ApoE-receptor of endothelial cells, which mediates crossing of this otherwise very tight barrier [43]. In addition, current discussion focuses on other transport pathways: clathrin-mediated endocytosis and non-clathrin-mediated endocytosis, the latter including internalization via caveolae [44–46]. Vesicular caveolae migrate from the luminal to the mucosal side of epithelial and endothelial cells [47]. After internalization caveolae are involved in endocytosis, transcytosis, and pinocytosis. In a recent study Rejman and coworkers showed that both pathways are clearly size-dependent for ligand-devoid polystyrene particles [48]. Using non-phagocytic cells, internalization of microspheres with a diameter < 200 nm involved clathrin-coated pits. With increasing size, a shift to a mechanism that relied on caveolae-mediated internalization became apparent, which became the predominant pathway of entry for particles of 500 nm in size. Under these conditions, delivery to the lysosomes was no longer apparent. The data indicate that the size itself of (ligand devoid) particles can determine the pathway of entry. The clathrin-mediated pathway of endocytosis shows an upper size limit for internalization of approx. 200 nm, and kinetic parameters may determine the almost exclusive internalization of such particles along this pathway rather than via caveolae. These studies on non-phagocytic melanoma cells B16 are supported by others using primary oral or esophageal epithelial cells [49, 50]. Likewise, endothelial cells are known to have bottleshape-like caveolaer invaginations of the plasma membrane [46]. Interestingly, studies aiming for cell transfection and gene delivery are focusing on these transport pathways [44–46].

Besides transcellular pathways para- or intercellular pathways are other routes by which nanoparticles can penetrate membranous cell barriers. Recently, Heckel and coworkers have shown in an inflammatory rabbit model induced by infused lipopolysaccharide (LPS) that intravenously administered 8 nm gold particles coated with autologous albumin were found on their way between endothelial as well as epithelial cells of the alveolar air–blood barrier [51]. The control rabbits showed the same pathway at a lower extent reflecting the transfer of albumin from the vascular to the luminal side of the air–blood barrier also under physiological conditions. Besides this paracellular pathway they also found albumin-coated gold particles in intracellular caveolae vesicles, indicative of transcellular pathways in both the control and the inflammatory model. At the same time it shows that nanoparticles coated with the appropriate serum protein can rapidly be translocated across a membrane, particularly when there is a protein gradient between the two sides of the barrier. It remains open whether these albumin-modulated para- and transcellular pathways are important in human lungs in the reverse direction, when an inhaled particle deposited on the alveolar epithelium is translocated into blood circulation.

4.3
Toxicological Plausibility of Health Effects Caused by Nanoparticles

Toxicological experiments can be categorized into *in vivo* and *in vitro* experiments. *In vivo* experiments investigate effects in living organisms such as experimental animals or healthy subjects and patients in clinical studies, whereas *in vitro* experiments are conducted in organs, tissues, cells or biomolecules isolated from the living organism. In general, toxicology studies on air pollutants are shorter-term experimental approaches that tend, for ethical reasons, not to study people but experimental animals. They often analyze the early events rather than waiting for final disease. In addition, in those studies high doses are required to detect significant effects; and they may use cells and biochemical systems rather than whole animals. To compensate for such shortcomings, models of susceptibility are an important experimental approach in which the biological system is predisposed prior to treatment with air pollutants. Predisposition of the system acts already like a change of the homeostatic balance; such systems may respond more vigorously than the normal system. If, in addition, the predisposition reflects or resembles human diseases or human predisposition, such a model may be a powerful analytical tool, providing insights on how air pollutants may change responses of the susceptible system.

While human clinical studies provide closer insight in nanoparticle related modes of action, they usually are limited in yielding firm dose-response-relationships because of ethical reasons. Such relationships, however, can be provided by toxicological studies on susceptible animal models as well as cell systems. Hence, animal studies are supposed to provide biological explanations and plausibility and create hypotheses that may then be proven in human clinical studies.

4.3.1
Pulmonary Inflammation Induced by Ultrafine Particles

Many of the effects that occur rapidly after ultrafine particles deposit on the respiratory epithelium are not fully understood. Cells in contact with ultrafine particles like macrophages, epithelial cells and neutrophilic granulocytes are activated and may synthesize compounds referred to as *reactive oxygen or nitrogen species* (including free radicals, hydrogen peroxide, and superoxide) that try to inactivate and eliminate the invading foreign material [52]. Within hours, cytokines and chemokines are synthesized and secreted into the affected area. These molecules are mediators that interact with specific receptors on the surfaces of many cell types and thereby activate cells in the surrounding environment as well as in the blood and other tissues. As a result, cells leave the bloodstream and enter the fluid filled interstitial spaces, where they can attack the foreign material. Consequently, particle-induced cell activation events in the airways frequently result in an *inflammatory response*. This response includes both the activation of cells of the epithelium (including the production of "pro-inflammatory" and reactive oxygen molecules described above) and the activation and migration of cells (particularly, neutrophilic granulocytes and eosinophilic granulocytes in the case of a specific immunological response) from the blood into the airways. While the inflammatory response may occur after interaction of both fine and ultrafine particles with respiratory tissues, the enhanced surface area of ultrafine and nanoparticles compared to fine particles (as the acting interface) suggests a more prominent reaction.

Airway nerve cells may also contribute to inflammation in the airways by synthesizing neurotransmitters [53]. In this *neurogenic inflammation*, the neurotransmitters can affect many types of white blood cells in the lung, as well as epithelial and smooth muscle cells. Inflammatory cytokines synthesized by white blood cells may also affect the nerve cells. The inflammatory response may damage the epithelial cell layer at the surface of the tissue and other cells in the airway (such as macrophages), which results in the loss of integrity of the tissue's defenses. One potential consequence may be increased exposure to and reduced capacity to defend against microorganisms.

Thus, particle deposition on the respiratory epithelium can trigger a cascade of events in many different cells, potentially resulting in changes in tissues and organs at sites progressively further from the initial stimulus. These defense mechanisms are normal responses in healthy individuals, but they may lead to deleterious changes in the host. Such changes may be rapid and temporary and may resolve quickly but, depending on the level and pattern of exposure and the agent to which the host is exposed to, they may last longer. It is not clear whether or how such modulations are relevant to the development of particle-induced adverse health effects at low levels of exposure. Yet, these changes are thought to have a greater impact on individuals whose respiratory, cardiac, or vascular tissues have been previously altered or damaged.

One possible consequence of damage to the airways is that the individual may become more susceptible to respiratory infections if exposed to viruses or bacteria, as discussed in Ref. [54]. A second possible consequence is that it may further decrease respiratory function in a person whose airways are already damaged by diseases such as bronchitis or asthma. As a result, symptoms of asthma, for example, may be exacerbated [55].

4.3.2
Systemic Inflammation and other Responses

Recent studies have suggested that exposure to particles results in systemic inflammatory effects within hours [56]. Pathways are discussed as either via direct particle translocation into circulation or via mediators released in the respiratory tract. In particular, the former pathway supports the concept of enhanced cardiovascular responses after ultrafine or nanoparticles exposure due to the higher likelihood of ultrafine particle translocation into circulation when compared with larger particles. Recently, a panel of cardiologists has reviewed the existing literature and compiled a statement in which possible pathways of interference of particulate matter are summarized [57]. They have developed a schematic (Fig. 4.4), which has been adopted in this chapter.

The panel link particle exposure with cardiovascular disease via four major routes: pulmonary inflammation, pulmonary reflexes and systemic translocation to circulation and to the heart. It is currently unclear whether the systemic response is a consequence of an inflammatory response in the respiratory tract, because some studies on systemic inflammation have detected little or no inflammatory lung response after exposure to PM. As described above, there are studies that indicate that either particles per se (ultrafine and nanoparticles in particular) or components that may detach or dissolve from particles may move rapidly into the circulation, triggering either oxidative stress or pro-thrombotic or acute-phase or other responses of the cardiovascular system [22, 56, 58]. Therefore, a direct systemic inflammatory response is possible via particle translocation into the circulation. This pathway of systemic inflammation is thought to be capable of triggering a cascade of responses, leading eventually to atherosclerotic plaque rupture and/or thrombosis as precursors of myocardial infarction. The other two pathways

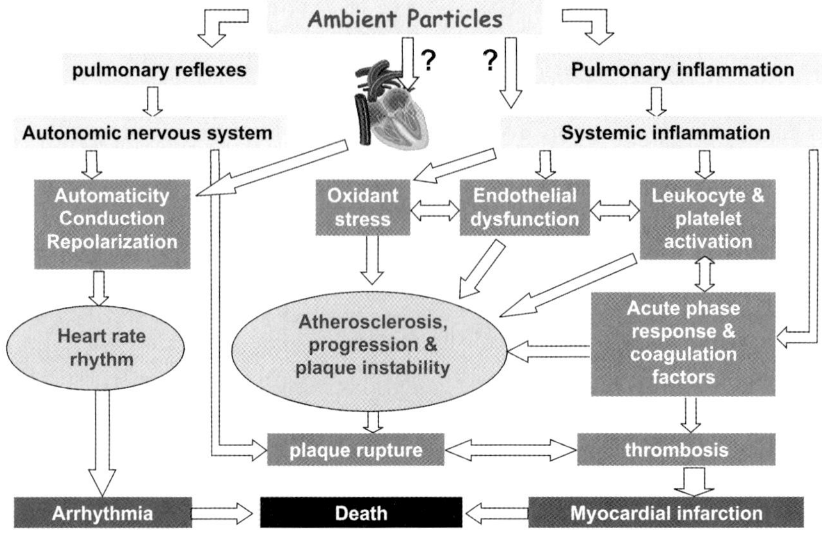

Fig. 4.4. Mechanisms involved in the link between cardiovascular disease and PM exposure. (After Ref. [57].)

of pulmonary reflexes and direct particle action on the heart are thought to interfere with heart rate functions, leading to arrhythmia. Both myocardial infarctions and arrhythmia are severe diseases with a substantial mortality rate.

4.3.3
Relevant Parameters in Nanoparticle Toxicology

Biologically, it appears unmeaningful to presume that low solubility ultrafine or nanoparticles are hazards per se, since adverse reactions result from the interaction between the ultrafine particle and biological tissues. Currently, certain parameters of ultrafine particles are considered to trigger or mediate a cascade of reactions, starting with the formation of free radicals, which lead to oxidative stress in extracellular matrix and cells with the subsequent onset of pro-inflammatory processes (outlined in Fig. 4.5) [2, 11, 59]. These particle parameters are now discussed below.

4.3.3.1 Number Concentration and Surface Area

Considering health effects initiated by the exposure to ultrafine and nanoparticles requires a change of the paradigm that effects are correlated with the mass of noxae accumulated during exposure. In ambient air the mass concentration of ultrafine particles is usually less than 10% of the mass concentration of $PM_{2.5}$. However, the number concentration of ultrafine particles dominates the number concentra-

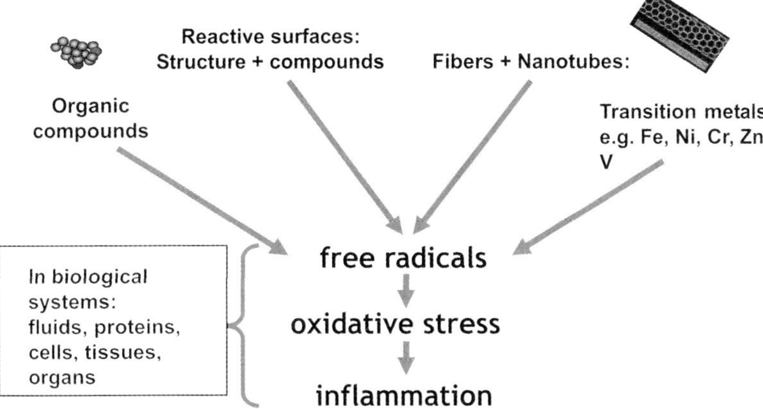

Fig. 4.5. Parameters of ultrafine particles considered to be involved in the initiation of oxidative stress and inflammatory processes. (Adapted from [3].)

tion of fine particles by >90% [60]. Therefore, surface area and number concentration appear to be the more reasonable metrics of ultrafine and nanoparticle exposure than mass concentration. Furthermore, exposure metrics may be inadequate since it may be the number of deposited particles per unit surface area or dose to a specific cell (e.g., alveolar macrophage) that determines response for specific regions. Therefore, the use of a metric depends on specific questions posed, requiring specifically defined metrics.

Oberdörster and co-workers have tested the relation between surface area of particles and inflammatory response [61]. Ultrafine TiO_2 with an average particle size of 20 nm and pigment grade (fine) TiO_2 with an average particle size of about 250 nm were used. Doses ranging from 30 to 2000 µg of TiO_2 were intratracheally instilled into rats and mice. When the deposited TiO_2 dose was expressed as particle surface area, there was a unique relationship to the inflammatory responses of these two different sizes of TiO_2 particles. The importance of particle surface area for eliciting inflammatory responses in the lung has been confirmed [62, 63]. This concept of particle surface area as the appropriate dose metric has been recognized as an important principle in particulate matter toxicology [11, 64, 65].

4.3.3.2 Particle Shape (Fibers and Nanotubes)

Newer materials or those under development, such as synthetic organic fibers and carbon nanotubes, may have different toxicology paradigms [3]. The existing paradigm for silicate fibers suggests that respirable fiber types vary in their ability to cause lung disease and that this can be understood on the basis of the length of the fibers and their biopersistence in the lungs. Because fibers are regulated on a fiber number basis and the hazard is understood on the basis of the number of long fibers, in fiber testing the dose should always be expressed as fiber number (not mass), and the length and diameter distribution also need to be known. Carbon nanotubes are long thin structures that can have diameters of a few nanometers but be many thousands of nanometers long. These could have very unusual toxicological properties, in that they share shape characteristics of both fibers and nanoparticles; such limited toxicology as presently exists supports the contention that these may be harmful to the lungs [3, 66]. Thus, the physiological relevance of these findings needs ultimately be determined by conducting inhalation toxicity studies.

4.3.3.3 Transition Metals

For more than 10 years transition metals have been suspected and proved to cause health effects. Convincing evidence has been provided by a combination of epidemiologic and subsequent toxicological studies. An epidemiological study showed reduced effects of several morbidity endpoints as well as mortality in the population of Utah Valley, Utah, during a one-year period when a steel mill had been shut down, resulting in considerable reduction of transition metals in ambient fine particles, while symptoms and metal-containing air pollution were high in the year prior and after the closure of the steel mill [67–69]. Dust sampled from each period was applied in a human clinical study [70] as well as in a mechanistic study demonstrating pulmonary injury, neutrophilic inflammation and increased airway responsiveness of the metal-rich samples using a sensitive animal model [71]. Hence, the clinical and toxicological animal study provided a better understanding of the modes of actions of the ambient particles that had been found to be associated with adverse health effects within the population of Utah Valley.

Similar combined studies are being applied to air pollution in the city of Hettstedt, in former Eastern Germany, which has a history of several centuries of nonferrous metal mining and smelting in comparison with the city of Zerbst, serving as a control in a nearby agricultural area. High $PM_{2.5}$ levels were shown to be associated with significant decline of lung function and with significant increase of prevalence for bronchitis, otitis media, frequent colds, and febrile infections in three surveys on children over a decade in the 1990s. At the same time, high levels of the transition metals zinc, lead, copper, and cadmium were associated with allergic responses [55, 72, 73]. Dusts from Hettstedt and Zerbst were studied in an allergic mouse model, showing increased allergic responses and increased allergic sensitization [74]. This dust also had the capacity to form radical species [75] and further clinical human studies demonstrated that dust obtained from Hettstedt induced distinct airway inflammation in healthy subjects with a selective influx of monocytes and increased generation of oxidant radicals [76].

Both series of combinatory "epi-tox" studies provide comprehensive evidence for the association of adverse heath effects in susceptible population groups and modes of actions of these ambient particles in either clinical and animal models studies or in *in vitro* studies. Importantly, $PM_{2.5}$ is the basic dose metric in these studies and not ultrafine particle parameters. Note that in the second series of epidemiological studies ultrafine particle number concentration was measured in addition and dominated the number concentration of the fine particle fraction by 90% [77]. Therefore, additional research is required to demonstrate whether the ultrafine particle fraction plays a role in the observed effects and modes of actions and whether they may even drive the effects because of the very peculiar properties of the ultrafine particle fraction.

4.3.3.4 Organic Compounds

Organic chemicals associated with ultrafine particles play a role in the pro-inflammatory effects of diesel exhaust particles (DEP), as demonstrated in several *in vitro* studies. For example, DEP caused modest stimulation of interleukin-8 (IL-8), granulocyte macrophage colony-stimulating factor (GM-CSF) and RANTES production by epithelial cells, and this activity was lost on extraction of the organic matter [78]; the benzene extracts contained most of the stimulatory activity seen in the whole DEP. Benzene extracts contained almost 90% of the benzo[*a*]pyrene, B(a)P, content and the authors concluded that poly aromatic hydrocarbons (PAHs) such as B(a)P were likely responsible for the stimulation of cytokine production by the epithelial cells. Boland et al. have demonstrated that DEP stimulated IL-8, GM-CSF, and IL-1β release from the bronchial epithelial cell-line 16HBE [79]. Furthermore, they contended, this was related to the amount of adsorbed organic compounds, because carbon black with virtually no adsorbed organic matter did not cause cytokine release. In support of this, exhaust gas post-treatments that diminished the adsorbed organic compounds also reduced the DEP-induced increase in GM-CSF release. Further studies with the organic extracts confirmed that most of the stimulating activity was in the organic fraction [80]. In another study, PAH extracted from DEP induced expression of IL-8 and RANTES in peripheral blood mononuclear cells [81], demonstrating that both macrophages and epithelial cells could be important in the pro-inflammatory effect induced by DEP in the lungs. Chin and coworkers [82] demonstrated that carbon-black treatment of the RAW264.7 mouse macrophage cell line had no effect on TNFα release but that the addition of B(a)P to the particles caused them to become stimulatory for TNFα.

More recently, investigators started to find evidence that amongst the large variety of organic compounds, particularly in the particulate fraction of ambient air originating from combustion processes, there are biologically highly reactive compounds like redox cycling quinones – oxidized and nitrated polyaromatic hydrocarbons, which can catalyze release of reactive oxygen species (ROS), leading to the induction of oxidative stress and inflammation. PM from the Los Angeles basin as well as organic extracts obtained from DEP induce a stratified oxidative stress response leading to Heme-oxygenase-1 expression, followed by activation of Jun kinase and pro-inflammatory interleukin-8 production, and culminated in cellular apoptosis in parallel with a sharp decline of antioxidant levels. These effects were more prominent in the fine particle fraction than the coarse, and they were positively correlated with higher contents of organic carbon and polyaromatic hydrocarbons [83]. Admittedly, the ultrafine particle fraction was not analyzed explicitly; however, the organic carbon load is highly associated with the ultrafine particle fraction since this fraction predominantly originates from combustion processes. Pro-inflammatory effects in the respiratory tract are related to the particle content of redox cycling chemicals and are involved in the adjuvant effect of DEP in atopic sensitization [84]. Cytotoxicity in epithelial cells and macrophages is the result of mitochondrial damage, which manifests as ultramicroscopic changes in organelle morphology, a decrease in the mitochondrial membrane potential, superoxide production, and ATP depletion.

4.3.3.5 Extrapolation of Health Effects Observed in Animals towards Human

Comparison between rodents and humans is rather difficult due to anatomical and physiological differences, which can result in considerably lower concentrations to sensitive regions of the respiratory tract of animals compared to similar regions in humans.

Laboratory animals used in toxicological studies are genetically very similar within specific strains, whereas human populations are heterogeneous. Thus, extrapolating results from animals to humans must not only take strain and species differences into account, but must also consider inter-individual variation among humans.

With animal studies, the presumed susceptible part of the human population is mimicked by inducing specific cardiopulmonary diseases or focusing on senescent animals. For instance, animal research has been done in the laboratory using a model of asthma, increased blood pressure in lung arteries, lung inflammation and general high blood pressure. An advantage of these models is that hypotheses on the mechanisms of action and biological plausibility can be examined. Using DEP, residual oil fly ash or ambient particles, it has been demonstrated in models for respiratory infection and allergy that symptoms can exacerbate. A drawback with these disease models in animals is that they are not completely equivalent to the human disease counterpart. Yet the development of susceptible animal models is a prerequisite to the search for biological plausibility of the higher vulnerability of susceptible and diseased individuals. A second issue is a certain degree of uncertainty as to whether the laboratory animal used is representative of the reactions of a human being, or whether the endpoints examined are sensitive enough and representative for human endpoints. An additional consideration is whether the timing of exposure and observation has been correctly chosen in the animal studies. However, studies in animals continue to be the main methodology on which to predict adverse health outcomes in humans and to clarify the relationship between exposure dose and toxic effects.

Since the development of ambient particle related health effects is a multifactorial process that may start with genetic predisposition and is propagated by lifelong exposure to air pollution and a history of diseases, acute effect studies in animal models seem not to be likely to mimic this progression, which would require long-term exposure studies to carefully controlled fractions of air pollutants. Whether long-term exposures to nanoparticles will occur and whether such precautions are required will depend on the use of future nanoparticles.

4.4
Integrated Concept of Risk Assessment of Nanoparticles

Nanoparticles are expected to be used in a wide range of new technologies. To highlight the importance of sustainable risk assessment, let us consider the use of nanoparticles in medicine: in this field nanoparticles are designed for therapeutic drug delivery and/or imaging techniques in diagnostics to be administered directly to the patient as multifunctional drug nanocarriers. This may require targeting across several membranes while the drug is sufficiently bound to the nanocarrier, and specific and controlled release of the drug at the target site, allowing for increased efficiency of the drug in target organs or cells, while side effects are minimized in sensitive but not-targeted organs and tissues [9]. This also includes non-toxic effects of the carrier nanoparticles. Although these may be future visions, nanomedicines are likely to represent the most challenging nanoparticles in terms of their safe and sustained application since patients predisposed by their disease are likely to be more susceptible to any such treatments, responding eventually more sensitively than healthy subjects. Hence, the rapid development of a multitude of nanoparticle applications needs to be complemented by assessing possible implications, assuring safe and sustainable handling of those nanoparticles. The challenge of an integrated application development and implication assessment is pro-active collaboration at the earliest stage to optimize functionality of the nanoparticle and to minimize its side effects without losses in terms of costs and time because of one-sided mismanagement or unfocussed or delayed initiation of risk assessment. While most other nanoparticles are not aimed for specific administration in human subjects, possible exposure scenarios still need to be considered.

Hence we propose an integrated conception to estimate the health hazards of newly generated nanoparticles during their development. For this risk analysis the whole life-cycle of the newly developed nanoparticle has to be considered, including its scientific or industrial generation, its storage and distribution, its anticipated application and possible abuse, and finally its disposal. That means potential human incorporation may vary at different stages of the life cycle of nanoparticles and different groups of the population may be exposed. During generation an occupational group of healthy adult workers may be exposed to nanoparticles, while its anticipated application may possibly lead to exposure of the whole population, including susceptible people, like infants, children, elderly and diseased individuals. The challenge is to produce a reasonable estimate that takes into account the widespread use of mass-wise produced nanoparticles versus those produced in small quantities.

As a result, possible incorporation pathways can be foreseen for the respective usage of a given nanoparticle, which will include organs of uptake, organs involved in the distribution and secondary target organs. In all these organs the nanoparticles or their metabolic products eventually accumulate in different ways. Within these organs, interaction of the nanoparticles will take place at the level of proteins of body fluids and on the cell surface and within cells; this will eventually be the beginning of a cascade of reactions, mediating or initiating adverse health effects, leading to disease, as outlined above. Therefore, only those proteins and cells need to be examined that interact with the nanoparticle. Nevertheless, these may be more than those within the organ of predominant uptake and anticipated target organs or tissues.

To assess the interaction between nanoparticles and biological materials, a strategic concept is necessary that (a) classifies nanoparticles according to their chemical compounds, physical structure and, particularly, such properties at their surface; (b) takes into account possible exposures in different phases of the nanoparticle's life cycle; (c) estimates delivered doses to the various biological systems; and (d) stratifies toxicological assessment from simple high-throughput screening methods towards more complex *in vivo* studies only when required and based on previous findings at a lower level of analysis. Initially, simple acellular tests may evaluate free radical formation, oxidative stress, antigen–antibody reactions, etc. followed by genetic (regulation of cytokines and mediators, nucleus-signaling) and proteomic (structural and functional modification of proteins) high-throughput methods. Actually, these methods, which aim to understand the underlying mechanisms, will predominantly make use of modern nanotechnology, such as gene- and protein-array chip technology specifically designed to screen nanoparticle–gene and –protein interactions. This genetic and proteomic information will provide guidance to the next step of assessment. As a result a second series of biologically and toxicologically more relevant tests aimed toward specific reactions may become necessary – first on cells, followed by multicell models and *in vivo* animal models, which finally may require clinical phase trials in the case of medicinal nanoparticles. Figure 4.6 shows a schematic of this integrated concept.

From this structured approach, life-cycle-specific recommendations for regulation will be derived for each nanoparticle. Such an approach would even allow the use of a nanoparticle that may have an elevated risk at some stage of its life as an

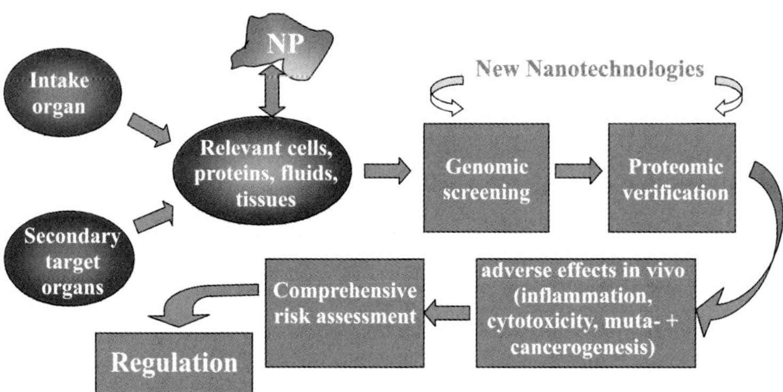

Fig. 4.6. Integrated concept for sustained toxicological risk assessment in relation to newly developed ultrafine particles and nanoparticles.

overall risk–benefit analysis may still be in support of its use while including specific prevention measures when they become necessary – e.g., in the application of nanoparticles in medicine. In fact, controlled application of new emerging nanoparticles will require knowledge-based development, production and application, including sustainable risk assessment prior to their widespread use.

Public perception is an important factor for the future development of nanosciences and nanotechnologies. Efforts need to be undertaken to convince the public of the beneficial potential of nanosciences and nanotechnologies. It would be a tragedy if a major accident jeopardized in part, or even largely, the development of this future technology with its splendid prognoses.

References

1 BSI-PAS71, British Standard Institution (BSI), **2005**, www.bsi-global.com/nano.

2 BORM, P. J., KREYLING, W., Toxicological hazards of inhaled nanoparticles – potential implications for drug delivery. *J. Nanosci. Nanotechnol.* **2004**, 4, 521–531.

3 DONALDSON, K., TRAN, C. L., An introduction to the short-term toxicology of respirable industrial fibres. *Mutat. Res.* **2004**, 553, 5–9.

4 KREYLING, W. G., SEMMLER, M., MÖLLER, W., Dosimetry and toxicology of ultrafine particles. *J. Aerosol Med.* **2004**, 17, 140–152.

5 The Royal Society (2004). Nanosciences and nanotechnologies: opportunities and uncertainties. The Royal Society and The Royal Academy of Engineering, London, UK. http://www.ost.gov.uk/policy/issues/introduction.htm.

6 SCHULZ, H., HARDER, V., IBALD-MULLI, A., KHANDOGA, A., KOENIG, W., KROMBACH, F., RADYKEWICZ, R., STAMPFL, A., THORAND, B., PETERS, A., Cardiovascular effects of fine and ultrafine particles. *J. Aerosol Med.* **2005**, 18, 1–22.

7 OBERDÖRSTER, G., OBERDÖRSTER, E., OBERDÖRSTER, J., Nanotoxicology: An emerging discipline evolving from studies of ultrafine particles. *Environ. Health Perspect.* **2005**, 113, 823–839.

8 SEATON, A., DONALDSON, K., Nanoscience, nanotoxicology, and the need to think small. *Lancet* **2005**, 365, 923–924.

9 ESF, European Science Foundation Policy Briefings., **2005**, http://www.esf.org/newsrelease/83/SPB23Nanomedicine.pdf.

10 WICHMANN, H. E., PETERS, A., Epidemiological evidence of the effects of ultrafine particle exposure. *Philos. Trans. Roy. Soc. A* **2000**, 358, 2751–2769.

11 DONALDSON, K., STONE, V., BORM, P. J., JIMENEZ, L. A., GILMOUR, P. S., SCHINS, R. P., KNAAPEN, A. M., RAHMAN, I., FAUX, S. P., BROWN, D. M., MACNEE, W., Oxidative stress and calcium signaling in the adverse effects of environmental particles (PM10). *Free Radic. Biol. Med.* **2003**, 34, 1369–1382.

12 POPE, C. A., III, Epidemiology of fine particulate air pollution and human health: biologic mechanisms and who's at risk? *Environ. Health Perspect.* **2000**, 108(Suppl. 4), 713–723.

13 US Environmental Protection Agency (EPA). Air quality criteria for particulate matter. EPA/600/p-99/022aD and bD. 2004. Research Triangle Park, NC, USEPA, National Center for Environmental Assessment.

14 OBERDOERSTER, G., FERIN, J., LEHNERT, B. E., Correlation between particle size, in vivo particle persistence, and lung injury. *Environ. Health Perspect.* **1994**, 102(Suppl 5), 173–179.

15 SCHULZ, H., BRAND, P., HEYDER, J., Particle deposition in the respiratory tract. In *Particle-Lung Interactions* (ed., P. GEHR, J. HEYDER), Marcel Dekker Inc., New York, Basel, **2000**, pp. 229–290.

16 ICRP Publication 66, Human respiratory tract model for radiological protection. A report of a Task Group of the International Commission on Radiological Protection. *Ann. ICRP* **1994**, 24, 1–482.

17 ASGHARIAN, B., HOFMANN, W., BERGMANN, R., Particle deposition in a multiple-path model of the human lung. *Aerosol Sci. Technol.* **2001**, 34, 332–339.

18 BROWN, J. S., ZEMAN, K. L., BENNETT, W. D., Ultrafine particle deposition and clearance in the healthy and obstructed lung. *Am. J. Respir. Crit. Care Med.* **2002**, 166, 1240–1247.

19 OBERDÖRSTER, G., Lung clearance of inhaled insoluble and soluble particles. *J. Aerosol Med.* **1988**, 1, 289–330.

20 OBERDÖRSTER, G., Lung dosimetry: pulmonary clearance of inhaled particles. *Aerosol Sci. Technol.* **1993**, 18, 279–289.

21 KREYLING, W. G., SCHEUCH, G., Clearance of particles deposited in the lungs. In *Particle-Lung Interactions* (ed. P. GEHR, J. HEYDER), Marcel Dekker Inc., New York, Basel, **2000**, pp. 323–376.

22 NEMMAR, A., HOET, P. H., VANQUICKENBORNE, B., DINSDALE, D., THOMEER, M., HOYLAERTS, M. F., VANBILLOEN, H., MORTELMANS, L., NEMERY, B., Passage of inhaled particles into the blood circulation in humans. *Circulation* **2002**, 105, 411–414.

23 WIEBERT, P., SANCHEZ-CRESPO, A., SEITZ, J., FALK, R., PHILIPSON, K., KREYLING, W. G., MÖLLER, W., SOMMERER, K., LARSSON, S., SVARTENGREN, M., High retention of 100 nm 99mTc-labeled carbonaceous particles in healthy and affected human lungs. *Eur. Respir. J.* **2006**, submitted.

24 OBERDÖRSTER, G., SHARP, Z., ATUDOREI, V., ELDER, A., GELEIN, R., LUNTS, A., KREYLING, W. G., COX, C., Extrapulmonary translocation of ultrafine carbon particles following whole-body inhalation exposure of rats. *J. Toxicol. Environ. Health A* **2002**, 65, 1531–1543.

25 TAKENAKA, S., KARG, E., ROTH, C., SCHULZ, H., ZIESENIS, A., HEINZMANN, U., SCHRAMEL, P., HEYDER, J., Pulmonary and systemic distribution of inhaled ultrafine silver particles in rats. *Environ. Health Perspect.* **2001**, 109, 547–551.

26 KREYLING, W. G., SEMMLER, M., ERBE, F., MAYER, P., TAKENAKA, S., SCHULZ, H., OBERDÖRSTER, G., ZIESENIS, A., Translocation of ultrafine insoluble iridium particles from lung epithelium to extrapulmonary organs is size dependent but very low. *J. Toxicol. Environ. Health A* **2002**, 65, 1513–1530.

27 SEMMLER, M., SEITZ, J., ERBE, F., MAYER, P., HEYDER, J., OBERDÖRSTER, G., KREYLING, W. G., Long-term clearance kinetics of inhaled ultrafine insoluble iridium particles from the rat lung, including transient translocation into secondary organs. *Inhal. Toxicol.* **2004**, 16, 453–459.

28 BODIAN, D., HOWE, H. A., The rate of progression of poliomyelitis virus in nerves. *Bull. Johns Hopkins Hospital* **1941**, 69, 79–85.

29 DE LORENZO, A. J. D., DARIN, J., The olfactory neuron and the blood-brain barrier. In *Taste and Smell in Vertebrates* (ed. G. E. W. WOLSTENHOLME, J. KNIGHT), Churchill, London, **1970**, pp. 151–176.

30 OBERDÖRSTER, G., SHARP, Z., ATUDOREI, V., ELDER, A., GELEIN, R., KREYLING, W., COX, C., Translocation of inhaled ultrafine particles to the brain. *Inhal. Toxicol.* **2004**, 16, 437–445.

31 HUNTER, D. D., UNDEM, B. J., Identification and substance P content of vagal afferent neurons innervating the epithelium of the guinea pig trachea. *Am. J. Respir. Crit. Care Med.* **1999**, 159, 1943–1948.

32 HOWE, H. A., BODIAN, D., Poliomyelitis in the chimpanzee: A clinical–pathological study. *Proc. Soc. Exp. Biol. Med.* **1940**, 43, 718–721.

33 OBERDÖRSTER, G., UTELL, M. J., Ultrafine particles in the urban air: to the respiratory tract – and beyond? *Environ. Health Perspect.* **2002**, 110, A440–441.

34 ROTH, C., FERRON, G. A., KARG, E., LENTNER, B., SCHUMANN, G., TAKENAKA, S., HEYDER, J., Generation of ultrafine particles by spark discharging. *Aerosol Sci. Technol.* **2004**, 38, 228–235.

35 SZYMCZAK, W., KREYLING, W. G., SEITZ, J., WITTMAACK, K., Mass spectrometric characterisation of pure and mixed ultrafine particles of iridium and carbon. *J. Aerosol Sci.* **2004**, 35, 37–38.

36 NEMMAR, A., HOYLAERTS, M. F., HOET, P. H., DINSDALE, D., SMITH, T., XU, H., VERMYLEN, J., NEMERY, B., Ultrafine particles affect experimental thrombosis in an in vivo hamster model. *Am. J. Respir. Crit. Care Med.* **2002**, 166, 998–1004.

37 NEMMAR, A., HOET, P. H., DINSDALE, D., VERMYLEN, J., HOYLAERTS, M. F., NEMERY, B., Diesel exhaust particles in lung acutely enhance experimental peripheral thrombosis. *Circulation* **2003**, 107, 1202–1208.

38 SILVA, V. M., CORSON, N., ELDER, A., OBERDÖRSTER, G., The rat ear vein model for investigating in vivo thrombogenicity of ultrafine articles (UFP). *Toxicol. Sci.* **2005**, 85, 983–989.

39 KAPP, N., KREYLING, W., SCHULZ, H., IM HOF, V., GEHR, P., SEMMLER, M., GEISER, M., Electron energy loss spectroscopy for analysis of inhaled ultrafine particles in rat lungs. *Microsc. Res. Technol.* **2004**, 63, 298–305.

40 GEISER, M., ROTHEN-RUTISHAUSER, B., KAPP, N., SCHÜRCH, S., KREYLING, W. G., SCHULZ, H., SEMMLER, M., IM HOF, V., HEYDER, J., GEHR, P., Ultrafine particles cross cellular membranes by non-phagocytic mechanisms in lungs and in cultured cells. *Environ. Health Perspect.* **2005**, 113, 1555–60.

41 MEIRING, J. J., BORM, P. J., BAGATE, K., SEMMLER, M., SEITZ, J., TAKENAKA, S., KREYLING, W. G., The influence of hydrogen peroxide and histamine on lung permeability and translocation of iridium nanoparticles in the isolated perfused rat lung. *Part. Fibre Toxicol.* **2005**, 2, 3.

42 BROOKING, J., DAVIS, S. S., ILLUM, L., Transport of nanoparticles across the rat nasal mucosa. *J. Drug Target.* **2001**, 9, 267–279.

43 KREUTER, J., SHAMENKOV, D., PETROV, V., RAMGE, P., CYCHUTEK, K., KOCH-BRANDT, C., ALYAUTDIN, R., Apolipoprotein-mediated transport of nanoparticle-bound drugs across the blood-brain barrier. *J. Drug Target.* **2002**, 10, 317–325.

44 NICHOLS, B., Caveosomes and endocytosis of lipid rafts. *J. Cell Sci.* **2003**, 116, 4707–4714.

45 PARTON, R. G., RICHARDS, A. A., Lipid rafts and caveolae as portals for endocytosis: new insights and common mechanisms. *Traffic* **2003**, 4, 724–738.

46 BATHORI, G., CERVENAK, L., KARADI, I., Caveolae – an alternative endocytotic pathway for targeted drug delivery. *Crit. Rev. Ther. Drug Carrier Syst.* **2004**, 21, 67–95.

47 GUMBLETON, M., Caveolae as potential macromolecule trafficking compartments within alveolar epithelium. *Adv. Drug Deliv. Rev.* **2001**, 49, 281–300.

48 REJMAN, J., OBERLE, V., ZUHORN, I. S., HOEKSTRA, D., Size-dependent internalization of particles via the pathways of clathrin- and caveolae-mediated endocytosis. *Biochem. J.* **2004**, 377, 159–169.

49 HOPWOOD, D., SPIERS, E. M., ROSS, P. E., ANDERSON, J. T., MCCULLOUGH, J. B., MURRAY, F. E., Endocytosis of fluorescent microspheres by human oesophageal epithelial cells: comparison between normal and inflamed tissue. *Gut* **1995**, 37, 598–602.

50 INNES, N. P., OGDEN, G. R., A technique for the study of endocytosis in human oral epithelial cells. *Arch. Oral Biol.* **1999**, 44, 519–523.

51 HECKEL, K., KIEFMANN, R., DORGER, M., STOECKELHUBER, M., GOETZ, A. E., Colloidal gold particles as a new in vivo marker of early acute lung injury.

Am. J. Physiol. Lung Cell Mol. Physiol. **2004**, 287, L867–L878.

52 NEL, A. E., DIAZ-SANCHEZ, D., LI, N., The role of particulate pollutants in pulmonary inflammation and asthma: evidence for the involvement of organic chemicals and oxidative stress. *Curr. Opin. Pulm. Med.* **2001**, 7, 20–26.

53 BARNES, P. J., Neurogenic inflammation in the airways. *Respir. Physiol.* **2001**, 125, 145–154.

54 GILMOUR, P. S., RAHMAN, I., HAYASHI, S., HOGG, J. C., DONALDSON, K., MACNEE, W., Adenoviral E1A primes alveolar epithelial cells to PM(10)-induced transcription of interleukin-8. *Am. J. Physiol. Lung Cell Mol. Physiol.* **2001**, 281, L598–606.

55 HEINRICH, J., HOELSCHER, B., FRYE, C., MEYER, I., WJST, M., WICHMANN, H. E., Trends in prevalence of atopic diseases and allergic sensitization in children in Eastern Germany. *Eur. Respir. J.* **2002**, 19, 1040–1046.

56 DONALDSON, K., STONE, V., SEATON, A., MACNEE, W., Ambient particle inhalation and the cardiovascular system: potential mechanisms. *Environ. Health Perspect.* **2001**, 109, 523–527.

57 BROOK, R. D., FRANKLIN, B., CASCIO, W., HONG, Y., HOWARD, G., LIPSETT, M., LUEPKER, R., MITTLEMAN, M., SAMET, J., SMITH, S. C., JR., TAGER, I., Air pollution and cardiovascular disease: a statement for healthcare professionals from the Expert Panel on Population and Prevention Science of the American Heart Association. *Circulation* **2004**, 109, 2655–2671.

58 NEMMAR, A., VANBILLOEN, H., HOYLAERTS, M. F., HOET, P. H., VERBRUGGEN, A., NEMERY, B., Passage of intratracheally instilled ultrafine particles from the lung into the systemic circulation in hamster. *Am. J. Respir. Crit. Care Med.* **2001**, 164, 1665–1668.

59 DONALDSON, K., The biological effects of coarse and fine particulate matter. *Occup. Environ. Med.* **2003**, 60, 313–314.

60 KREYLING, W. G., TUCH, T., PETERS, A., PITZ, M., HEINRICH, J., STÖLZEL, M., CYRYS, J., HEYDER, J., WICHMANN, H. E., Diverging long-term trends in ambient urban particle mass and number concentrations associated with emission changes caused by the German unification. *Atmos. Environ.* **2003**, 37, 3841–3848.

61 OBERDÖRSTER, G., Toxicology of ultrafine particles: in vivo studies. *Philos. Trans. Roy. Soc. A* **2000**, 358, 2719–2739.

62 LI, X. Y., GILMOUR, P. S., DONALDSON, K., MACNEE, W., Free radical activity and pro-inflammatory effects of particulate air pollution (PM10) in vivo and in vitro. *Thorax* **1996**, 51, 1216–1222.

63 FAUX, S. P., TRAN, C. L., MILLER, B. G., JONES, A. D., MONTEILLER, C., AND DONALDSON, K. (2003). In vitro determinants of particulate toxicity: The dose-metric for poorly soluble dusts. Health and Safety Executive, Crown, Norwich, UK.

64 OBERDÖRSTER, G., Significance of particle parameters in the evaluation of exposure-dose-response relationships of inhaled particles. *Inhal. Toxicol.* **1996**, 8, 73–89.

65 DONALDSON, K., LI, X. Y., MACNEE, W., Ultrafine (nanometre) particle mediated lung injury. *J. Aerosol Sci.* **1998**, 29, 553–560.

66 WARHEIT, D. B., LAURENCE, B. R., REED, K. L., ROACH, D. H., REYNOLDS, G. A., WEBB, T. R., Comparative pulmonary toxicity assessment of single-wall carbon nanotubes in rats. *Toxicol. Sci.* **2004**, 77, 117–125.

67 POPE, C. A., III, Respiratory disease associated with community air pollution and a steel mill, Utah Valley. *Am. J. Public Health* **1989**, 79, 623–628.

68 RANSOM, M. R., POPE, C. A., III, Elementary school absences and PM10 pollution in Utah Valley. *Environ. Res.* **1992**, 58, 204–219.

69 POPE, C. A., III, SCHWARTZ, J., RANSOM, M. R., Daily mortality and PM10 pollution in Utah Valley. *Arch. Environ. Health* **1992**, 47, 211–217.

70 GHIO, A. J., DEVLIN, R. B., Inflammatory lung injury after bronchial instillation of air pollution particles. *Am. J. Respir. Crit. Care Med.* **2001**, 164, 704–708.

71 Dye, J. A., Lehmann, J. R., McGee, J. K., Winsett, D. W., Ledbetter, A. D., Everitt, J. I., Ghio, A. J., Costa, D. L., Acute pulmonary toxicity of particulate matter filter extracts in rats: coherence with epidemiologic studies in Utah Valley residents. *Environ. Health Perspect.* **2001**, 109, 395–403.

72 Heinrich, J., Hoelscher, B., Wichmann, H. E., Decline of ambient air pollution and respiratory symptoms in children. *Am. J. Respir. Crit. Care Med.* **2000**, 161, 1930–1936.

73 Heinrich, J., Hoelscher, B., Frye, C., Meyer, I., Pitz, M., Cyrys, J., Wjst, M., Neas, L., Wichmann, H. E., Improved air quality in reunified Germany and decreases in respiratory symptoms. *Epidemiology* **2002**, 13, 394–401.

74 Gavett, S. H., Bishop, L. R., Haykal-Coates, N., Heinrich, J., Gilmour, M. I., Effects of particles from two German cities on allergic responses in mice. *Am. J. Respir. Crit. Care Med.* **2001**, 163, A50.

75 Shi, T., Schins, R. P., Knaapen, A. M., Kuhlbusch, T. A. J., Pitz, M., Heinrich, J., Borm, P. J. A., Hydroxyl radical generation by electron paramagnetic resonance as a new method to monitor ambient particulate matter composition. *J. Environ. Monitor.* **2003**, 5, 550–556.

76 Schaumann, F., Borm, P. J., Herbrich, A., Knoch, J., Pitz, M., Schins, R. P., Luettig, B., Hohlfeld, J. M., Heinrich, J., Krug, N., Metal-rich ambient particles (particulate matter 2.5) cause airway inflammation in healthy subjects. *Am. J. Respir. Crit. Care Med.* **2004**, 170, 898–903.

77 Pitz, M., Kreyling, W. G., Holscher, B., Cyrys, J., Wichmann, H. E., Heinrich, J., Change of the ambient particle size distribution in East Germany between 1993 and 1999. *Atmos. Environ.* **2001**, 35, 4357–4366.

78 Kawasaki, S., Takizawa, H., Takami, K., Desaki, M., Okazaki, H., Kasama, T., Kobayashi, K., Yamamoto, K., Nakahara, K., Tanaka, M., Sagai, M., Ohtoshi, T., Benzene-extracted components are important for the major activity of diesel exhaust particles: effect on interleukin-8 gene expression in human bronchial epithelial cells. *Am. J. Respir. Cell Mol. Biol.* **2001**, 24, 419–426.

79 Boland, S., Baeza-Squiban, A., Fournier, T., Houcine, O., Gendron, M. C., Chévrier, M., Jouvenot, G., Coste, A., Aubier, M., Marano, F., Diesel exhaust particles are taken up by human airway epithelial cells in vitro and alter cytokine production. *Am. J. Physiol.* **1999**, 276, L604–613.

80 Boland, S., Bonvallot, V., Fournier, T., Baeza-Squiban, A., Aubier, M., Marano, F., Mechanisms of GM-CSF increase by diesel exhaust particles in human airway epithelial cells. *Am. J. Physiol.* **2000**, 278, L25–32.

81 Fahy, O., Tsicopoulos, A., Hammad, H., Pestel, J., Tonnel, A. B., Wallaert, B., Effects of diesel organic extracts on chemokine production by peripheral blood mononuclear cells. *J. Allergy Clin. Immunol.* **1999**, 103, 1115–1124.

82 Chin, B. Y., Choi, M. E., Burdick, M. D., Strieter, R. M., Risby, T. H., Choi, A. M., Induction of apoptosis by particulate matter: role of TNF-alpha and MAPK. *Am. J. Physiol.* **1998**, 275, L942–949.

83 Li, N., Kim, S., Wang, M., Froines, J., Sioutas, C., Nel, A., Use of a stratified oxidative stress model to study the biological effects of ambient concentrated and diesel exhaust particulate matter. *Inhal. Toxicol.* **2002**, 14, 459–486.

84 Li, N., Wang, M., Oberley, T. D., Sempf, J. M., Nel, A. E., Comparison of the pro-oxidative and proinflammatory effects of organic diesel exhaust particle chemicals in bronchial epithelial cells and macrophages. *J. Immunol.* **2002**, 169, 4531–4541.

5

Impact of Ceramic and Metallic Nano-scaled Particles on Endothelial Cell Functions *in Vitro*

Kirsten Peters, Ronald E. Unger, Antonietta M. Gatti,
Enrico Sabbioni, Andrea Gambarelli, and C. James Kirkpatrick

5.1
Introduction

5.1.1
Origin of Particles in the Human Environment

The term "particles" is defined as "very small pieces of solid or liquid matter, light enough to be suspended in the air". The human body is exposed to many types of particles during its lifetime. These particles can be of natural origin or they can develop as (by-)products of industrial processes, and technical or pharmaceutical engineering. Thus, these particles vary largely in composition, size, shape, surface property and habitat. Dependent on these characteristics, the internalization of particles and their dissemination within the body is variable and may take place by ingestion (by polluted food or food additives) [1], inhalation (smoking, diesel soot, medical aerosols), via the skin (e.g., cosmetics, pharmaceutics), implantation (e.g., wear from implants), and also injection (e.g., drug delivery and cancer therapy) [2]. Particulate air pollution is associated with enhanced mortality from respiratory and cardiovascular diseases [3]. The World Health Organisation (WHO) estimates that inhalation of particulate matter is responsible for at least 500 000 deaths each year worldwide [4].

Particles can be divided into those that are internalized accidentally (e.g., by air or food pollution) and those that are administered intentionally (e.g., for drug delivery, diagnostic agents). Among the naturally occurring particles in the air are pollen grains and their fragments and other vegetable particles, such as starch granules, which might, for example, carry mold spores with high allergenic potential. Furthermore, particles from volcanic eruptions occur naturally in the atmosphere. Particles evolved by pollution are from industry, motor vehicles, and other sources of thermodegradation. Furthermore, man-made/engineered particles have become relevant in recent years, e.g., due to the enormous progress in industrial use (e.g., in automotive, electronic, textile, household, and chemical industries)

Nanotechnologies for the Life Sciences Vol. 5
Nanomaterials – Toxicity, Health and Environmental Issues. Edited by Challa S. S. R. Kumar
Copyright © 2006 WILEY-VCH Verlag GmbH & Co. KGaA, Weinheim
ISBN: 3-527-31385-0

and in pharmaceutical development of nanometer-scaled particles for drug delivery in asthma or cancer diagnosis and therapy [2, 5]. However, this rapid technical development has led to concerns about unknown risks of engineered nanometer-scaled materials [6].

5.1.1.1 Evidence for Size-dependent Toxicity of Particles

Among the most abundant air pollutants in urban areas is particulate matter with a mean diameter of ≤10 µm (also called PM_{10}, defined as particulate matter 10 µm in diameter and smaller, by environmental toxicologists). Over the years it has become clear that particles with very low sizes (especially those below 100 nm) are more noteworthy than larger particles since they induce more severe effects [7, 8]. There are different reasons for this phenomenon:

1. Particle size as a limiting factor of accessibility to the body's organs and tissues.
   For example, intact pollen grains, >10 µm in diameter, are too large to enter the lower airways. They are eliminated by mucociliary clearance. However, the naturally occurring break-up of pollen grains (e.g., by osmotic shock) produces smaller pollen fragments that can reach the lower parts of the airways and exert asthma-inducing effects in sensitized patients [9, 10].
2. Particle size as a limiting factor of accessibility to cells and cell compartments.
   Recent studies indicate that particle size alone can strongly affect the efficiency of cellular uptake [11].
3. The surface/size-ratio increases exponentially with decreasing particle sizes, leading to increased surface reactivity. This increased surface reactivity might lead to greater biological activity per given mass compared to larger particles, which in turn might have effects on, for example, particle internalization into tissues, cells and organelles, toxicity, or the induction of oxidative stress [2, 12].

This leads to the question: Are materials that are generally recognized as safe as bulk materials by the accredited standard tests also safe as (nano-scaled) particles? To date no standards exist for testing the safety of nanoparticles.

As mentioned above, particles with sizes below 100 nm are especially problematic. By definition, particles smaller than 100 nm are called nanoparticles or ultrafine particles. Nanoparticles are at least 100-fold smaller than mammalian cells and are mostly smaller than viruses. Since the diameter of DNA molecules is 2 nm and atoms have diameters between 0.1 and 0.4 nm, concern about the interference of small nanoparticles with cellular structures at the molecular level is legitimate.

5.1.1.2 Dissemination and Interferences of Nanoparticles within the Body

Owing to the minute size of nanoparticles, internalization into the body's tissues appears to be extremely easy. This has been shown by experiments in human volunteers with radioactive-labeled carbon nanoparticles (i.e., "Technegas") that passed rapidly into the systemic circulation after inhalation. Radioactivity could already be detected in blood after 1 min of inhalation [13]. Furthermore, animal studies revealed that inhaled nanoparticles were translocated into the liver [14] and into the brain [15]. Thus, nanoparticles seem to be able to circumvent the tight blood–brain barrier; the movement of nanoparticles across the blood–placenta barrier has also been discussed [16, 17].

The incidence of higher asthma frequency during severe air pollution episodes has long been known [18]. Recent studies indicate that the ultrafine particles in air pollution are especially important in the course of asthma and chronic obstructive pulmonary disease (COPD) [19, 20]. Furthermore, other tissues in addition to the lung seem to be affected by nanoparticle exposure: In mice exposed to nanoparticles in ambient air the levels of pro-inflammatory cytokines were increased in brain tissue so that a coherency between inhaled particulate matter and the development of neurodegenerative diseases was suggested [21]. Moreover, nanoparticles are suggested to be involved in thrombus formation in the blood [22, 23]. There is evidence that fine and ultrafine particles are involved in the pathogenesis of Crohn's disease, a transmural inflammation of the gastrointestinal tract [24].

5.1.1.3 Endothelial Cells and Nanoparticle Exposure

As the sources of internalized nanoparticles (food, air, etc.) and the location of particle detection are generally far apart, a distribution via the blood stream must have occurred. Thus, endothelial cells, which line the inner surface of blood vessels, will have direct contact with the particles. Endothelial cells are important in inflammation and wound healing. Upon pro-inflammatory stimulation of the endothelium, adhesion molecules are expressed on the cell surface, thus mediating leukocyte attachment (e.g., E-selectin and intercellular adhesion molecule-1/ICAM-1). Furthermore, endothelial cells are able to release cytokines, such as interleukin-8 (IL-8, a key factor in neutrophil chemotaxis). Thus, these features contribute to the pro-inflammatory endothelial phenotype that permits the transmigration of leukocytes from the blood into the perivascular space [25]. Activation of IL-8, E-selectin and ICAM-1 is regulated by the same transcription factors, NF-κB (nuclear factor-κB) and AP-1 (activator protein-1) [26–28].

5.1.1.4 Testing of Nanoparticle-induced Effects on Human Endothelial Cells *In Vitro*

Little is known about the effects of nanoparticles on endothelial cell functions. Therefore, their effects on human endothelial cells have been studied *in vitro* and are reported here. Ceramic nanoparticles of TiO_2 and SiO_2 and metallic nanoparticles of Co and Ni were examined with respect to cellular internalization and their influence on cell viability, proliferative activity, and the pro-inflammatory endothelial phenotype. Moreover, due to the effects of the metallic nanoparticles they were compared with metal ion treatment. Endothelial cells *in vitro* were able to internalize many particles and reacted differentially in response to the internalization, dependent on the composition of the different nanoparticles. Furthermore, divergent effects of metallic nanoparticles vs. metal ions were observed. The link between these results and the possible risk of nanoparticles to human health is also discussed.

5.2
Materials and Methods

All chemicals were obtained from Sigma if not otherwise indicated.

5.2.1
Cell Culture

Human dermal microvascular endothelial cells (HDMEC) were isolated from juvenile foreskin as described before [29] and cultured in Endothelial Cell Basal Medium MV (PromoCell) supplemented with 15% fetal calf serum (Invitrogen), basic fibroblast growth factor (bFGF, 2.5 ng mL^{-1}), sodium heparin (10 µg mL^{-1}), penicillin/streptomycin (10 000 units penicillin per mL, 10 000 µg streptomycin sulfate per mL, Invitrogen), cultivated in a humidified atmosphere at 37 °C (5% CO_2) and used in passage 4.

5.2.2
Particles

SiO_2 and TiO_2 particles were produced by flame spray pyrolysis (TAL Materials Inc.). The size spectrum of SiO_2 particles was between 4 and 40 nm with 14 nm mean particle size. The TiO_2 particles were between 20 and 160 nm with 70 nm mean particle size. The mean size of Co particles was 28 nm (Nanoamor) and the Ni particles had a mean size of 62 nm (Nanoamor). Particles were added to the cell culture medium and tested at three different concentrations (0.5, 5, and 50 µg per mL of culture medium).

Particles were analyzed by means of an Environmental Scanning Electron Microscope (ESEM-Quanta, FEI-Company). This instrument is called "Environmental" by the Manufacturer, as it can analyze samples in many different modes: at high and medium vacuum, and also at environmental conditions. It can also accept wet or oily samples, which is ideal for biological specimens.

5.2.3
Transmission Electron Microscopy (TEM)

Cells were seeded onto fibronectin-coated Thermanox coverslips (Nunc). Exposure to particles was performed two days after seeding (50 µg particles per mL medium). After 48 h incubation cells were fixed in cacodylate-buffered glutaraldehyde (2.5%) and embedded in Agar100 (Plano). Ultrathin sections were made with an Ultracut E microtome (Leica). TEM was performed with a Phillips 410 EM (Phillips).

5.2.4
Cytotoxicity Assay

To evaluate cytotoxicity the CellTiter 96® AQueous non-radioactive assay (Promega) was performed according to the manufacturer's instructions. This assay gives a measure of the enzymatic conversion of a tetrazolium salt (MTS reagent) by mitochondrial dehydrogenase and thus presents indirect evidence for cell viability.

5.2.5
Detection of Ki67 Expression

Cells were seeded onto fibronectin-coated 96-well microtiter plates (6500 cells per well) and grown to subconfluence. Afterwards, cells were exposed to particles (0.5, 5, and 50 µg mL^{-1} culture medium) and cultivated for an additional 24 h. The cells were fixed with methanol:ethanol (2:1, 15 min, room temperature) and permeabilized with buffered 0.1% Triton X-100 (5 min, room temperature).

Ki67, a protein expressed in the nucleus of proliferating cells, was detected with mouse-anti human Ki67-antibody (Dako). The secondary antibody was a peroxidase-conjugated rabbit-anti mouse-antibody (Dako). The staining reaction was performed by addition of the peroxidase substrate (*o*-phenylenediamine dihydrochloride) for 15 min at 37 °C. The staining reaction was stopped with 3 M HCl. Light extinction was determined with a microtiter plate photometer (ThermoLab Systems) at 492 nm.

5.2.6
Quantification of IL-8 Release in Cell Culture Supernatant

Cells were seeded onto fibronectin-coated microtiter plates (13 500 cells per well), grown for 24 h and exposed to particles (0.5, 5, and 50 µg per mL of culture medium for different samples) and TNFα (300 U mL^{-1}; inflammatory control). Cell culture supernatants were collected 24 h after substance or particle exposure. The IL-8 content in supernatants was assayed using human IL-8 immunoassay/ELISA (Hiss Diagnostics) according to the manufacturer's instructions.

5.2.7
Quantification of E-selectin Cell Surface Protein Expression

This cell surface antigen is generally only expressed in inflammatory-stimulated endothelial cells and can be detected by using an enzyme-linked immunoassay based on a peroxidase staining reaction and subsequent dye quantification by a microplate reader. Therefore, cells were seeded onto fibronectin-coated 96-well microtiter plates (13 500 cells per well) and grown to confluence. Cells were then subjected to specific cell culture conditions (different particles and TNFα as a positive control, 300 U mL^{-1}) and cultivated for an additional 4 h. The cells were fixed with methanol:ethanol (2:1, 15 min, room temperature).

E-selectin was detected with mouse-anti human E-selectin-antibody (Bender MedSystems). The secondary antibody was a biotinylated goat-anti mouse-antibody (Amersham). Afterwards, the streptavidin–horseradish peroxidase conjugate was added (Amersham). The staining reaction was performed by addition of the peroxidase substrate (*o*-phenylenediamine dihydrochloride) for 15 min at 37 °C. The staining reaction was stopped with 3 M HCl. Light extinction was determined with a microtiter plate spectrophotometer (ThermoLab Systems) at 492 nm.

5.2.8
Fluorescence Staining

HDMEC were seeded onto fibronectin-coated glass chamber-slides (Nunc). After 48 h, cells were exposed to particles (50 µg mL^{-1}) or CoCl$_2$ (0.7 mM), incubated for an additional 24 h and fixed with buffered 3.7% paraformaldehyde (15 min, room temperature). Staining for Hypoxia-inducible factor-1α (HIF-1α) was performed with the HIF-1α-antibody (IgG1, BD Transduction Laboratories). Nuclear staining was performed with Hoechst 33342. Fluorescence-labeled cells were covered with GelMount (Biomeda/Natutec).

5.2.9
Statistical Analysis

All results are shown as means $\pm$ standard deviations (SD). Statistical analysis was carried out with Microsoft Excel software. According to the results of variance ratio analysis (F-test $p < 0.05$) an unpaired t-test for either homoscedastic or heteroscedastic variances was performed ($p < 0.05$ or $p < 0.001$ as indicated in the figures).

5.3
Results

Analysis of the different nanoparticles acquired by Environmental Scanning Electron Microscopy (ESEM) revealed a relative homogenous particle size for the ceramic nanoparticles (TiO$_2$, Fig. 5.1a; SiO$_2$, Fig. 5.1b). According to the manufacturer's specification the size spectrum of TiO$_2$ particles was between 20 and 160 nm with 70 nm mean particle size and that of SiO$_2$ particles between 4 and 40 nm with 14 nm mean particle size. In contrast, the metallic nanoparticles of Co and Ni were more inhomogeneous (Co, Fig. 5.1c; Ni, Fig. 5.1d). Both particle types possess nanoparticle character since the specified mean sizes of Co and Ni particles were 28 and 62 nm, respectively.

Ultrastructural studies (TEM) from perpendicular sections of endothelial cell monolayers demonstrated a flattened cell phenotype. Cytoplasm of the untreated control cells contained numerous organelles and vacuoles (Fig. 5.2a, arrowhead: vacuole with autophagic function containing cellular debris). When HDMEC were exposed to the different particles, internalization of the nanoparticles occurred that was independent of particle composition. The particles were localized within cytoplasmic vacuoles, partially containing cellular debris. Both the TiO$_2$ and SiO$_2$ particles were partially detectable as large aggregations and partially as smaller particulate matter. However, striking ultrastructural changes were not observed (Fig.

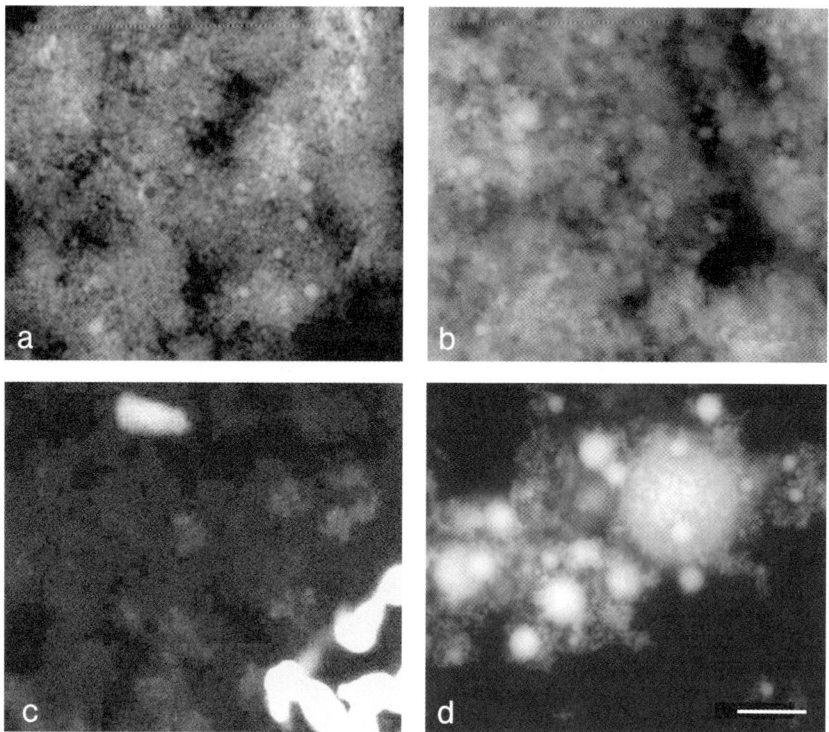

Fig. 5.1. ESEM images of the different nanoparticles used in this study. (a) TiO$_2$, (b) SiO$_2$, (c) Co, (d) Ni (scale bar: 2 μm).

5.2b/c). The exposure to Co (Fig. 5.2d) and Ni (Fig. 5.2e) particles led to the enlargement of vacuoles; simultaneously, the number of vacuoles appeared to decrease. The Co- and Ni-particle-induced vacuoles partially filled the complete height of the cells (Fig. 5.2d/e). Interestingly, Co-particle exposure induced some annular-shaped, electron-dense material within the vacuoles (Fig. 5.2d, asterisk). Beside this electron-dense material the vacuoles of the Co-particle exposed cells contained a large amount of cellular debris.

Exposure of TiO$_2$ particles to HDMEC did not induce an effect on cell number within 24 h (Fig. 5.3a; detected by a DNA staining with crystal violet; 0.5, 5 and 50 μg mL^{-1} were each tested). Also, the staining for the proliferation marker Ki67 after 24 h (Fig. 5.3b) and metabolic activity after 72 h (shown by the MTS conversion assay, Fig. 5.3c) did not show significant changes after exposure to TiO$_2$ particles. High amounts of SiO$_2$ particles (50 μg mL^{-1}) induced a slight decrease in cell number after 24 h (Fig. 5.3a). This SiO$_2$-particle-induced decrease is also reflected by a slight decrease of Ki67 protein expression after 24 h (Fig. 5.3b). However, the MTS conversion assay showed no significant reduction after 72 h (Fig.

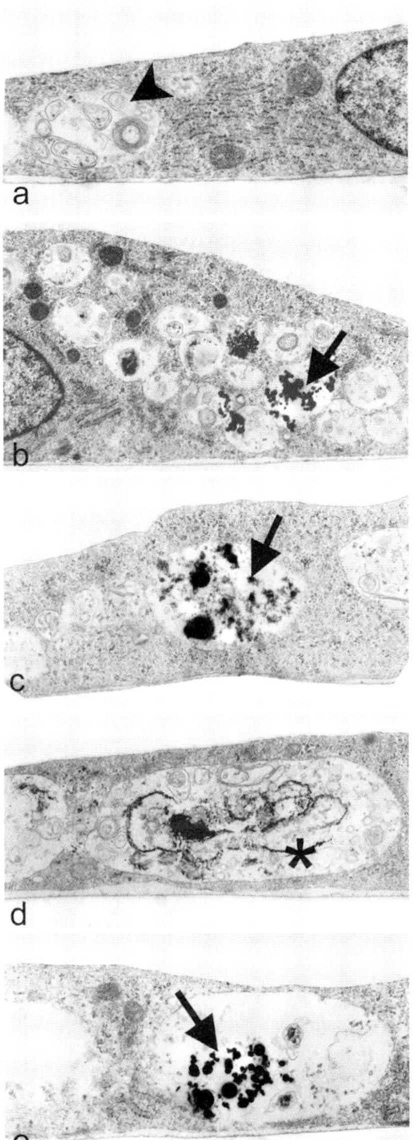

Fig. 5.2. Perpendicular sections of HDMEC monolayers: (a) Nontreated HDMEC (control) and HDMEC exposed to particles of (b) TiO_2, (c) SiO_2, (d) Co and (e) Ni (TEM, magnification 26000×).

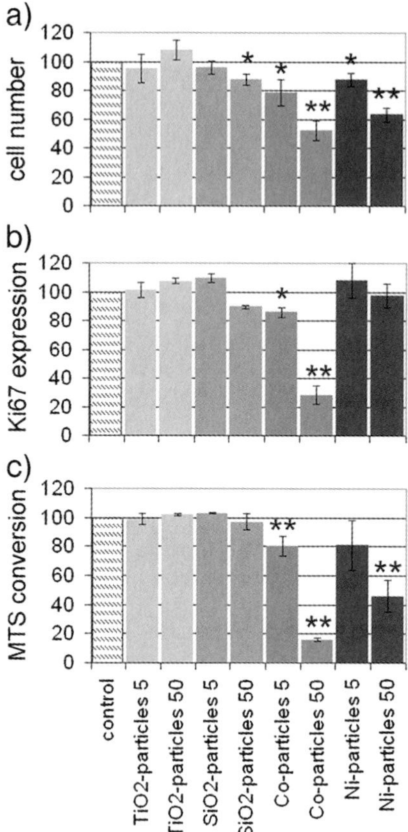

Fig. 5.3. Cytotoxicity of different nanoparticles. Tests were for (a) cell number after exposure of 24 h (crystal violet staining), (b) proliferation after 24 h particle exposure (Ki67 detection), and (c) metabolic activity after 72 h (MTS conversion). Particle amounts are in µg mL^{-1}, untreated control set as 100%; $n = 4$, means $\pm$ SDs, significantly different from normoxia: $^*p < 0.05$, $^{**}p < 0.001$.

5.3c). Co and Ni particles induced a significant, concentration-dependent decrease in cell number after 24 h (Fig. 5.3a). Also, the reduction of Ki67 expression and MTS conversion in Co-particle treated cells was significantly reduced after Co-particle exposure (Fig. 5.3b/c). Interestingly, the Ni particles, which also induce a concentration-dependent decrease in cell number, showed no significant deviations in protein expression of Ki67 (Fig. 5.3b). Nickel particles led to a significant, concentration-dependent reduction of MTS conversion after 72 h (Fig. 5.3c). Thus, protein expression of the proliferation marker Ki67 was not reduced after Ni particle exposure, although cell number and metabolic activity were decreased.

Four hours after exposure of the different particles the cells did not show

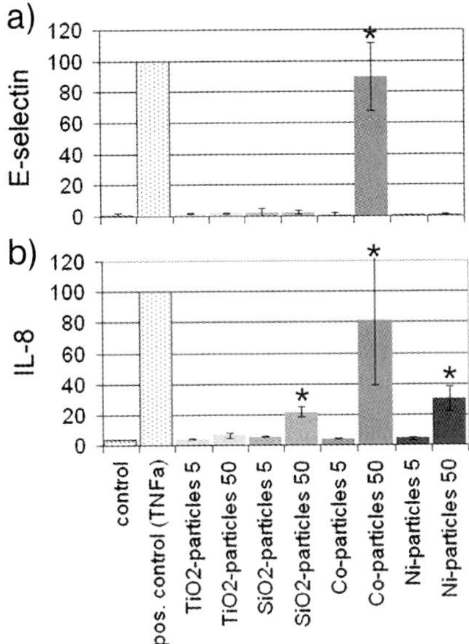

Fig. 5.4. Detection of pro-inflammatory effects induced by particle exposure. Detection of (a) E-selectin after 4 h and (b) IL-8 after 24 h particle exposure; TNFα-stimulated cells serve as positive control and set as 100%, $n = 4$, means $\pm$ SDs, significantly different from the untreated control: $*p < 0.05$.

E-selectin protein expression except for high Co particle amounts, which induced a significant increase in E-selectin protein expression (Fig. 5.4a, TNFα-stimulated cells served as the positive control and were set as 100%). In addition, ICAM-1 cell surface protein expression after 24 h was induced by high Co-particle amounts only; whereas all other particles did not induce an increase (data not shown). IL-8 release in the cell culture supernatant was stimulated by high amounts of SiO_2, Co and Ni particles. High amounts of TiO_2 particles induced only a minor, non-significant increase (Fig. 5.4b).

An important protein in cell signaling is the hypoxia-inducible factor HIF-1α. This protein is not detectable in normoxic cells (controls, Fig. 5.5a/b). Under oxygen deficiency (hypoxia) the protein is stabilized within the cells by complex mechanisms and transported into the nuclei. In this context it was important that divalent cobalt and nickel ions (Co^{2+} and Ni^{2+}) were able to stabilize and induce a translocation of the protein into the nuclei comparable to hypoxia conditions (results for Co^{2+}-treatment shown in Fig. 5.5c/d). Comparable with the effects of Co^{2+}, the exposure of Co particles induced the stabilization and translocation of HIF-1α protein into the nuclei of endothelial cells *in vitro* (Fig. 5.5e/f). These HIF-1α effects also occurred upon exposure to Ni-ions and -particles (data not shown).

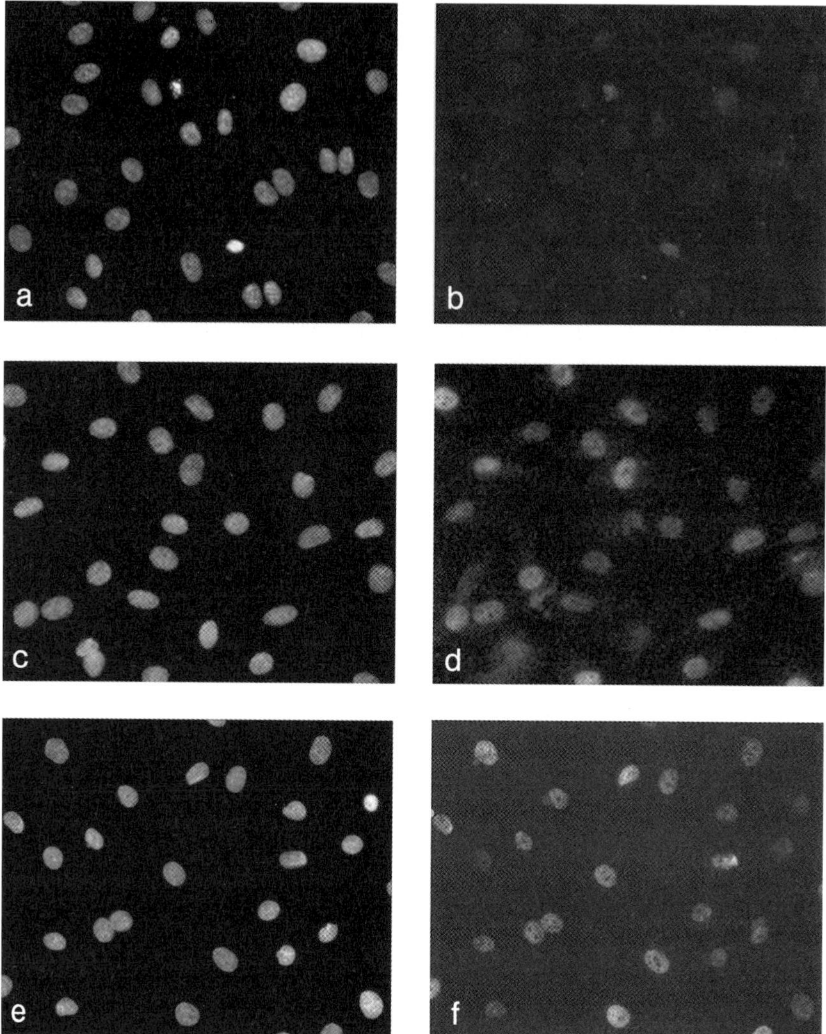

Fig. 5.5. Nuclear- and HIF-1α-staining in HDMEC. (a/b)
Control: (a) nuclear and (b) HIF-1α staining images the same
section; (c/d) Co^{2+}-treated HDMEC; 24 h, 0.7 mM, (c) nuclear
and (d) HIF-1α staining; (e/f) Co-particle-treated HDMEC; 24
h, 50 μg mL^{-1}, (e) nuclear and (f) HIF-1α staining.

The major changes after particle exposure, induced by the metallic nanoparticles
and the Co^{2+}-induced comparable stabilization of HIF-1α induced after Co particle
exposure, suggested that these effects were triggered by the release of metal ions
from the particles. Therefore we compared the effects of the respective ions in con-

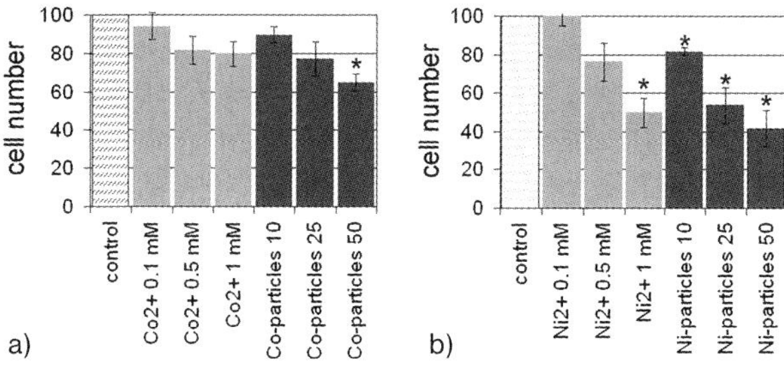

Fig. 5.6. Analysis of HDMEC number after exposure to Co and Ni ions and Co and Ni particles: (a) Co-ions and -particles, (b) Ni-ions and -particles; untreated control set as 100%, $n = 4$, means $\pm$ SDs, significantly different from normoxia: $^*p < 0.01$, crystal violet staining after 24 h exposure.

centrations that were equivalent to the molarity of the utilized particles (e.g., 50 µg Co-particles per mL corresponds to 0.85 mM cobalt, 25 µg mL^{-1} corresponds to 0.42 mM, and 10 µg mL^{-1} to 0.17 mM; the calculations depend on the assumption that pure Co- and Ni-particles were present and are, therefore, only an approximation).

The cell number was concentration-dependently decreased by Co-/Ni-ions and -particles after 24 h of exposure (Fig. 5.6a/b). Nickel showed more pronounced effects than cobalt (for both ions and particles). Interestingly, both types of particles induced a stronger reduction of cell number than the ions, indicating a higher cytotoxicity of particles than the respective ions: 1 mM of CoCl$_2$ induced a 20% decrease in cell number, whereas 50 µg mL^{-1} ($\sim$0.85 mM) of cobalt particles induced a nearly 40% decrease (Fig. 5.6a), 1 mM of NiCl$_2$ induced a nearly 50% reduction of cell number whereas 50 µg mL^{-1} of Ni particles (also $\sim$0.85 mM) induced a 60% reduction (Fig. 5.6b).

We also compared the pro-inflammatory capacity of the Co and Ni ions with the respective particles. Cobalt ions and particles were effective in inducing the cell surface protein expression of E-selectin (Fig. 5.7a, effect after 4 h). Interestingly, only Ni ions showed a concentration-dependent increase in E-selectin expression, whereas the Ni particles failed to induce E-selectin cell surface expression (Fig. 5.7b). The same effect was detectable in the expression of ICAM-1 after 24 h; again Co-particles and -ions induced ICAM-1 protein expression on the cell surface, whereas only Ni ions were able to induce ICAM-1 expression, with the Ni particles eliciting no change (data not shown). Moreover, an increase in IL-8 release after 24 h was effectively induced by Co-ions and -particles (Fig. 5.7c). Nickel ions were also able to induce IL-8 release. Contrary to the absence of pro-inflammatory stimulation of Ni particles in the expression of the cell surface adhesion molecules E-selectin and ICAM-1, release of the pro-inflammatory chemokine IL-8 was induced by Ni particles (Fig. 5.7d).

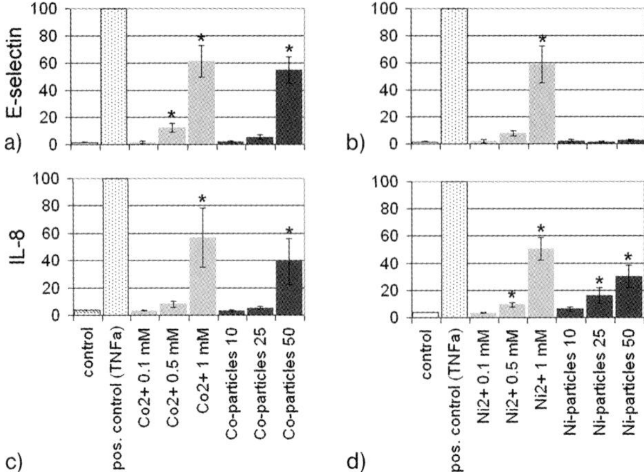

Fig. 5.7. Detection of pro-inflammatory stimulation of HDMEC after exposure to Co and Ni ions and Co and Ni particles: (a) E-selectin protein expression after Co-ion and -particle exposure (4 h), (b) E-selectin expression after Ni-ion and -particle exposure (4 h), (c) IL-8 release after Co-ion and -particle exposure (24 h), and (d) IL-8 release after Ni-ion and -particle exposure (24 h, TNFα-treated cells serve as positive control and set as 100%, $n = 4$, means $\pm$ SDs, significantly different from normoxia: $^*p < 0.01$).

5.4
Discussion

The role of particulate matter of nano-scaled size has received increasing attention in recent years. As mentioned before, this attention evolved due to the accidental internalization of particles with (partially) described effects on human health and also due to the intended administration of pharmaceutical, therapeutic, and diagnostic agents in nano-scaled sizes, which today is an important research field [30, 31]. Independent of the pathway of nanoparticle internalization, the distribution of nanoparticles within the body must occur in large part via the blood stream. Therefore, the endothelium will have contact with the nanoparticles during their passage throughout the body. The endothelium is an important cell population in the development of a multitude of diseases (e.g., tumor growth, atherosclerosis, and inflammatory diseases). Therefore, we focused on the effects of nanoparticles on viability and inflammatory status in human endothelial cells *in vitro*.

Endothelial cells *in vitro* maintain several features that their originals *in vivo* express under physiological situations. Furthermore, endothelial cells *in vitro* can be induced by pro-inflammatory compounds to synthesize and release factors that play an important role in the development of diseases. Thus, endothelial cells *in vitro* are a model system suitable for the examination of physiological and pathophysiological situations. Due to the availability of different recombinant growth factors and highly defined cell culture media several types of human endothelial cell types can be more or less easily propagated *in vitro* (e.g., macrovascular endothelial cells from arteries or veins and microvascular cells derived from the capillaries of different tissues) [32, 33].

5.4.1
Particle Internalization

Our study has shown that human endothelial cells possess a large capacity for the internalization of nanoparticles. All nanoparticles tested were taken up by the endothelial cells and to a major extent into vacuoles. Endothelial cells are able to internalize particles by different mechanisms. A large portion of the endothelial cell population possesses a prominent vesicular system that is called the vesiculovacuolar organelle (VVO) that other cells do not have with this specificity. The VVO is, together with specific plasma membrane compartments, the caveolae (invaginations of the plasma membrane), involved in the regulated transendothelial cell passage of macromolecules and particles [34]. Since, primarily, endothelial cells possess such a distinct vacuole system this way of particle internalization might be specific for a part of the endothelial cell population. There is also evidence for other means of particle internalization: In endothelial cells, particle internalization is also suggested to occur via specific receptors (e.g., low-density lipoprotein/LDL-receptor, platelet-derived growth factor/PDGF-receptor, albumin-receptor) since nanoparticles covered with LDL [35], PDGF [36], and bovine serum albumin [37] were shown to be internalized via coated pits/vesicles. It is not yet known if both pathways work independently in endothelial cells or if there is a mechanistic link.

There is evidence that a particle's size also affects the pathway of internalization, e.g., beads of 200 nm diameter were internalized into murine melanoma cells (cell line B16-F10) via clathrin-coated pits, whereas 500 nm beads enter the cells by a clathrin-independent pathway [11]. Whether endothelial cells show these size-dependent differences in internalization mechanisms is not yet known and a particle internalization pathway for the endothelial cells used in this study cannot be defined.

Nanoparticles of the biodegradable compound D,L-lactide-*co*-glycolide with a mean size of about 300 nm (larger than the particles used in this study, with mean sizes between 14 and 120 nm) are rapidly internalized by endothelial cells *in vitro* (as early as at 30 min) [38]. Moreover, *in vivo* experiments showed rapid particle internalization by endothelial cells; carbon nanoparticles (~50 nm) injected intravenously into tumor-bearing guinea pigs were detected in vacuoles and in the subendothelial space of tumor blood vessels 60 min after administration. However, the endothelial cells of non-tumorous blood vessels did not show carbon particle internalization in this animal model [34].

In our *in vitro* model, particles were detectable within vacuoles. The occurrence of vacuoles containing cellular debris together with the internalized particles was partially detectable. This co-localization of cellular debris and particles indicates an overlap between endocytosis and autophagocytosis [39, 40]. Interestingly, these "mixed content" vacuoles (called amphisomes [41]) appeared very distinct in the case of the Co and Ni particle treatment, whereas TiO_2 and SiO_2 particle filled vacuoles were mostly free of cellular debris. The vacuole enlargement after Co and Ni particle treatment might be connected directly with the possible endocytotic/ autophagocytic overlap and may be regulated by different mechanisms depending on the type of vacuole that evolves. The sizes of autophagic vacuoles can be regulated by vacuole fusion or enlargement of the vacuole [42]. The regulation of amphisomal sizes is unknown. We have suggested a connection between the release of divalent metal ions from metallic nanoparticles and the development of amphisomes in endothelial cells. This is supported by the fact that HIF-1α was stabilized after Co- and Ni-particle exposure, as described for Co- and Ni-ion exposure [43, 44].

5.4.2
Particle Cytotoxicity

Whereas the ceramic particles of TiO_2 and SiO_2 showed no significant cytotoxic effects, the nanoparticles of Co and Ni induced a significant, concentration dependent impairment of cellular viability. This impairment is obvious at different levels of cellular function (i.e., decrease of cell number, protein expression of the proliferation marker Ki67, and the metabolic activity).

Due to the stabilization of HIF-1α (see above) it appeared that the observed metal particle cytotoxicity was induced by the release of divalent metal ions from the particles. Therefore we compared the effects of the nanoparticles with those of the metal ions Co^{2+} and Ni^{2+} (as chloride salts) in concentrations similar to the solid matter utilized. The particles exerted a higher cytotoxicity than the corresponding ions (e.g., 1 mM Co^{2+} and ca. 0.85 mM Co particles induced a reduction in cell number after 24 h of ca. 20% and ca. 35%, respectively). These findings are in agreement with the results of a study on the cytotoxicity of Co nanoparticles and Co^{2+} ions in mouse fibroblast Balb/3T3 cells [45].

At this point we can only speculate about the higher cytotoxicity of transition metal nanoparticles compared to the corresponding metal ions. The presence of metal-ion-specific transporters is one possible explanation: Transition metal ions like Co^{2+} and Ni^{2+} (which are essential trace elements and also toxic when present in excess) are transported into cells against concentration gradients with ion selectivity. Furthermore, mammalian cells have intracellular mechanisms to deliver the metal ions to specialized proteins [46]. This specified delivery system is necessary due to the toxicity of metal ions such as Co^{2+} and Ni^{2+} [47, 48]. The vacuolar system of eukaryotic cells is important in metal ion homeostasis because it provides several organelles for storage of metal ions and also provides the proper amounts of transporters in the various cellular membranes through the secretory pathway. Thus, the concentration of metal ions is regulated [46]. Therefore, it might be suggested that an excess of metal ions can be regulated via specific transporters or that an excess of metal ions is transferred into cellular organelles possessing mechanisms to avoid damage of sensitive cellular compartments. This would, however, imply that metal nanoparticles are translocated to regions of the cell that are more sensitive to the release of metal ions than the regions that are attained via the specific metal ion transporters.

Severe problems may not only be a result of ion release but also due to the development of free radicals or reactive oxygen species (ROS) that occur from both transition metal ions [49, 50] and (nano-)particles [51–53]. Furthermore, it has been shown that nanosized particulate matter together with transition metals induces ROS production that exhibits a higher response than the single compounds (tested were carbon black particles with a mean size of 14 nm plus exposure of different iron and cupric salts). This increased ROS production suggested a synergistic effect of nanoparticles together with transition metals [54]. The metal particles used in this study, which are made of transition metals, contained a combination of both of these characteristics, i.e., nano-scaled particles and transition metals. Interestingly, Co particles might produce ROS by mechanisms different from ROS production by Co^{2+} ions [55, 56]. In addition, the binding of metal ions to extracellular and intracellular proteins such as transferrin and albumin, which are also known to bind Co^{2+} and Ni^{2+} efficiently [57–59], might influence the cytotoxicity of metal ions. A recent *in vitro* biokinetic study on simultaneous exposure of mouse fibroblast Balb/3T3 cells to Co particles and Co^{2+} ions showed different abilities of the two Co compounds to penetrate cells and cellular organelles. In addition, the study confirmed that serum components, particularly albumin and histidine, play a crucial role in determining the eventual toxic effects [60].

Interestingly, Ni particles did not show a reduction of Ki67 protein although the cell number and the metabolic activity were significantly decreased. In contrast, Co particles showed a very distinct down-regulation of Ki67 protein expression, in accordance with the reduced cell number and metabolic activity. Since we have previously shown that Co ions induced apoptosis whereas Ni ions induced cell death that was most likely due to necrosis (no signs of apoptosis were detectable) [48], we suggest that the absence of down-regulation of Ki67 expression after Ni particle exposure indicates a missing regulation of proliferative activity in the course of cell death. This is absent in necrotic cell death [61].

5.4.3
Pro-inflammatory Activation

A pro-inflammatory effect in HDMEC occurred after exposure to SiO_2, Co, and Ni particles and was apparent by an enhanced release of IL-8. Only higher particle concentrations (25 and 50 µg mL^{-1}) induced this increase in IL-8 release. E-selectin protein expression was enhanced by high amounts of Co particles, whereas Ni particles induced no protein expression of E-selectin. In contrast to the particles, divalent Co and Ni ions induced the expression of all pro-inflammatory markers tested (i.e., IL-8, E-selectin, ICAM-1).

The enhancement of IL-8 release in response to high amounts of SiO_2 particles indicated a pro-inflammatory state. Since SiO_2 particles can cause chronic inflammatory lung disease by inhalation [62] a comparable effect might be present in nanoparticle-induced pro-inflammatory activation of endothelial cells *in vitro*. The SiO_2-induced inflammation is mediated by the activation of the transcription factors NF-κB and AP-1, which are both involved in the regulation of inflammation. This transcription factor activation appears to be triggered by adverse biologic reactions such as the generation of ROS [63]. Moreover, SiO_2-induced AP-1 activation plays an important role in neoplastic transformation and tumor promotion [64].

The observed pro-inflammatory activation after Co-particle exposure may be attributed to a release of divalent Co-ions by the particles, since the exposure of endothelial cells with these ions leads to impaired endothelial viability and pro-inflammatory stimulation [65]. In addition, with Co-ions the concerted activation of the above-mentioned transcription factors NF-κB and AP-1 has been demonstrated [66, 67] and the Co particles used in this study induced pro-inflammatory activation with Co-ion comparable dimensions.

This contrasts with the effects of the Ni particles. Here, the suggestion of a Ni ion release by the particles, resulting in an induced pro-inflammatory stimulation, is not congruent with the pro-inflammatory effects induced by the respective ions, since Ni-ions induced both an increase in the release of IL-8 and the protein expression of endothelial cell adhesion molecules (i.e., E-selectin and ICAM-1), whereas Ni-particles induced only an increased IL-8 release and the expression of adhesion molecules was not initiated. This indicates an activation mechanism for the Ni particles that deviates from the Ni-ion-induced activation shown to occur via a cooperation of the above-mentioned transcription factors NF-κB and AP-1 [66, 67]. Such differential activation of IL-8 and ICAM-1 was also shown by the treatment of endothelial and epithelial cells with H_2O_2 [68, 69]: Whereas H_2O_2 induced an IL-8 expression in epithelial cells lines without the expression of ICAM-1, endothelial cells expressed ICAM-1 after H_2O_2 treatment without the expression of IL-8. There is evidence that this cell-type-specific differential induction of IL-8 gene expression by H_2O_2 (and thus oxidative stress) is by a differential binding of NF-κB and AP-1 to the IL-8 promoter [68]. Since oxidative stress is also a relevant aspect in the mechanisms of (Ni-) particulate-matter-induced effects [52] this mechanism of differential activation of pro-inflammatory gene promoters might play a role. Thus, it can be suggested that Ni-ion release by the nanoparticles remains under the critical limit for pro-inflammatory activation but further Ni-nanoparticle-induced effects (possibly oxidative stress) are responsible for the enhanced IL-8 release. However, this hypothesis requires further examination and is the subject of a separate study.

5.4.4
Conclusions and Consideration of the Risk of Nanoparticles to Human Health

This study has shown that nanoparticles exert effects that deviate from the effects of bulk materials and also from possible corrosion products. The experiments were performed with a cell type highly relevant for nanoparticle transmigration from the blood into tissues, i.e., primary human endothelial cells. Whether the described effects shown *in vitro* are of relevance *in vivo* remains unanswered. However, nanoparticles clearly exert effects that are not easily interpreted. If a pro-inflammatory stimulation of endothelial cells by nanoparticles occurs *in vivo*, chronic inflammation (such as granulomatosis) could be a possible consequence.

Although much progress has been made in recent years to understand the effects of nano-scaled particulate matter, knowledge about its risk in humans is clearly limited. This lack of knowledge is due to the complexity of the chemical and physical characteristics of nanoparticles combined with the difficulty of predicting reactivity in biological systems. Furthermore, nanoparticles possess an impressively high accessibility to different types of tissues and cells, which large-scaled particulate matter usually does not have (reviewed in Ref. [2]). Thus, two different risk aspects are combined in nanoparticles: The ability of foreign materials to enter biological compartments that are usually not easily accessible and the potential for the exertion of effects on these locations, which are difficult to control.

The above-mentioned characteristic behavior of nanoparticles could be useful for medical applications such as in diagnostic and therapeutic devices. The endothelium is an important target for therapy due to its role in several physiological and pathological conditions. Also, the penetration of physiological barriers with endothelial participation such as, for example, the blood–brain barrier is of interest. Therefore, several diagnostic and therapeutic approaches that involve nanoparticulate compounds are under consideration (reviewed in Refs. [5, 70]). However, the utilization of nanoparticle-formulated drugs appears to be afflicted by a similar complexity described above, e.g., the chemotherapeutic agent paclitaxel showed different rates of clearance and different tissue distributions when formulated as nanoparticles compared to the conventional formulation as emulsion [71]. Thus, both accidental and intended exposure of humans to nanoparticles is connected to a certain risk that could be influenced by the type of particle and the state of health of the affected person. Therefore, detailed toxicological research should be carried out to improve the scientific basis of the risk assessment at all stages of the life cycle of nanotechnology [72].

Acknowledgments

This work was supported by the Deutsche Forschungsgemeinschaft (Priority Programme Biosystem 322 1100) and the European Commission (QOL-2002-147).

The authors thank Susanne Barth, Marianne Müller, and Karin Molter for their excellent technical assistance.

References

1 LOMER, M. C., THOMPSON, R. P., COMMISSO, J., KEEN, C. L., POWELL, J. J., Determination of titanium dioxide in foods using inductively coupled plasma optical emission spectrometry. *Analyst* **2000**, 125, 2339–2343.

2 OBERDORSTER, G., OBERDORSTER, E.,

OBERDORSTER, J., Nanotoxicology: An emerging discipline evolving from studies of ultrafine particles. *Environ. Health Perspect.* **2005**, 113, 823–839.

3 POPE, C. A., III, Epidemiology of fine particulate air pollution and human health: Biologic mechanisms and who's at risk?, *Environ. Health Perspect.* **2000**, 108(Suppl 4), 713–723.

4 Anonymous, Air pollution in the world's megacities. *Environment* **1994**, 36, 4–37.

5 BRIGGER, I., DUBERNET, C., COUVREUR, P., Nanoparticles in cancer therapy and diagnosis. *Adv. Drug Deliv. Rev.* **2002**, 54, 631–651.

6 COLVIN, V. L., The potential environmental impact of engineered nanomaterials. *Nat. Biotechnol.* **2003**, 21, 1166–1170.

7 DONALDSON, K., STONE, V., Current hypotheses on the mechanisms of toxicity of ultrafine particles. *Ann. Ist Super Sanita* **2003**, 39, 405–410.

8 OBERDORSTER, G., Pulmonary effects of inhaled ultrafine particles. *Int. Arch. Occup. Environ. Health* **2001**, 74, 1–8.

9 SUPHIOGLU, C., Thunderstorm asthma due to grass pollen. *Int. Arch. Allergy Immunol.* **1998**, 116, 253–260.

10 SALVAGGIO, J. E., Inhaled particles and respiratory disease. *J. Allergy Clin. Immunol.* **1994**, 94, 304–309.

11 REJMAN, J., OBERLE, V., ZUHORN, I. S., HOEKSTRA, D., Size-dependent internalization of particles via the pathways of clathrin- and caveolae-mediated endocytosis. *Biochem. J.* **2004**, 377, 159–169.

12 DUFFIN, R., CLOUTER, A., BROWN, D. M., TRAN, C. L., MACNEE, W., STONE, V., DONALDSON, K., The importance of surface area and specific reactivity in the acute pulmonary inflammatory response to particles. *Ann. Occup. Hyg.* **2002**, 46, 242–245.

13 NEMMAR, A., HOET, P. H., VANQUICKENBORNE, B., DINSDALE, D., THOMEER, M., HOYLAERTS, M. F., VANBILLOEN, H., MORTELMANS, L., NEMERY, B., Passage of inhaled particles into the blood circulation in humans. *Circulation* **2002**, 105, 411–414.

14 OBERDORSTER, G., SHARP, Z., ATUDOREI, V., ELDER, A., GELEIN, R., LUNTS, A., KREYLING, W., COX, C., Extrapulmonary translocation of ultrafine carbon particles following whole-body inhalation exposure of rats. *J. Toxicol. Environ. Health A* **2002**, 65, 1531–1543.

15 OBERDORSTER, G., SHARP, Z., ATUDOREI, V., ELDER, A., GELEIN, R., KREYLING, W., COX, C., Translocation of inhaled ultrafine particles to the brain. *Inhal. Toxicol.* **2004**, 16, 437–445.

16 REICHRTOVA, E., DOROCIAK, F., PALKOVICOVA, L., Sites of lead and nickel accumulation in the placental tissue. *Hum. Exp. Toxicol.* **1998**, 17, 176–181.

17 KAIGLOVA, A., REICHRTOVA, E., ADAMCAKOVA, A., WSOLOVA, L., Lactate dehydrogenase activity in human placenta following exposure to environmental pollutants. *Physiol. Res.* **2001**, 50, 525–528.

18 LEVY, D., GENT, M., NEWHOUSE, M. T., Relationship between acute respiratory illness and air pollution levels in an industrial city. *Am. Rev. Respir. Dis.* **1977**, 116, 167–173.

19 FRAMPTON, M. W., UTELL, M. J., ZAREBA, W., OBERDORSTER, G., COX, C., HUANG, L. S., MORROW, P. E., LEE, F. E., CHALUPA, D., FRASIER, L. M., SPEERS, D. M., STEWART, J., Effects of exposure to ultrafine carbon particles in healthy subjects and subjects with asthma. *Res. Rep. Health Eff. Inst.* **2004**, 1–47; discussion 49–63.

20 MACNEE, W., DONALDSON, K., Mechanism of lung injury caused by PM10 and ultrafine particles with special reference to COPD. *Eur. Respir. J.* **2003**, 40(Suppl.), 47s–51s.

21 CAMPBELL, A., OLDHAM, M., BECARIA, A., BONDY, S. C., MEACHER, D., SIOUTAS, C., MISRA, C., MENDEZ, L. B., KLEINMAN, M., Particulate matter in polluted air may increase biomarkers of inflammation in mouse brain. *Neurotoxicology* **2005**, 26, 133–140.

22 NEMMAR, A., HOYLAERTS, M. F., HOET, P. H., DINSDALE, D., SMITH, T., XU, H., VERMYLEN, J., NEMERY, B.,

Ultrafine particles affect experimental thrombosis in an in vivo hamster model. *Am. J. Respir. Crit. Care Med.* **2002**, 166, 998–1004.

23 GATTI, A. M., MONTANARI, S., MONARI, E., GAMBARELLI, A., CAPITANI, F., PARISINI, B., Detection of micro- and nano-sized biocompatible particles in the blood. *J. Mater. Sci. Mater. Med.* **2004**, 15, 469–472.

24 LOMER, M. C., THOMPSON, R. P., POWELL, J. J., Fine and ultrafine particles of the diet: Influence on the mucosal immune response and association with Crohn's disease. *Proc. Nutr. Soc.* **2002**, 61, 123–130.

25 COOK-MILLS, J. M., DEEM, T. L., Active participation of endothelial cells in inflammation. *J. Leukoc. Biol.* **2005**, 77, 487–495.

26 MONTGOMERY, K. F., OSBORN, L., HESSION, C., TIZARD, R., GOFF, D., VASSALLO, C., TARR, P. I., BOMSZTYK, K., LOBB, R., HARLAN, J. M., et al., Activation of endothelial-leukocyte adhesion molecule 1 (ELAM-1) gene transcription. *Proc. Natl. Acad. Sci. U.S.A.* **1991**, 88, 6523–6527.

27 ROEBUCK, K. A., RAHMAN, A., LAKSHMINARAYANAN, V., JANAKIDEVI, K., MALIK, A. B., H2O2 and tumor necrosis factor-alpha activate intercellular adhesion molecule 1 (ICAM-1) gene transcription through distinct cis-regulatory elements within the ICAM-1 promoter. *J. Biol. Chem.* **1995**, 270, 18 966–18 974.

28 MUKAIDA, N., OKAMOTO, S., ISHIKAWA, Y., MATSUSHIMA, K., Molecular mechanism of interleukin-8 gene expression. *J. Leukoc. Biol.* **1994**, 56, 554–558.

29 PETERS, K., SCHMIDT, H., UNGER, R. E., OTTO, M., KAMP, G., KIRKPATRICK, C. J., Software-supported image quantification of angiogenesis in an in vitro culture system: Application to studies of biocompatibility. *Biomaterials* **2002**, 23, 3413–3419.

30 KUBIK, T., BOGUNIA-KUBIK, K., SUGISAKA, M., Nanotechnology on duty in medical applications. *Curr. Pharm. Biotechnol.* **2005**, 6, 17–33.

31 REYNOLDS, A. R., MOEIN MOGHIMI,

S., HODIVALA-DILKE, K., Nanoparticle-mediated gene delivery to tumour neovasculature. *Trends Mol. Med.* **2003**, 9, 2–4.

32 KIRKPATRICK, C. J., UNGER, R. E., KRUMP-KONVALINKOVA, V., PETERS, K., SCHMIDT, H., KAMP, G., Experimental approaches to study vascularization in tissue engineering and biomaterial applications. *J. Mater. Sci. Mater. Med.* **2003**, 14, 677–681.

33 KIRKPATRICK, C. J., KRUMP-KONVALINKOVA, V., UNGER, R. E., BITTINGER, F., OTTO, M., PETERS, K., Tissue response and biomaterial integration: The efficacy of in vitro methods. *Biomol. Eng.* **2002**, 19, 211–217.

34 FENG, D., NAGY, J. A., DVORAK, H. F., DVORAK, A. M., Ultrastructural studies define soluble macromolecular, particulate, and cellular trans-endothelial cell pathways in venules, lymphatic vessels, and tumor-associated microvessels in man and animals. *Microsc. Res. Technol.* **2002**, 57, 289–326.

35 HANDLEY, D. A., ARBEENY, C. M., CHIEN, S., Sinusoidal endothelial endocytosis of low density lipoprotein-gold conjugates in perfused livers of ethynyl-estradiol treated rats. *Eur. J. Cell Biol.* **1983**, 30, 266–271.

36 ROSENFELD, M. E., BOWEN-POPE, D. F., Ross, R., Platelet-derived growth factor: Morphologic and biochemical studies of binding, internalization, and degradation. *J. Cell Physiol.* **1984**, 121, 263–274.

37 GEOFFROY, J. S., BECKER, R. P., Endocytosis by endothelial phagocytes: Uptake of bovine serum albumin-gold conjugates in bone marrow. *J. Ultrastruct. Res.* **1984**, 89, 223–239.

38 DAVDA, J., LABHASETWAR, V., Characterization of nanoparticle uptake by endothelial cells. *Int. J. Pharm.* **2002**, 233, 51–59.

39 LIOU, W., GEUZE, H. J., GEELEN, M. J., SLOT, J. W., The autophagic and endocytic pathways converge at the nascent autophagic vacuoles. *J. Cell Biol.* **1997**, 136, 61–70.

40 GORDON, P. B., HOYVIK, H., SEGLEN,

P. O., Prelysosomal and lysosomal connections between autophagy and endocytosis. *Biochem. J.* **1992**, 283 (Pt 2), 361–369.

41 BERG, T. O., FENGSRUD, M., STROMHAUG, P. E., BERG, T., SEGLEN, P. O., Isolation and characterization of rat liver amphisomes. Evidence for fusion of autophagosomes with both early and late endosomes. *J. Biol. Chem.* **1998**, 273, 21 883–21 892.

42 DUNN, W. A., JR., Studies on the mechanisms of autophagy: Maturation of the autophagic vacuole. *J. Cell Biol.* **1990**, 110, 1935–1945.

43 HIRSILA, M., KOIVUNEN, P., XU, L., SEELEY, T., KIVIRIKKO, K. I., MYLLYHARJU, J., Effect of desferrioxamine and metals on the hydroxylases in the oxygen sensing pathway. *FASEB J.* **2005**.

44 PETERS, K., SCHMIDT, H., UNGER, R. E., KAMP, G., PROLS, F., BERGER, B. J., KIRKPATRICK, C. J., Paradoxical effects of hypoxia-mimicking divalent cobalt ions in human endothelial cells in vitro. *Mol. Cell Biochem.* **2005**, 270, 157–166.

45 SABBIONI, E., GATTI, A. M., HARTUNG, T., Pathology of new diseases induced by nanomaterials and in vitro toxicology research. *Pathol. Int.* **2004**, 54, S141–148.

46 NELSON, N., Metal ion transporters and homeostasis. *EMBO J.* **1999**, 18, 4361–4371.

47 ERMOLLI, M., MENNE, C., POZZI, G., SERRA, M. A., CLERICI, L. A., Nickel, cobalt and chromium-induced cytotoxicity and intracellular accumulation in human hacat keratinocytes. *Toxicology* **2001**, 159, 23–31.

48 PETERS, K., UNGER, R. E., BARTH, S., GERDES, T., KIRKPATRICK, C. J., Induction of apoptosis in human microvascular endothelial cells by divalent cobalt ions. Evidence for integrin-mediated signaling via the cytoskeleton. *J. Mater. Sci. Mater. Med.* **2001**, 12, 955–958.

49 STOHS, S. J., BAGCHI, D., Oxidative mechanisms in the toxicity of metal ions. *Free Radic. Biol. Med.* **1995**, 18, 321–336.

50 GINSBURG, I., SADOVNIC, M., VARANI, J., TIROSH, O., KOHEN, R., Hemolysis of human red blood cells induced by the combination of diethyldithio-carbamate (DDC) and divalent metals: Modulation by anaerobiosis, certain antioxidants and oxidants. *Free Radic. Res.* **1999**, 31, 79–91.

51 PETIT, A., MWALE, F., TKACZYK, C., ANTONIOU, J., ZUKOR, D. J., HUK, O. L., Induction of protein oxidation by cobalt and chromium ions in human U937 macrophages. *Biomaterials* **2005**, 26, 4416–4422.

52 DICK, C. A., BROWN, D. M., DONALDSON, K., STONE, V., The role of free radicals in the toxic and inflammatory effects of four different ultrafine particle types. *Inhal. Toxicol.* **2003**, 15, 39–52.

53 DONALDSON, K., BESWICK, P. H., GILMOUR, P. S., Free radical activity associated with the surface of particles: A unifying factor in determining biological activity?, *Toxicol. Lett.* **1996**, 88, 293–298.

54 WILSON, M. R., LIGHTBODY, J. H., DONALDSON, K., SALES, J., STONE, V., Interactions between ultrafine particles and transition metals in vivo and in vitro. *Toxicol. Appl. Pharmacol.* **2002**, 184, 172–179.

55 LISON, D., CARBONNELLE, P., MOLLO, L., LAUWERYS, R., FUBINI, B., Physicochemical mechanism of the interaction between cobalt metal and carbide particles to generate toxic activated oxygen species. *Chem. Res. Toxicol.* **1995**, 8, 600–606.

56 LISON, D., DE BOECK, M., VEROUG-STRAETE, V., KIRSCH-VOLDERS, M., Update on the genotoxicity and carcinogenicity of cobalt compounds. *Occup. Environ. Med.* **2001**, 58, 619–625.

57 AISEN, P., AASA, R., REDFIELD, A. G., The chromium, manganese, and cobalt complexes of transferrin. *J. Biol. Chem.* **1969**, 244, 4628–4633.

58 BAR-OR, D., CURTIS, G., RAO, N., BAMPOS, N., LAU, E., Characterization of the Co(2+) and Ni(2+) binding amino-acid residues of the N-terminus of human albumin. An insight into the mechanism of a new assay for myocardial ischemia. *Eur. J. Biochem.* **2001**, 268, 42–47.

59 SADLER, P. J., TUCKER, A., VILES, J. H., Involvement of a lysine residue in the N-terminal Ni2+ and Cu2+ binding site of serum albumins. Comparison with Co2+, Cd2+ and Al3+. *Eur. J. Biochem.* **1994**, 220, 193–200.

60 DEL TORCHIO, R. (**2005**). Nanotossicologia in vitro mediante tecniche analitiche avanzate. Uno studio di Co nanoparticelle in fibroblasti di topo (Balb/3T3). MSc thesis, University of Milan, Milan.

61 CORCORAN, G. B., FIX, L., JONES, D. P., MOSLEN, M. T., NICOTERA, P., OBERHAMMER, F. A., BUTTYAN, R., Apoptosis: Molecular control point in toxicity. *Toxicol. Appl. Pharmacol.* **1994**, 128, 169–181.

62 MCDONALD, J. C. (**1996**). Silica and lung cancer. In *Silica and Silica-induced Lung Diseases* (ed. V. CASTRANOVA, V. VALLYATHAN, W. E. WALLACE), CRC Press, Boca Raton, FL.

63 CASTRANOVA, V., Signaling pathways controlling the production of inflammatory mediators in response to crystalline silica exposure: Role of reactive oxygen/nitrogen species. *Free Radic. Biol. Med.* **2004**, 37, 916–925.

64 DING, M., CHEN, F., SHI, X., YUCESOY, B., MOSSMAN, B., VALLYATHAN, V., Diseases caused by silica: Mechanisms of injury and disease development. *Int. Immuno-pharmacol.* **2002**, 2, 173–182.

65 KIRKPATRICK, C. J., BARTH, S., GERDES, T., KRUMP-KONVALINKOVA, V., PETERS, K., [Pathomechanisms of impaired wound healing by metallic corrosion products]. *Mund Kiefer Gesichtschir* **2002**, 6, 183–190.

66 WAGNER, M., KLEIN, C. L., VAN KOOTEN, T. G., KIRKPATRICK, C. J., Mechanisms of cell activation by heavy metal ions. *J. Biomed. Mater. Res.* **1998**, 42, 443–452.

67 WAGNER, M., KLEIN, C. L., KLEINERT, H., EUCHENHOFER, C., FORSTERMANN, U., KIRKPATRICK, C. J., Heavy metal ion induction of adhesion molecules and cytokines in human endothelial cells: The role of NF-kappaB, I kappaB-alpha and AP-1. *Pathobiology* **1997**, 65, 241–252.

68 LAKSHMINARAYANAN, V., DRAB-WEISS, E. A., ROEBUCK, K. A., H2O2 and tumor necrosis factor-alpha induce differential binding of the redox-responsive transcription factors AP-1 and NF-kappaB to the interleukin-8 promoter in endothelial and epithelial cells. *J. Biol. Chem.* **1998**, 273, 32670–32678.

69 LAKSHMINARAYANAN, V., BENO, D. W., COSTA, R. H., ROEBUCK, K. A., Differential regulation of interleukin-8 and intercellular adhesion molecule-1 by H2O2 and tumor necrosis factor-alpha in endothelial and epithelial cells. *J. Biol. Chem.* **1997**, 272, 32910–32918.

70 EMERICH, D. F., THANOS, C. G., Nanotechnology and medicine. *Expert Opin. Biol. Ther.* **2003**, 3, 655–663.

71 YEH, T. K., LU, Z., WIENTJES, M. G., AU, J. L., Formulating paclitaxel in nanoparticles alters its disposition. *Pharm. Res.* **2005**, 22, 867–874.

72 European Commission (**2004**). Towards a European strategy for nanotechnology. http://www.cordis.lu/nanotechnology

1996 for the discovery of fullerene. In 2000, this federal effort was raised by President Clinton to the level of a federal initiative, which was known as the National Nanotechnology Initiative (NNI) [2]. The NNI is promoting nanotechnology research and development to lead the United States to the next industrial revolution [3]. The National Science Foundation [4] predicted that nanotechnology will drive prodigious nanoscience research and engineering development efforts in materials science, physics, chemistry, biology, medicine, and biotechnology, and will generate, in 10 to 15 years' time, an annual business and economic impact of close to $1 trillion. One of the major objectives of the NNI is "developing materials that are 10 times stronger than steel, but a fraction of the weight for making all kinds of land, sea, air and space vehicles lighter and more fuel efficient." The materials implicated in the initiative are CNTs.

6.3
Manufactured Carbon Nanotubes: Their Synthesis, Properties, and Potential Applications

6.3.1
Discovery and Synthesis

CNTs are the most important and most-studied nanomaterials. They are a new allotropic form of carbon similar to fullerene (Fig. 6.1). Buckminsterfullerene was first synthesized in 1985 by a laser ablation process developed by Richard Smalley, a Nobel Prize laureate at Rice University (Houston, TX), and his colleagues [5]. CNTs were discovered in 1991 by Sumio Iijima during his investigation of fullerene formation from atomized carbon dissociated from heated graphite in an arc-discharge process [6, 7]. This Japanese electron microscopist observed CNTs, predominately multiwalled (Fig. 6.1), and other nanoparticles deposited at the graphite cathode. Ebbesen and Ajayan of Iijima's laboratory showed that CNTs could be produced in bulk quantities by varying the arc-evaporation conditions [8]. Iijima found that the synthetic yield of single-wall carbon nanotubes (SWCNTs; Fig. 6.1) could be increased by incorporating cobalt or other catalytic transition metals with the graphite source in the arc vaporization process [9]. Smalley's group at Rice University also succeeded in synthesizing SWCNTs, by adapting the laser ablation process used to make C_{60} [10, 11]. In the arc vaporization and laser ablation processes, solid or powdered graphite is used as the carbon source, but in the chemical vapor deposition method, carbon-bearing gaseous compounds such as methane, acetylene, or other hydrocarbons are the source [12, 13]. Using carbon monoxide as a feedstock, Nikolaev in Smalley's laboratory developed a gas-phase catalytic growth of SWCNTs from carbon atoms generated from a stream of continuous-flow high-pressure carbon monoxide [14]. This patented synthesis method is referred to by Smalley's group as the HiPco^{TM} process.

All these synthetic processes involve formation of nanotubes from carbon atoms thermally generated from the carbon-bearing sources. CNT synthesis is generally carried out in an argon or other inert atmosphere at 600–1200 °C [15]. Typically,

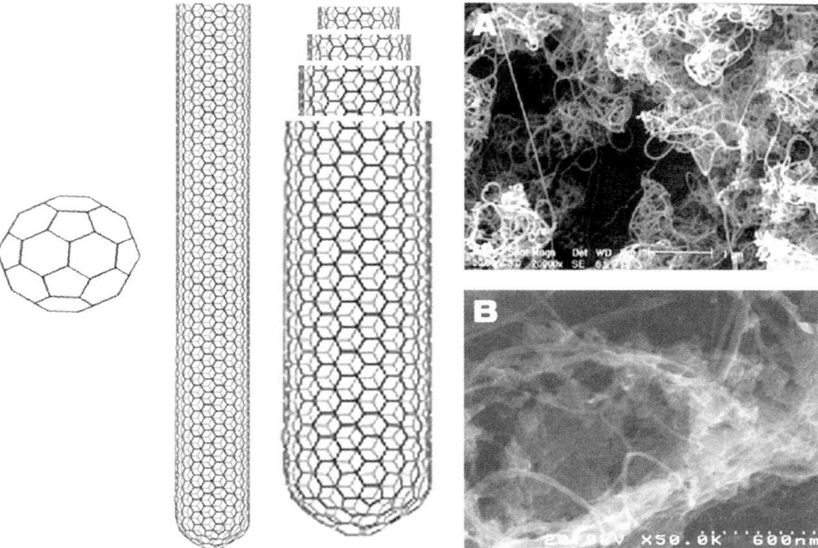

Fig. 6.1. Drawings of a C_{60} fullerene, a single-wall carbon nanotube (SWCNT), and a multiwall carbon nanotube (MWCNT); also shown are scanning electron micrographs of SWCNT ropes (A) and MWCNT ropes (B). ((B) is courtesy of Jordi Rodriguez of University of Barcelona, Spain.)

the carbon atoms are formed on the surface of the catalytic metal; they then dissolve in molten metal nanoparticles. From the molten metal, CNTs grow [15]; when they reach a certain length, they drop off from the metal particles. At the temperature of synthesis, the metal(s) needs to be catalytically active and remain in the molten state, allowing dissolution of carbon atoms in the metal(s); these requirements limit the metals that can be used, of which cobalt, nickel, iron, and molybdenum are the most common. All unprocessed SWCNT or MWCNT products contain residual metal(s). Generally metal(s) accounts for up to 30% of a raw SWCNT product; the metal content in a MWCNT product is much less (Table 6.1 below). The metal impurities are undesirable; some products on the market are sold in purified form after removal of metal. CNT products may contain other impurities that consist of non-nanotube carbon. Product purity depends on both manufacturing processes and post-manufacturing treatments.

6.3.2
Physical and Chemical Properties

Structurally, a SWCNT is a rolled-up sheet of graphene or graphite with its carbon atoms arranged in hexagonal and pentagonal patterns. It is about 1 nm in diameter and several micrometers long. MWCNTs contain multiple concentric layers with various diameters and lengths. Immense efforts have been devoted to making CNTs longer for practical uses. Because MWCNTs are more heterogeneous, characterization of CNT physicochemical properties has been carried out predominately on SWCNTs.

Theoretical calculations and experimental results have shown that SWCNTs have highly desirable mechanical, thermal, photochemical, and electrical properties. SWCNTs are both strong and stiff, yet flexible; in fact, they are the strongest of all synthetic fibers [16]. According to Smalley, "… calculations show [they] should be somewhere between 30 and 100 times stronger than steel." [17, 18]. The van der Waals force attraction of SWCNTs causes the tubes to bundle into microscopic ropes, which in turn aggregate to form loose clumps. SWCNTs are among the best electrical conductors and can conduct electricity twice as well as copper [19]. SWCNTs have unique electron-transport properties; they may be either metallic or semiconducting, determined by the chiral vector of the tubes [20]. In a typical batch of synthesized material, one-third of the SWCNTs are metal conductors and two-thirds are semiconductors [21, 22]. MWCNTs are made up of concentrically rolled-up graphene sheets. Each rolled-up sheet is like a SWCNT, and MWCNTs probably have many physicochemical properties similar to those of SWCNTs.

6.3.3
Applications

Both SWCNTs and MWCNTs are very light; they are very strong and stiff and yet flexible [16]. These very desirable mechanical properties make them ideal materials, by themselves or in composites, for potential wide applications in engineering structures. Even composite materials containing CNTs may be strong enough for building such things as spacecraft structures, space elevators, artificial muscles, combat jackets, and land and sea vehicles [23]. Commenting on the SWCNTs that have metal-like properties, Smalley stated [24], "… they will conduct electricity better than copper. Membranes made from arrays of these nanotubes are expected to have a revolutionary impact on the technology of rechargeable batteries and fuel cells, perhaps giving us all-electric vehicles within the next 10–20 years." Smalley [24] further predicted that, several decades from now, CNT-based nano-electronics having vastly greater performance and scope will supplant our current silicon-based microelectronics. CNTs also have many other potential applications.

6.4
Occurrence of Carbon Nanotubes in the Environment

6.4.1
Potential Occupational Exposures and Environmental Impact of Manufactured Carbon Nanotubes

Unprocessed CNTs are very light and can become airborne (Fig. 6.2), and if they entered the environment as fine suspended particulates they could reach the lungs

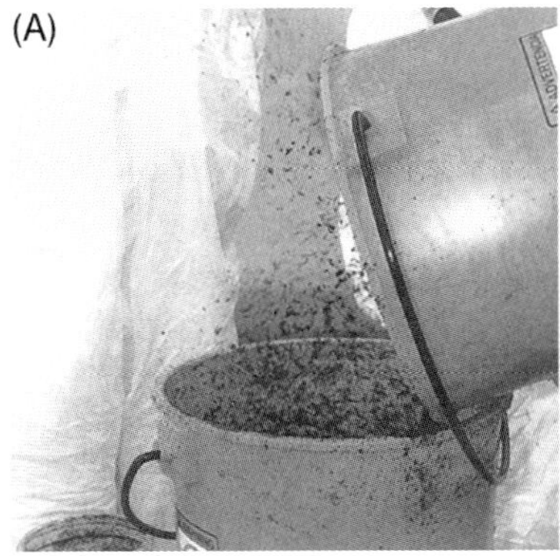
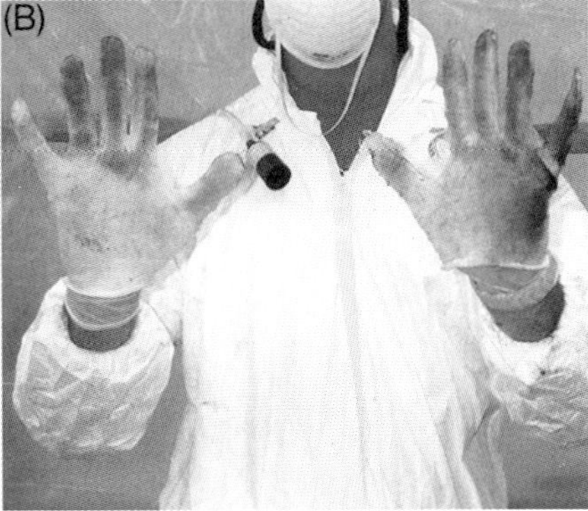

Fig. 6.2. (A) SWCNT particles became airborne when the raw material (HiPco CNTs) was poured between containers. (B) Contaminated gloves showing SCWNT particles appearing like black soot. (Courtesy of Dr. Andrew Maynard of NIOSH [27].)

of exposed workers. If CNTs exist in nanosize fibers, they might penetrate unprotected skin (Fig. 6.2). The CNT manufacturing industry is still in its infancy and the products remain expensive. In an interview conducted in 2003 [25], Smalley stated that the price of SWCNTs was hundreds of thousands of dollars per pound. Currently (2005), raw SWCNTs and MWCNTs are sold by BuckyUSA (Houston) at >\$100 per gram (or >\$50 000 lb^{-1}) [26]. In 2003, Baron et al. [27]. of the aerosol group of the National Institute of Occupational Safety and Health (NIOSH) visited CNT synthesis laboratories at Rice University and NASA's Johnson Space Center and the CNT manufacturing facility at Carbon Nanotechnologies Incorporated (Houston) where SWCNT are produced by the HiPco and laser processes. They observed the recovery of CNTs from synthetic ovens and reported that handling of the collected samples was gentle, and losses of this expensive material were minimized. Occupational exposure in the facilities that make these expensive materials is expected to be minimal, with very little CNT industrial waste contaminating the environment.

Smalley [28] predicted, however, that "… in time, millions of tonnes of nanotubes will be produced worldwide every year." The Department of Energy's 2010 target goal for the price of CNTs is \$8 kg^{-1} (or <\$20 lb^{-1}). The extent of industrial and commercial applications would depend on the price of CNT products. If millions of tons of CNTs could be produced annually and if the CNT industry achieves the goal of "a couple dollars a pound," [25] occupational exposures to airborne dusts of these lightweight materials during synthesis, processing, and product manufacturing would be very substantial. Because of the potential for rapid merging or incorporation of CNTs into fabrics, plastic, lubricants, composite materials, and household commodities that are used worldwide, concerns have arisen about the adverse impact of CNTs on human health and the environment (Fig. 6.3). If

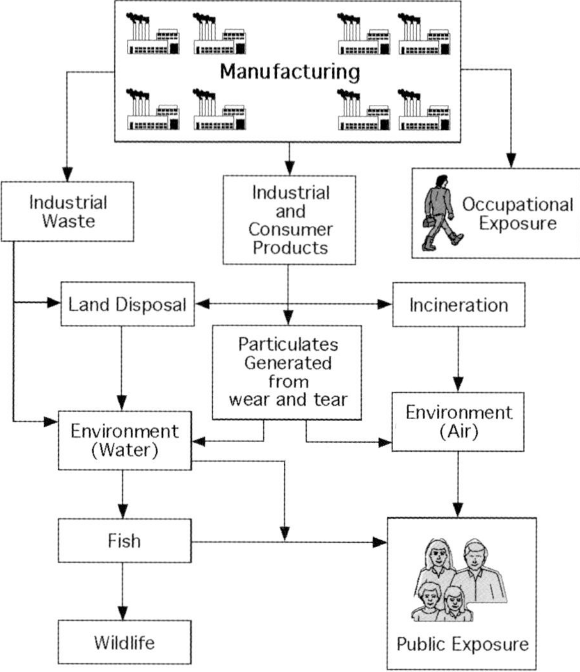

Fig. 6.3. Potential environmental impact and health concern when the production volume of carbon nanotubes becomes very large.

they are incorporated into materials such as plastic and composite structures, the CNT materials will need to be pulverized; working with pulverized CNTs would pose an inhalation exposure risk. In the U.S. we have also witnessed the use of asbestos in automotive brake shoes until it was found that asbestos particles generated from abrasion contribute to environmental pollution, leading the U.S. Environmental Protection Agency to ban the use of this carcinogenic material in automobiles [29]. CNTs are light and strong, and if the price of CNTs drops to a few dollars a pound, a potential exists for finding this type of application for CNTs and for them to contribute to environmental pollution. It is important that the chronic toxicity of CNT dusts be known before CNTs become more widespread.

6.4.2
Combustion-generated Carbon Nanotubes in the Environment

The fact that thermal processes are used to synthesize fullerenes and MWCNTs from atomized carbon generated from heated graphite [5, 9] and the finding that carbon nanotubes can be produced by chemical vapor deposition methods involving thermally dissociated methane, acetylene, and other hydrocarbons [30] have triggered interest in looking for these carbon allotropes in the environment.

6.4.2.1 MWCNT Formation from Natural Gas Combustion Indoors

Samples of airborne particulate material (PM) collected from combustion streams of methane, propane, or natural gas (containing 96% methane) of typical kitchen gas-ranges were found to contain aggregates of MWCNTs and other carbonaceous nanoforms [31–33]. The aggregates, which had aerodynamic diameters of about 0.4 to 2 μm, were essentially pure carbon or graphene. They contained several hundred to several thousand individual MWCNTs and other carbon nanocrystals of about 20 nm diameter; individual MWCNTs were ca. 3 to 30 nm in diameter [34]. The authors [34] concluded that CNTs and fullerene nanoparticles are ubiquitous in homes with gas cooking ranges. Their findings suggest that MWCNTs and other carbonaceous nanoparticles may also be produced by water heaters, furnaces, and other appliances that use natural gas.

6.4.2.2 MWCNTs in Metropolitan Outdoor Air

Outdoor airborne PM collected in El Paso (TX) by Murr et al. and analyzed using a transmission electron microscope showed the presence of MWCNTs aggregated with other forms of carbon nanocrystals (shells, spheres, and other structures) (Fig. 6.4a and b) [34]. The structure of these aggregates collected outdoors was similar to that of MWCNTs collected indoors, except that the outdoor PM included agglomerates of other mineral nanocrystals, such as silica, that are common in the atmosphere [34]. The carbon nanocrystals, largely MWCNTs, accounted for 15% of the weight of the dust. Diesel-related aggregates accounted for 5% of the dust collected. According to Murr et al., their laboratory had collected hundreds of samples in El Paso over several years and had previously characterized dusts as silica or other nanocrystals with carbonaceous materials [35]; reexamination of these samples indicated that 90% of them contained MWCNTs and other carbon nanocrystals [33]. Environmental samples collected recently from areas close to a heavy-traffic road in Houston were also shown to contain complex aggregates of MWCNTs and other carbon nanocrystals intermixed with silica nanocrystals (Fig. 6.4c and d) [34]. Murr et al. concluded that MWCNTs and carbonaceous nanoparticles are ubiquitous in the environment, and they speculated that MWCNTs make up a significant portion of airborne PM both indoors and outdoors [32].

6.4.2.3 MWCNTs in Ancient Ice

Nanoparticulates were found in an ice sample from a core drilled into a Greenland ice cap to a depth of 1646 feet and dated at roughly 10 000 years old [36]. MWCNTs, fullerene-like nanocrystal forms, and silica nanoparticles were observed in complex mixtures of nanoaggregates in this sample. The aggregates were less than 1 μm in diameter. The authors reported that the particulate pattern was similar to that they observed in samples collected from metropolitan air. These findings showed that MWCNTs and other carbonaceous nanoparticles were present in respirable particles in the air in prehistoric times.

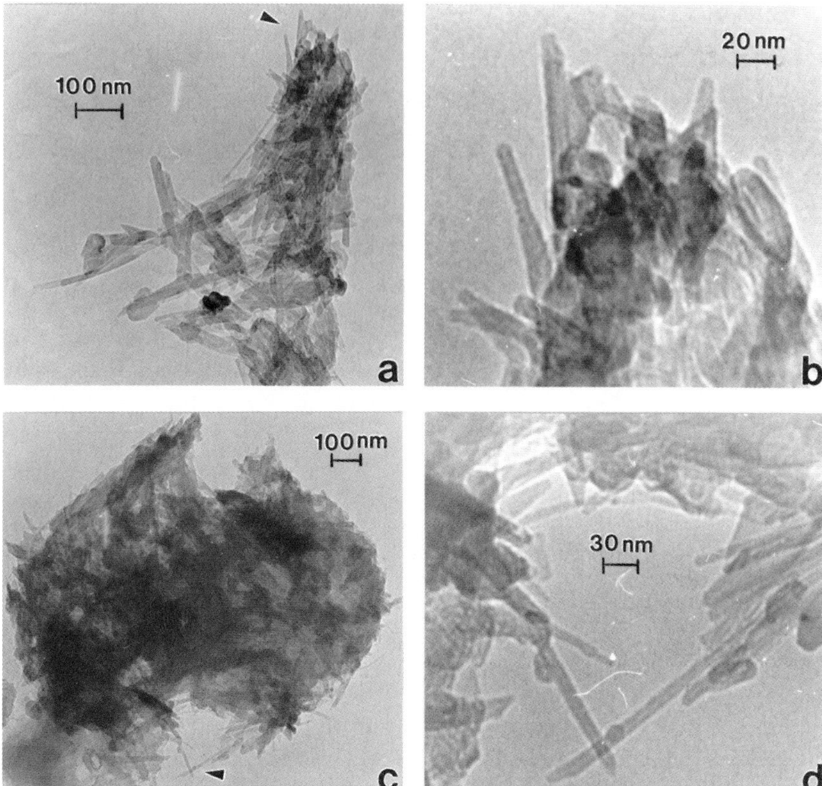

Fig. 6.4. Transmission electron microscope images of environmental particulate matter collected from El Paso, TX (a, b) and from Houston, TX (c, d), showing the presence of multi-wall carbon nanotubes [33] (Courtesy of Dr. L. Murr of University of Texas at El Paso, TX, and with permission to print from Springer Publishing Company, New York, NY.).

6.4.2.4 Concern about Combustion-generated MWCNTs in the Environment

MWCNTs are components of airborne particulate aggregates smaller than 2.5 μm in diameter ($PM_{2.5}$) [32, 33]. Natural gas is considered an environmentally clean fuel. Combustion of natural gas produces only 7 pounds of PM per billion BTU compared with 84 pounds for oil and 2774 pounds for coal combustion [37]. The numbers may be underestimated if nanosize particulates were not captured. Even though natural gas produces a relatively small amount of PM, global fuel-gas consumption is very large, making the contribution of MWCNTs to air pollution very substantial. In 1999, 22 096 billion BTU of natural gas was consumed in the U.S. [38]; this was 27% of the global consumption. The consumption of natural gas, especially in indoor activities, is expected to increase, and this can be expected to increase the contribution of MWCNTs to air pollution.

Other well-established sources of $PM_{2.5}$ include products of fuel combustion by automobiles, power plants, wood burning, industrial processes, and diesel-powered vehicles such as buses and trucks [39]. Combustion processes typically generate very fine particles of from 0.01 to 2.5 μm [40]; combustion of fossil fuels is the major contributor of fine particulates [41]. The PM contains elemental carbon, organic carbon, trace elements, and common ions [42]. Commenting on CNTs and fullerenes, Richard Smalley stated [23, 43], "They're also made in every candle flame and in forest fires." If CNTs are formed in such mundane places as candle flames and forest fires, MWCNTs are likely to be produced by combustion of other fuels in addition to natural gas. Because airborne MWCNTs are ubiquitous and are present in substantial amounts in the environment, all people are exposed to this newly identified environmental pollutant.

6.4.3
Comparison of Physical Structures of Manufactured and Non-manufactured Carbon Nanotubes

Manufactured CNTs are produced in ovens that allow them to form bundles, ropes, and clumps. Under optimal conditions, long fibers of high purity are preferably produced for practical applications such as spinning into threads or ropes. Most of the particles produced by CNT manufacture (such as by the HiPco and laser processes) are probably larger than respirable sizes [27]. In the environment outside the laboratory, where fuels are heterogeneous, combustion conditions are various, the reactions are not confined, and metal catalysts are generally not present, combustion-generated CNTs, which are exclusively multi-walled, are expected to be highly irregular in size and quality [34]. This probably decreases the effectiveness of van der Waals forces; MWCNTs thus produced are less orderly, shorter in length, and fewer in number than manufactured CNTs, and are intermingled with other nanoparticles (Fig. 6.4). This may explain why manufactured CNTs generally exist in larger dust aggregates whereas aggregates containing MWCNTs in the environment are often found in particles of respirable size [34].

6.5
Toxicological Studies and Toxicity of Manufactured CNTs

Toxicological studies on CNTs were conducted primarily to investigate the potential toxicity of manufactured CNTs in the lungs for occupational risk assessment. The first study of SWCNT toxicity, in which lung histopathology in exposed mice was examined, was triggered by NASA's concern that workers in occupational settings could be exposed to the airborne dust of this light material of unknown toxicity [44]. The study was also supported by the Center of Nanoscale Science and Technology of Rice University; both organizations have facilities that make SWCNTs. Particle characterization of the HiPco and laser CNTs by NIOSH aerosol scientists showed that respirable particles can be generated, albeit with difficulty, from bulk

SWCNT materials [27]. Inhalation would be the primary route of occupational exposure. Parallel to the NASA study [44], Warheit et al. [45] of Du Pont Company (Wilmington, DE) conducted a toxicity study in rats of a SWCNT product made by their company. Although the two studies yielded some similar histopathological findings, the two groups reached different conclusions about the toxicity of SWCNTs, which have been a subject of wide debate. Shvedova et al. [46] of NIOSH conducted a very comprehensive study in mice "to resolve this conflict...." The toxicity of MWCNTs had drawn little attention until very recently a Belgian group led by Muller et al. [47] published their study. All these toxicological studies are outlined in Table 6.1 and are the subjects of this chapter.

The portion of dust that could reach the pulmonary region (where air exchange takes place) of the respiratory system is termed the respirable fraction (or dust). For humans, a dust particle of respirable size is generally 10 μm or less, depending on its density and geometry; for rodents, the diameter of a respirable dust particle is a few micrometers less. The extent (fraction) of deposition of respirable dust in different locations in the respiratory system depends on the particle size, density, and geometry. To assess the toxicity of a dust in the lung, the respirable fraction is generally isolated or prepared from the bulk material.

The action of the van der Waals force causes CNTs to have a great tendency to bundle to form ropes, each containing a few hundred parallel tubes [48]. These secondary structures, in turn, aggregate into loose clumps. As shown by the NIOSH aerosol scientists for the HiPco and laser products [27], it would be difficult to isolate and collect enough fine CNT particles or clumps of respirable sizes from the bulk materials to use in an inhalation study, which would require a large amount of fine dust of this expensive material. A controlled CNT concentration would have to be generated in a chamber, and particle sizes and the actual exposure level would have to be monitored. This is difficult even with more workable dusts or powders of other compounds. Because it is technically difficult and costly to conduct inhalation toxicity experiments, investigators often assess the effects of dusts and aerosols in the lungs by intratracheal instillation (ITI) [49–51].

A dust is administered by the ITI route by injecting a suspension of a fine dust into the trachea of a small animal and allowing the dust to be pulled deeper into the lung during breathing. Dust administered by inhalation is inhaled continuously; when dust is administered by ITI it is generally instilled into the trachea as a bolus dose. Administration by the ITI route does not allow investigators to assess the effects of a test dust in the upper respiratory tract, and the distribution of dust in the lung is unnatural and less even than with administration by inhalation. In an ITI study, dust aggregates suspended in an aqueous solution usually need to be dissociated or broken down to respirable sizes by ultrasonication with or without a non-toxic dispersion agent. A dose of dust particles administered by ITI can swamp the respiratory system's dust clearance mechanisms, and ITI results are often exaggerated. However, an ITI study does allow investigators to determine the relative toxicity of the test material by giving reference compounds of known inhalation toxicities to control groups of animals [52, 53]. The ITI administration

Tab. 6.1. Pulmonary toxicity studies of SWCNTs in animals and characteristics of these materials.

Test materials and characteristics	Maker of test materials	Synthetic process	Metal content (%)[a]	Animal species	Ref.
Soot containing CNTs	Toyo Tanso Co. Ltd., Japan	Electric arc	Co/Ni No info on %	Guinea pig	Huczko et al., 2001 [54]
SWCNTs	Rice University, Houston, TX	Laser[b]	Ni: 10[c]	Mouse	Lam et al., 2000 (unpublished report)
SWCNTs	Rice University, Houston, TX	HiPco[b]	Fe: 26.9 Mo: 0.95 Ni: 0.8	Mouse	Lam et al., 2004 [44]
SWCNTs, purified	Rice University, Houston, TX	HiPco[b]	Fe: 2.1	Mouse	Lam et al., 2004 [44]
SWCNTs	CarboLex Inc., Lexington, KY	Electric arc	Ni: 26.0 Y: 5.0 Fe: 0.5	Mouse	Lam et al., 2004 [44]
SWCNTs	DuPont Co., Wilmington, DE	Laser[b]	Ni: 5 Co: 5	Rat	Warheit et al., 2004 [45]
SWCNTs, purified[d]	Carbon Nano-Technologies, Inc. Houston, TX	HiPco[b]	Fe: 0.23	Mouse	Shvedova et al., 2005 [46]
SWCNTs	Facultés Universitaires Notre-Dame de la Paix in Namur, Belgium	CVD	Co: 0.95 Fe: ~1	Rat	Muller et al., 2005 [47]

[a] Percent by weight in final products.
[b] Process developed by Rice University or originally developed by Rice University.
[c] Information provided by Smalley's group at Rice University.
[d] Purified by NASA Johnson Space Center Nanotechnology Laboratory.

route is acceptable for screening dusts for pulmonary toxicity, if the investigators or risk assessors recognize the limitations of this technique [51].

6.5.1
Study of SWCNTs in Guinea Pigs by Huczko et al. of Warsaw University

Huczko et al. conducted the first animal study on the toxicity of CNTs [54]. They intratracheally instilled two groups of five guinea pigs each with a bolus dose of 0 or 25 mg of CNT-containing soot in saline solution. They found no difference between the two groups in tidal volume, breathing frequency, or pulmonary resistance. Analysis of bronchoalveolar lavage fluid (BALF) obtained from the CNT-treated and control animals indicated that treatment with CNTs had no effect on cell differentials or total protein concentration. The authors concluded that "the soot with a high content of CNTs does not induce any abnormalities of pulmonary function or measurable inflammation in guinea pigs treated with carbon nanotubes." However, examination of lung pathology, which is the most critical toxicological endpoint of any pulmonary toxicity study with dust, was not included in the study. Lung pathology was examined in the other studies reviewed in this section.

6.5.2
Study of SWCNTs in Mice by Lam et al. of NASA-JSC Toxicology Laboratory

Lam et al. conducted a pilot 7-day ITI study to determine pulmonary toxicity of an early experimental SWCNT sample made at Rice University by the laser evaporation process and containing nickel. The ITI method used was a modified version of a method previously used in our laboratory [54]. Granulomas were observed in mice (C57/BL/6J) treated with 1 mg SWCNT per mouse. Because the test sample contained 10% (by weight) residual nickel, the lung lesions could not be attributed to the effect of CNTs.

A "core" study was conducted to assess the intrinsic toxicity of CNTs and the influence of the residual metals in the toxicological manifestation of the test compounds. The three types of CNTs we studied had been manufactured by different processes and contained different types and/or amounts of residual metals (Table 6.1) [44]. These materials were (1) unprocessed iron-containing SWCNTs made by the HiPco process, (2) a purified HiPco product that had been vigorously treated with concentrated acid to remove metal residues [56], and (3) a CarboLex SWCNT sample, made by an arc-discharge process, that contained nickel and yttrium. The core study included two standard reference materials: carbon black (Printex 90®, a dust with relatively low toxicity) and quartz (Min-U-Sil 5®, a fibrogenic dust). Groups of B6C3F$_1$ mice (4 mice per group for 7 d, 5 mice per group for 90 d) were intratracheally instilled with a suspension of the test dust (0, 0.1, or 0.5 mg 50 µL^{-1} per mouse, equal to about 0, 3.3, or 16.5 mg kg^{-1}, respectively). Mice were euthanized 7 or 90 d after the single treatment, and their lungs were excised, fixed, and stained for histopathological study [44].

Lung histopathology results showed that CNTs induced lesions, chiefly interstitial granulomas (Fig. 6.5), in the lungs of the 7-d and 90-d groups of mice [44]. These microscopic nodules, located beneath the bronchial epithelium, were present throughout most of the microscopic fields of lung tissue. The lesions were similar to the ones we saw in the pilot study. The granulomas contained macrophages that had engulfed CNTs, but they contained very few inflammatory cells. The severity of lesions was dose-dependent. Prominent granulomas were found in the lungs of all the mice that received 0.5 mg CNTs per mouse. Granulomas were less prominent but were still observed in the mice each treated with 0.1 mg of HiPco-synthesized CNTs. The lung lesions in the 90-d high-dose groups were generally more pronounced than those in the 7-d high-dose groups. The lungs of some animals in the 90-d groups showed peribronchial and interstitial inflammation, fibrosis, and necrosis that had extended into the alveolar septa. Mice in the group treated with carbon black had black particles in alveolar regions, but tissue reactions were minimal (Fig. 6.5a). The lungs of mice treated with the high dose of quartz had mild to moderate inflammation. The lesions induced by quartz were considered much less severe than those produced by CNTs. Similar results were obtained for all three types of SWCNTs. The lungs of mice in the serum control groups were normal. Lam et al. concluded that SWCNTs are intrinsically toxic to the lungs; these authors advise caution in allowing exposure to the dust and advocate implementation of strategies to minimize human exposures [44].

6.5.3
Study of SWCNTs in Rats by Warheit et al. of DuPont Company

Warheit et al. also presented evidence that CNTs produce granulomas in the lungs of treated animals [45]. These authors instilled a suspension containing a laser-synthesized SWCNT product (containing nickel and cobalt) into the trachea of Sprague–Dawley (Crl:CD(SD)IGS BR) rats. In addition to examining lung histopathology, Warheit's group assessed biomarkers of toxicity in bronchoalveolar lavage fluid (BALF) obtained from the treated rats. The animals were given the CNTs (suspended in saline containing 1% Tween 80) at 0, 1, or 5 mg kg^{-1} (0, 0.25, or 1.25 mg per rat) and were euthanized at 1, 7, 30, or 90 d after the single treatment. The high dose in this rat study was comparable to the low dose in the NASA mouse study. Histopathological results revealed multifocal granulomas, which became evident 1 month after the treatment (Fig. 6.6). The authors found that instillation of CNT dust produced granulomatous lesions in the lungs of treated rats, and the lesions were non-dose-dependent, non-uniform, and non-progressive. Quartz was given as a positive control, and was observed to produce cytotoxicity, inflammation, and fibrosis in a dose-dependent manner. Study of the BALF showed that SWCNTs induced only a transient increase in the concentration of lactate dehydrogenase (LDH; marker of cytotoxicity) in the 1-d group. Quartz at a high dose produced increases in LDH and protein concentrations at all time points. Unable to find a dose-dependent and time-dependent granulomatous response, or prolonged inflammation in the lungs, coupled with the observations

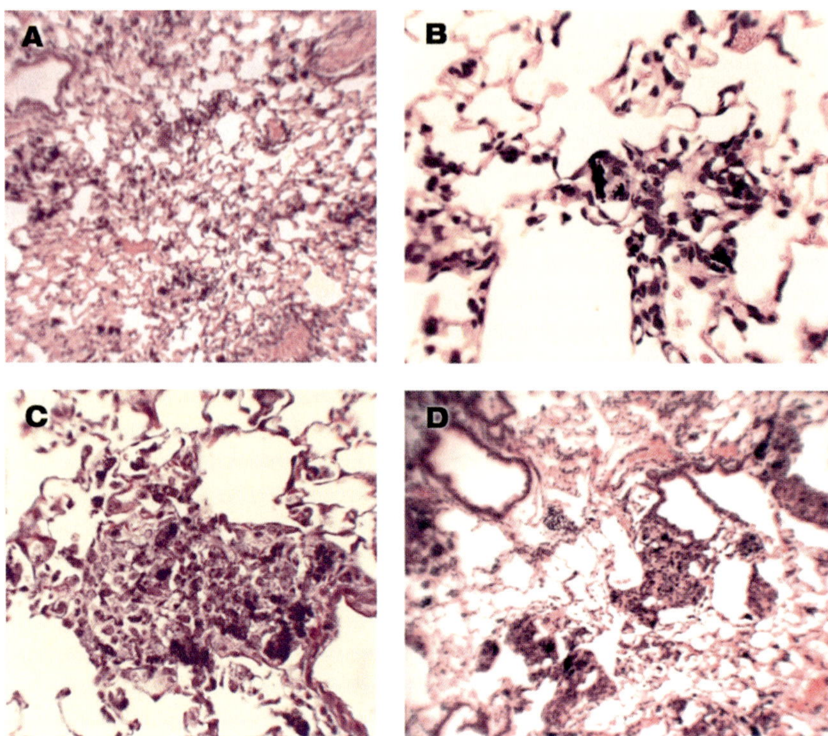

Fig. 6.5. Lung tissues from mice intratracheally instilled with 0.1 or 0.5 mg per mouse of a test material and euthanized 7 or 90 days (d) after the single treatment. (A) Carbon black, 0.5 mg, 90 d. Particles were scattered in alveoli, and no tissue reaction was observed. (B) Purified HiPco CNTs, 0.1 mg, 7 d. The figure shows a low-grade granuloma. (C) CarboLex arc-produced CNTs, 0.5 mg, 7 d. A well-defined granuloma is shown. (D) Unprocessed HiPco CNTs, 0.5 mg, 90 d. Granulomas, alveolar wall thickening, and some fibrotic tissue are shown. Magnifications 40–200×.

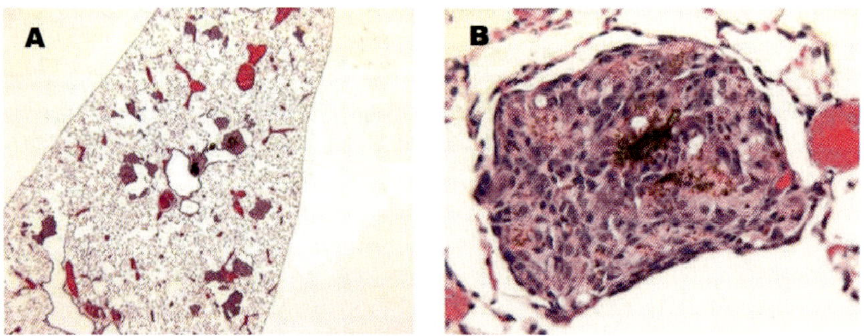

Fig. 6.6. Granulomas in lung tissues from rats 1 month after they were intratracheally instilled with 1 mg kg^{-1} of laser-synthesized CNTs. (Courtesy of Dr. David Warheit.)

that CNTs have a great tendency to aggregate, Warheit et al. concluded that the granulomatous reaction was a nonspecific response to instilled aggregates of SWCNTs and the results may not be relevant to human exposures [45]. Warheit showed that some instilled CNTs stuck in the major airway and the mechanical blockage suffocated 15% of the high-dose group. The proportion of the dose that clogged the airway must have been substantial. The inability of the dose to reach the alveolar region, where CNT-induced granulomas could occur, may partially explain why the results were both non-dose- and non-time-dependent.

6.5.4
Study of SWCNTs in Mice by Shvedova et al. of NIOSH

As pointed out above, to address the differences in conclusions about the potential hazard of exposures to SWCNTs drawn by Lam et al. and Warheit et al., Shvedova et al. [46] carried out a comprehensive pulmonary toxicity study in mice (C57CL/ 6), testing a purified HiPco CNT product (>99% SWCNTs) that was exhaustively subjected to purification to remove metals (final iron content 0.23% by weight). The animals were given a single treatment of CNTs, carbon black, or quartz at a dose of 0, 10, 20, or 40 µg per mouse (about 0, 0.5, 1, or 2 mg kg^{-1}, respectively). Aqueous suspensions of test dusts were aspirated at the pharyngeal area of mice, allowing droplets to be pulled into the lung during inspiration. The mice were then killed 1, 3, 7, 28, or 60 d after treatment. Histopathological examination of the lungs showed an acute inflammation, early onset of formation of granulomas, and progressive fibrosis. The histopathology was characterized by SWCNT-induced granulomas mainly associated with hypertrophied epithelial cells surrounding the dust aggregates, and diffusive interstitial fibrosis and alveolar wall thickening likely associated with dispersed SWCNTs. The total mass of granulomas in the lungs of mice in the 60-d group increased with an increase in the CNT dose. In general, lung lesions were dose-dependent and progressive, like those reported by Lam et al. [44]. Pulmonary function tests showed increases in functional respiratory deficiencies with increased concentrations of CNTs, a finding consistent with fibrosis. Compared with saline-treated controls, CNT-treated mice showed slower bacterial clearance assessed 7 days after bacterial inoculation. The test doses of quartz and carbon black did not induce granulomas or fibrosis.

Like Warheit et al., Shvedova et al. [46] examined the biomarkers of toxicity in BALF from CNT-treated animals. The results showed increases in total protein concentration, cell counts, concentration of transforming growth factor beta (TGF-β), and LDH and γ-glutamyltranspeptidase activities; these biomarkers of inflammation or cytotoxicity in the lungs were dose-dependent. Shvedova et al. concluded that crystalline silica caused less cytotoxicity than CNTs (compared on an equal-weight basis) and recruited fewer polymorphonuclear leucocytes into the lungs. These biomarker results from mice treated with quartz and SWCNTs differ from those of a similar study in rats reported by Warheit et al. and discussed above [45]. Like Lam et al. [44], Shvedova et al. demonstrated that CNTs were intrinsically toxic and cautioned that exposures of workers to high concentrations of respirable SWCNT particles may pose a risk of developing some lung lesions.

6.5.5
Study of MWCNTs by Muller et al. of Belgium

MWCNTs have been shown to produce lung lesions similar to those observed in studies with SWCNTs. Muller et al. tested two forms of MWCNTs, unprocessed (unground) MWCNTs and MWCNTs that had been ground [47]. They reported that 60 days after rats (Sprague–Dawley) were each given a single ITI dose of 0.5, 2, or 5 mg MWCNTs (sonicated and suspended in a normal saline solution containing a dispersing agent, Tween 80) their lungs showed inflammation, granulomas, and fibrosis. The unground CNTs remained in the bronchial lumen and produced collagen-rich granulomas. The bronchial lumen was partially or completely blocked, as in the study by Warheit et al. [45]; very few CNT particles were seen in the parenchymal (alveolar) region. The ground CNTs were "better dispersed" in the parenchyma and in the interstitium induced granulomas consisting of macrophages laden with particles, multinuclear giant cells, and some inflammatory cells, like those reported by Lam et al. [44]. Muller et al. [47] also showed that hydroxyproline and soluble collagen, two biomarkers of fibrosis, increased in the lung tissues in a dose-dependent fashion. BALF obtained from rats 3 days after the CNT treatment showed dose-dependent increases in LDH activity, total protein concentration, and neutrophil number. Muller et al. also concluded that CNTs are potentially toxic and advocated strict industrial hygiene [47].

6.6
Health Risk Implications

6.6.1
Toxicity Summary of CNTs and Occupational Exposure Risk

The results of four histopathology studies reported previously in the literature and reviewed here collectively showed that SWCNTs and MWCNTs themselves were capable of inducing granulomas and other lesions in the lungs, regardless of the process by which they were synthesized and the types and amounts of metals they contained. The metal residues and other impurities played only a small role in the formation of these pulmonary lesions (see Lam et al. [1] for detailed discussion). Dust of respiratory size is difficult to separate from bulk materials because CNT bundles or ropes tend to stick to each other. Moreover, all four studies were conducted similarly by ITI to expose a rat or mouse to a fine-particle suspension of a CNT dust ultrasonicated in an aqueous system containing a dispersing agent, or by pharyngeal aspiration of a CNT suspension obtained by boiling and sonicating CNTs. These instillation or aspiration studies, involving CNT particles mechanically dispersed in an aqueous system containing a dispersion agent, are considered screening assays of CNTs for potential pulmonary toxicity [44, 45]. However, even though these studies do not answer the important health risk question of whether airborne CNT particles can reach the lungs, they do reveal the intrinsic toxicity of CNTs. The findings convey the important message that if a CNT product contains a substantial amount of respirable dust that reaches the lung at a high enough concentration it is likely to produce the types of serious lung lesions seen in rodents. Certainly, it would be very important to conduct inhalation studies to confirm these pathology findings; data from inhalation studies are also needed for setting occupational exposure limits. Because it is difficult to conduct an inhalation study on CNTs, inhalation toxicity data will not be available for some time.

Lam et al. [44] and Shvedova et al. [46], who used carbon black and quartz as references in their comparative toxicity studies, concluded that if CNTs reach the lungs, under the test conditions described here and on an equal-weight basis they are much more toxic than carbon black and can be more toxic than quartz, chronic inhalation of which is considered a serious occupational health hazard. Study with MWCNTs led Muller et al. [47] to reach the same conclusion, i.e., that CNTs are intrinsically toxic. Therefore, it is prudent to assume that if significant amounts of airborne fine CNT particles were present in a workplace, occupational exposures to CNTs could produce substantial injury in the lungs and potentially the upper respiratory tract (see Lam et al. [1] for a detailed risk assessment of occupational exposures). If CNT dust is present in a work environment, strategies to minimize human exposure to it should be implemented.

6.6.2
Impact of SWCNTs on Environmental Health

The manufacture of CNTs is still on a small scale and the products remain expensive; the current impact of manufactured CNTs on environmental health is non-existent or very minimal. However, if millions of tons of CNTs are produced annually some day, as predicted by Richard Smalley [28], and if the CNT industry achieves the goal of charging "a couple dollars a pound" for CNTs [25], then CNTs will likely be incorporated or formulated into fabric, plastic, composite materials, and household commodities that will touch all aspects of human life. Then CNT-containing industrial wastes and degraded CNT-containing materials will probably appear in the environment (Fig. 6.2). Will CNTs bioaccumulate in the environment? Will ingested SWCNTs in the alimentary canal reach internal organs to produce toxicity? Studies will need to address these environmental health issues.

6.6.3
Toxicity of MWCNTs and Impact on Environmental Health

Murr and colleagues found MWCNTs in fine particulate matter generated from combustion of natural gas in typical kitchen ranges [31–33]. Finding MWCNTs in PM collected indoors and outdoors, they concluded [32] that MWCNTs and carbonaceous nanoparticles are ubiquitous in the environment. They further speculated that MWCNTs are a major component of indoor and outdoor airborne PM. Because MWCNTs are ubiquitously present in fine airborne particulate aggregates in our environments [34], it is reasonable to postulate that all humans are exposed to low levels of MWCNTs.

Dockery et al. found a positive correlation between fine particulate air pollution and excess mortality in six U.S. cities [56]. In a large-scale epidemiological study with 1.2 million adults, Pope et al. [57, 58] found that fine PM in ambient air is a risk factor associated with cardiopulmonary mortality and cardiovascular and pulmonary diseases. The underlying mechanisms by which exposure to fine PM might play a role in the pathogenesis of cardiopulmonary diseases are not known [59]. Pollutants may produce oxidative lung damage and inflammation [60]. Seaton et al. have proposed that fine particles deposited in the lung provoke alveolar inflammation, which causes potentially harmful cytokines to be released [61].

The pulmonary toxicity of SWCNTs and MWCNTs are similar; collectively, CNTs can produce lung lesions and biomarkers of toxicity, such as inflammation, fibrosis, granulomas, harmful cytokine release, and oxidative biochemical toxicological changes. Shvedova et al. also showed that SWCNTs impaired pulmonary functions and bacterial clearance. In a cell culture study, Monteiro-Riviere et al. have found that MWCNTs caused release of the proinflammatory cytokine interleukin 8 [62].

Gauderman et al. have found that adverse effects on the growth of lung functions in teenagers were associated with exposures to NO_2, acid vapor, fine PM, and elemental carbon, which had the highest correlation ($p = 0.007$) [63]. Fine PM, derived primarily from combustion, contains elemental carbon; MWCNTs, which are also produced by combustion, were found in fine PM collected in outdoor air in El Paso and Houston. The results of the four studies reviewed here showed that manufactured MWCNTs and SWCNTs were much toxic than ultrafine carbon black (used as a negative control by Muller et al. [47] and Shvedova et al. [46]) and graphite (used in the study by Warheit et al. [45]). It is probably true that environmental MWCNTs are a minor component in fine PM, but the unique toxicity of CNTs, which has not been seen with other elemental forms of carbon, raises a concern about these combustion-generated fibrous MWCNTs. Very possibly, they play a significant role in the pathogenesis of pollution-related cardiopulmonary diseases. Confirmation of this postulation requires results from toxicity studies on this newly-identified environmental pollutant.

Acknowledgment

The authors thank Dr. J. Krauhs for technical editing and Dr. D. Warheit of DuPont Company, Dr. A. Maynard of NIOSH, Dr. L. Murr of University of Texas at El Paso, TX, Dr. J. Rodriguez of University of Barcelona, Spain, and Springer Publishing Company, New York, NY, for granting permission to use their figures.

References

1 C.-w. Lam, J. T. James, R. McCluskey, S. Arepalli, R. L. Hunter, A review of carbon nanotube toxicity and assessment of potential occupational and human health, *Crit. Rev. Toxicol.* **2006**, in press.

2 White House Press Secretary, National Nanotechnology Initiative: Leading to the next industrial revolution. Press release January 21, 2000, in *What's New at the White House*, U. S. Government, Washington, DC, **2000**, http://clinton4.nara.gov/textonly/WH/New/html/20000121_4.html, accessed May 25, 2005.

3 National Science and Technology Council, *National Nanotechnology Initiative: Research and Development Supporting the Next Industrial Revolution. Supplement to the President's FY 2004 Budget*, National Science and Technology Council, Washington, DC, **2003**, http://nano.gov/html/res/fy04-pdf/fy04-main.html, accessed May 25, 2005.

4 M. Roco, Government nano-technology funding: An international outlook, *J. Minerals, Metals Mater. Soc.* **2002**, *54*, 22–23.

5 H. W. Kroto, J. R. Heath, S. C. O'Brien, R. F. Curl, R. E. Smalley, C60: Buckminsterfullerene, *Nature* **1985**, *318*, 162–163.

6 S. Iijima, Helical microtubules of graphitic carbon, *Nature* **1991**, *354*, 56–58, http://www.nature.com/nature/journal/v354/n6348/abs/354056a0.html, accessed May 25, 2005.

7 S. Iijima, Carbon nanotube, NEC Laboratories, **2004**, http://www.labs.nec.co.jp/Eng/innovative/E1/top.html, accessed June 10, 2005.

8 T. W. Ebbesen, P. M. Ajayan, Large-scale synthesis of carbon nanotubes, *Nature* **1992**, *358*, 220–222.

9 S. Iijima, T. Ichihashi, Single-shell carbon nanotubes of 1-nm diameter, *Nature* **1993**, *363*, 603.

10 T. Guo, P. Nikolaev, A. Thess, D. T. Colbert, R. E. Smalley, Catalytic growth of single-walled nanotubes by laser vaporization, *Chem. Phys. Lett.* **1995**, *243*, 49–54.

11 A. Thess, R. Lee, P. Nikolaev, H. J. Dai, P. Petit, J. Robert, C. H. Xu, Y. H. Lee, S. G. Kim, A. G. Rinzler, D. T. Colbert, G. E. Scuseria, D. Tomanek, J. E. Fischer, R. E. Smalley, Crystalline ropes of metallic carbon nanotubes, *Science* **1996**, *273*, 483–487, http://www.sciencemag.org/cgi/content/abstract/273/5274/483, accessed May 25, 2005.

12 J. Kong, A. M. Cassell, H. Dai, Chemical vapor deposition of methane for single-walled carbon nanotubes, *Chem. Phys. Lett.* **1998**, 292, 567–574.

13 Z. F. Ren, Z. P. Huang, J. W. Xu, J. H. Wang, P. Bush, M. P. Siegal, P. N. Provencio, Synthesis of large arrays of well-aligned carbon nano-tubes on glass, *Science* **1998**, *282*, 1105–1107.

14 P. Nikolaev, M. J. Bronikowski, R. K. Bradley, F. Rohmund, D. T. Colbert, K. A. Smith, R. E. Smalley, Gas-phase catalytic growth of single-walled carbon nanotubes from carbon monoxide, *Chem. Phys. Lett.* **1999**, *313*, 91–97.

15 A. Moisala, A. G. Nasibulin, E. I. Kauppinen, The role of metal nanoparticles in the catalytic production of single-walled carbon nanotubes – a review, *J. Phys.: Condensed Matter* **2003**, *15*, S3011–S3035.

16 P. Ball, Focus carbon nanotubes, *Nat. Scienceupdate* **1999**, http://www.nature.com/nsu/991202/991202-1.html, accessed May 25, 2005.

17 S. Arepalli, P. Nikolaev, W. Holmes, B. S. Files, Production and measurements of individual single-wall nanotubes and small ropes of carbon, *Appl. Phys. Lett.* **2001**, *78*, 1610–1612.

18 P. Ball, Roll up for the revolution, *Nature* **2001**, *414*, 142–144, http://www.phys.washington.edu/~cobden/

P600/Philip_Ball_tube_news.pdf, accessed May 25, 2005.

19 Rice University, Small is big @ Rice! The world's largest nanotube model, in *blogs@Rice*, Rice University, **2005**, http://blogs.rice.edu/blog/templates/blog_274/nanotube/files/nano-program.pdf, accessed June 13, 2005.

20 A. Jensen, J. R. Hauptmann, J. Nygård, J. Sadowski, P. E. Lindelof, Hybrid devices from single wall carbon nanotubes epitaxially grown into a semiconductor heterostructure, *Nano Lett.* **2004**, *4*, 349–352, http://pubs.acs.org/cgi-bin/jcen?nalefd/4/i02/html/nl0350027.html, accessed June 14, 2005.

21 H. Stahl, *Electronic Transport in Ropes of Single Wall Carbon Nanotubes* (Doctoral Thesis Dissertation), Aachen University of Technology, **2000**, http://sylvester.bth.rwth-aachen.de/dissertationen/2000/28/00_28.pdf, accessed September 7, 2005.

22 P. C. P. Watts, W. K. Hsu, D. P. Randall, H. W. Kroto, D. R. M. Walton, Non-linear current-voltage characteristics of electrically conducting carbon nanotube-polystyrene composites, *Phys. Chem. Chem. Phys.* **2002**, *4*, 5655–5662.

23 Wikipedia, Carbon nanotube, **2005**, http://en.wikipedia.org/wiki/Carbon_nanotube, accessed August 30, 2005.

24 R. E. Smalley, Nanotechnology. Prepared written statement and supplemental material of R. E. Smalley, Rice University, June 22, 1999, in *U.S. House of Representatives Committee on Science, Basic Research Subcommittee Hearings*, U. S. Government, Washington, DC, **1999**, pp. 1–17, http://www.house.gov/science/smalley_062299.htm, accessed September 7, 2005.

25 T. Bearden, Interview: Nobel Prize Winner Dr. Richard Smalley, in *Online NewsHour*, Public Broadcasting System, **2003**, http://www.pbs.org/newshour/science/hydrogen/smalley.html, accessed June 10.

26 BuckyUSA, BuckyUSA – price list, **2005**, http://home.flash.net/~buckyusa/page17.html, accessed October 6, 2005.

27 P. A. Baron, A. D. Maynard, M. Foley, *Evaluation of Aerosol Release during the Handling of Unrefined Single Walled Carbon Nanotube Material*, Tech. Report No. NIOSH DART-02-191, National Institute of Occupational Safety and Health, Cincinnati, OH, **2003**.

28 G. Taubes, Nanotechnology. An interview with Dr. Richard Smalley, *ISI Essential Sci. Indicators Special Top.* **2002**, http://www.esi-topics.com/nano/interviews/Richard-Smalley.html, accessed May 25, 2005.

29 Anonymous, 40 CFR Part 763, *Fed. Regist.* **1989**, *54*.

30 B. O. Boskovic, V. Stolojan, R. U. A. Khan, S. Haq, S. R. P. Silva, Large-area synthesis of carbon nanofibres at room temperature, *Nat. Mater.* **2002**, *1*, 165–168.

31 W. Merchan-Merchan, A. V. Saveliev, L. A. Kennedy, Carbon nanostructures in opposed-flow methane oxy-flames, *Combust. Sci. Technol.* **2003**, *175*, 2217–2236.

32 J. J. Bang, P. A. Guerrero, D. A. Lopez, L. E. Murr, E. V. Esquivel, Carbon nanotubes and other fullerene nanocrystals in domestic propane and natural gas combustion streams, *J. Nanosci. Nanotechnol.* **2004**, *4*, 716–718.

33 L. E. Murr, J. J. Bang, D. A. Lopez, P. A. Guerrero, E. V. Esquivel, A. R. Choudhuri, M. Subramanya, M. Morandi, A. Holian, Carbon nanotubes and nanocrystals in methane combustion and the environmental implications, *J. Mater. Sci.* **2004**, *39*, 2199–2204.

34 L. E. Murr, J. J. Bang, E. V. Esquivel, P. A. Guerrero, D. A. Lopez, Carbon nanotubes, nanocrystal forms and complex nanoparticle aggregates in common fuel-gas combustion source and ambient air, *J. Nanoparticle Res.* **2004**, *5*, 1–11.

35 J. J. Bang, L. E. Murr, Collecting and characterizing atmospheric nanoparticles, *J. Minerals, Metals Mater. Soc.* **2002**, *54*, 28–30.

36 E. V. Esquivel, L. E. Murr, A TEM analysis of nanoparticulates in a polar ice core, *Mater. Characterization* **2004**, *52*, 15–25.

37 Natural Gas Supply Association, Natural gas and the environment, **2004**, http://www.naturalgas.org/environment/naturalgas.asp, accessed August 30, 2005.

38 National Energy Foundation, Fuel consumption stats, **2002**, http://www.nef1.org/ea/eastats.html, accessed September 8, 2005.

39 U.S. Environmental Protection Agency, PM2.5. Objectives and history, in *Laboratory and Field Operations – PM 2.5. Ambient Air Quality Monitoring Program*, U.S. Environmental Protection Agency, Washington, DC, **2003**, http://www.epa.gov/region4/sesd/pm25/p2.htm, accessed September 9, 2005.

40 F. E. Huggins, G. P. Huffman, W. P. Linak, C. A. Miller, Quantifying hazardous species in particulate matter derived from fossil-fuel combustion, *Environ. Sci. Technol.* **2004**, *38*, 1836–1842.

41 J. Schwartz, G. Norris, T. Larson, L. Sheppard, C. Claiborne, J. Koenig, Episodes of high coarse particle concentrations are not associated with increased mortality, *Environ. Health Perspect.* **1999**, *107*, 339–342, http://ehp.niehs.nih.gov/members/1999/107p339-342schwartz/schwartz-full.html, accessed September 12, 2005.

42 U.S. Environmental Protection Agency, Research opportunities: Epidemiologic research on health effects of long-term exposure to ambient particulate matter and other air pollutants, **2003**, http://es.epa.gov/ncer/rfa/current/2003_pm_epi.html, accessed September 7, 2005.

43 I. Amato, The soot that could change the world, *Fortune Mag.* **2001**, *June 25*, http://www.nano-lab.com/fortune.html, accessed August 30, 2005.

44 C.-W. Lam, J. T. James, R. McCluskey, R. L. Hunter, Pulmonary toxicity of single-wall carbon nanotubes in mice 7 and 90 days after intratracheal instillation, *Toxicol. Sci.* **2004**, *77*, 126–134.

45 D. B. Warheit, B. R. Laurence, K. L. Reed, D. H. Roach, G. A. M. Reynolds, T. R. Webb, Comparative pulmonary toxicity assessment of single-wall carbon nanotubes in rats, *Toxicol. Sci.* **2004**, *77*, 117–125.

46 A. A. Shvedova, E. R. Kisin, R. Mercer, A. R. Murray, V. J. Johnson, A. I. Potapovich, Y. Y. Tyurina, O. Gorelik, S. Arepalli, D. Schwegler-Berry, A. F. Hubbs, J. Antonini, D. E. Evans, B. K. Ku, D. Ramsey, A. Maynard, V. E. Kagan, V. Castranova, P. Baron, Unusual inflammatory and fibrogenic pulmonary responses to single walled carbon nanotubes in mice, *Am. J. Physiol. Lung Cell. Mol. Physiol.* **2005**, *289*, L698–L708.

47 J. Muller, F. Huaux, N. Moreau, P. Misson, J. F. Heilier, M. Delos, M. Arras, A. Fonseca, J. B. Nagy, D. Lison, Respiratory toxicity of multi-wall carbon nanotubes, *Toxicol. Appl. Pharmacol.* **2005**, *207*, 221–231.

48 J.-P. Salvetat, J.-M. Bonard, N. Thomson, A. Kulik, L. Forró, Mechanical properties of carbon nanotubes, *Appl. Phys. A* **1999**, *69*, 255–260.

49 C. P. Sabaitis, B. K. Leong, D. A. Rop, C. S. Aaron, Validation of intratracheal instillation as an alternative for aerosol inhalation toxicity testing, *J. Appl. Toxicol.* **1999**, *19*, 133–140.

50 B. K. Leong, J. K. Coombs, C. P. Sabaitis, D. A. Rop, C. S. Aaron, Quantitiative morphometric analysis of pulmonary deposition of aerosol particles inhaled via intratracheal nebulization, intratracheal instillation or nose-only inhalation in rats, *J. Appl. Toxicol.* **1998**, *18*, 149–160.

51 K. E. Driscoll, D. L. Costa, G. Hatch, R. Henderson, G. Oberdorster, H. Salem, R. B. Schlesinger, Intratracheal instillation as an exposure technique for the evaluation of respiratory tract toxicity: Uses and limitations, *Toxicol. Sci.* **2000**, *55*, 24–35.

52 C.-W. Lam, J. T. James, J. N. Latch, R. F. Hamilton, A. Holian, Pulmonary toxicity of simulated lunar and Martian dusts in mice. II. Biomarkers of acute responses after intratracheal instillation, *Inhal. Toxicol.* **2002**, *14*, 917–928.

53 C.-W. Lam, J. T. James, R. McCluskey, S. Cowper, C. Muro-Cacho, Pulmonary toxicity of simulated lunar and Martian dusts in mice. I. Histopathology 7 and 90 days after intratracheal instillation, *Inhal. Toxicol.* **2002**, *14*, 901–916.

54 A. Huczko, H. Lange, E. Całko, H. Grubek-Jaworska, P. Droszcz, Physiological test of carbon nanotubes: Are they asbestos-like? *Fullerenes, Nanotubes, Carbon Nanostruct.* **2001**, *9*, 251–254.

55 A. G. Rinzler, J. Liu, H. Dai, P. Nikolaev, C. B. Huffman, F. J. Rodriguez-Macias, P. J. Boul, A. H. Lu, D. Heymann, D. T. Colbert, R. S. Lee, J. E. Fischer, A. M. Rao, P. C. Eklund, R. E. Smalley, Large-scale purification of single-wall carbon nanotubes: Process, product, and characterization, *Appl. Phys. A* **1998**, *67*, 29–37.

56 D. W. Dockery, C. A. Pope, 3rd, X. Xu, J. D. Spengler, J. H. Ware, M. E. Fay, B. G. Ferris, Jr, F. E. Speizer, An association between air pollution and mortality in six U.S. cities, *N. Engl. J. Med.* **1993**, *329*, 1753–1759.

57 C. A. Pope, R. T. Burnett, M. J. Thun, E. E. Calle, D. Krewski, K. Ito, G. D. Thurston, Lung cancer, Cardiopulmonary mortality, and long-term exposure to fine particulate air pollution, *JAMA* **2002**, *287*, 1132–1141.

58 C. A. Pope, 3rd, R. T. Burnett, G. D. Thurston, M. J. Thun, E. E. Calle, D. Krewski, J. J. Godleski, Cardiovascular mortality and long-term exposure to particulate air pollution: Epidemiological evidence of general pathophysiological pathways of disease, *Circulation* **2004**, *109*, 71–77.

59 C. A. Pope, Epidemiology of fine particulate air pollution and human health: Biologic mechanisms and who's at risk? *Environ. Health Perspect.* **2000**, *108(Suppl. 4)*, 713–723.

60 C. A. Pope, D. W. Dockery, R. E. Kanner, G. M. Villegas, J. Schwartz, Oxygen saturation, pulse rate, and particulate air pollution: A daily time-series panel study, *Am. J. Respir. Crit. Care Med.* **1999**, *159*, 354–356.

61 A. Seaton, W. MacNee, K. Donaldson, D. Godden, Particulate air pollution and acute health effects, *Lancet* **1995**, *345*, 176–178.

62 N. A. Monteiro-Riviere, R. J. Nemanich, A. O. Inman, Y. Y. Wang, J. E. Riviere, Multi-walled carbon nanotube interactions with human epidermal keratinocytes, *Toxicol. Lett.* **2005**, *155*, 377–384.

63 W. J. Gauderman, E. Avol, F. Gilliland, H. Vora, D. Thomas, K. Berhane, R. McConnell, N. Kuenzli, F. Lurmann, E. Rappaport, H. Margolis, D. Bates, J. Peters, The effect of air pollution on lung development from 10 to 18 years of age, *N. Engl. J. Med.* **2004**, *351*, 1057–1067.

7
Toxicity of Nanomaterials – New Carbon Conformations and Metal Oxides

Harald F. Krug, Katrin Kern, Jörg M. Wörle-Knirsch, and Silvia Diabaté

7.1
Introduction

Nanomaterials are on the same scale as most elements of living cells, including proteins, nucleic acids, lipids and even cellular organelles. When considering nanoparticles it must be asked how man-made nanostructures can interact with or influence biological systems. On the one hand, nanosystems are specifically engineered to interact with biological systems for particular medical or biological applications. On the other hand, the large-scale production of nanoparticles for either non-medical applications or as side-product of combustion processes may affect a wide range of organisms throughout the environment. Since the 1970s, an increasing number of investigations concerning the use of nanoscale structures, e.g., liposomes, for drug transport and comparable applications have been undertaken [1–9]. In addition to liposomes, nanoparticles produced from other materials came into the focus of physicians for various treatments of diseases [10–16]. This work aims to design inert auxiliary accompanying materials and to use body-friendly and biodegradable excipients. However, dependent on their target organ and functionality, not all of these materials are degradable and some stay in the body for long periods. In light of this, side effects and foreign body reactions may be detectable and a good local and systemic tolerance during and after medication should be a condition *sine qua non*.

Nanostructured materials come into contact with biological systems not only through their use in drug delivery systems or for gene transfer. They are also produced for food and cosmetic chemistry and many other technical applications (Tab. 7.1). The increasing production, particularly of metal oxide nanoparticles and new carbon materials, will enhance the possible exposure at work places, packing stations and during application of the products [17]. In addition, waste treatment and containment at the end of a products life cycle must be considered. For all these reasons, it is of great interest to determine how these materials, when coming in contact with living organisms, are taken up, transported in or through cell layers, and affect biological functions. To cover these questions we review the latest

Nanotechnologies for the Life Sciences Vol. 5
Nanomaterials – Toxicity, Health and Environmental Issues. Edited by Challa S. S. R. Kumar
Copyright © 2006 WILEY-VCH Verlag GmbH & Co. KGaA, Weinheim
ISBN: 3-527-31385-0

Tab. 7.1. Examples of metal oxides and carbon modifications manufactured as nanomaterials of commercial interest.

Type	Examples for use
Metal oxides	
• Silica (SiO_2)	• Additives for polymer composites
• Titania (TiO_2)	• UV-A protection
• Alumina (Al_2O_3)	• Solar cells
• Iron oxide (Fe_3O_4, Fe_2O_3)	• Pharmacy/medicine
• Zirconia (ZrO_2)	• Additives for scratch resistance coatings
• Zinc dioxide (ZnO_2)	
Carbon modifications	
• Carbon black	• Tires, printer, copier
Fullerenes	
• Buckminsterfullerene (C_{60})	• Mechanical and tribological applications/additives to grease
Carbon nanotubes	
• Single-wall carbon nanotubes	• Additives for polymer composites
• Multi-wall carbon nanotubes	• Electronic field emitters Batteries Fuel cells
Carbon nanofibers	
• Various conformations	• Mechanical and tribological applications Carrier for catalysts Additives for polymer composites Elastic foams

results from various studies on the biological effects of nanoparticles that may be the basis for adverse effects, especially in humans. Because metal oxides are the most prominent produced variants of nanoparticles and new carbon modifications are the most promising ones we focus on these two types with respect to their cellular uptake and possible influence on important cellular mechanisms *in vitro*. The effects of ambient particulate matter or particulate emissions from combustion devices such as diesel engines or oil burners are not reviewed. These particles, produced unintentionally with a complex chemical composition, are released into the environment and affect the general population. Although these particles stimulated much concern over their health effects, this chapter concentrates on the effects of intentionally produced nanoparticles of low solubility with well-known chemical composition, form and size to provide condensed information for a possible occupational exposure. Our short overview on biological hazards ends with a more general aspect of the risks connected with the production and use of nanomaterials and, on the other hand, the opportunities that are an important component in all long-term considerations.

7.1.1
Nanoscale Materials and Adverse Health Effects: Precautionary Measures

Although nanoparticles have been used in various products for several decades, the expected increase of production and use of newly developed materials makes the question of their safety to life and the environment increasingly important. However, classical risk regulation is not adequate because the risk can not be quantified. At this early stage, where most of the materials are under development and produced only in small amounts for research laboratories, precautionary measures can be taken to keep exposure below particular thresholds and avoid possible adverse effects by employing the "as low as reasonably achievable – ALARA" principle. As soon as the conditions required for the risk management approach are no longer fulfilled, controversies and ambivalent situations result. This is especially the case if fundamental knowledge concerning the toxicity of these materials is missing, controversial, or based on not sufficiently validated experimental models. In past cases, severe adverse effects resulting from the implementation of new materials or technologies were not been detected at an early stage (e.g., asbestos) and the resulting health, environmental, and economic damage has spurred calls for stronger regulatory measures. These debates resulted in the implementation of the precautionary principle in the European Union that reached wide international agreement during the Earth Summit (United Nations Conference on Environment and Development, UNCED) in Rio de Janeiro 1992 and became part of Agenda 21. Based on this, several demands have to be made for nanotechnology and its products:

1. Without knowledge of possible adverse effects, nanoparticle exposure should be avoided at work places as well as in the population and the environment.
2. Multiple studies are necessary to clarify the biological effects of nanoparticles, with the caveat that different materials, sizes and surface characteristics often behave differently.
3. As with normal chemicals, extrapolation from *in vitro* system and/or animal experiments with regards to a specific nanoparticle is reasonable for judging human exposure to the nanoparticle in question, but can not be the base for a fundamental evaluation or assessment of nanoparticles in general.

As past experience has illustrated, precautionary measures are needed for nanotechnological developments, new materials, and nanoparticles [18, 19]. However, the call for a moratorium on nanotechnology is unrealistic because a moratorium for the chemical, physical and pharmaceutical industries (new substances and new techniques are possible new hazards and risks) would logically follow. From the point of view of toxicologists, the database of biological effects of nanoparticles must be increased by intensifying research on adverse health effects of these new

Risk Assessment

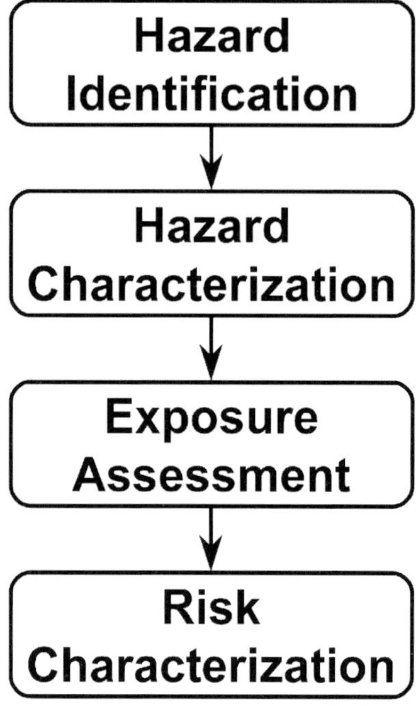

Fig. 7.1. Sequence of events leading to risk assessment.

materials. Most of the presented data in this chapter on nanosized metal oxides and carbonaceous materials are either preliminary or poorly confirmed by other research groups. So far, few nanomaterials have been investigated intensely. These two nanomaterial categories are in production for commercial applications and, therefore, several studies on their health effects have already been performed, whereas information on the impact of nanomaterials on the environment is rare and hence considered marginally.

7.1.2
Hazard Identification and Exposure Estimation

The health and safety issues related to metal oxides and carbon nanomaterials are in a very early phase. Hence, it seems premature to draw far-ranging conclusions regarding the potential hazards related to exposures to these materials. Since the toxicology database for inhalation or dermal exposure to these nanomaterials is rather sparse, efforts to obtain this information must be intensified. Most important is the development of methodologies and protocols concomitant with the implementation of hazard/toxicity studies, as well as workplace exposure assessments, to better ascertain the impact of nanomaterials on human health. For a valid risk assessment both parts of the following equation must be taken into account:

$$\text{Risk} = \text{Exposure} \times \text{Hazard}$$

Within the risk assessment procedure, hazard identification is the first step (Fig. 7.1). Because of the lack of knowledge, it is an open question as to whether established mechanisms of risk analysis and risk regulation may be applied to nanotechnology. Regardless, both hazard characterization and exposure assessment are fundamental pre-requisites, leading to risk characterization.

7.2
Production and Use of "New Carbon Modifications" and Metal Oxides

Nanoscaled insoluble metal oxides are used in applications in almost all fields of technology and industry (Tab. 7.1). They are used as additives in sun screens and textiles to block UV light (TiO_2), to clear apple juice and beer (SiO_2) as well as to degrade toxic chemical waste very efficiently (MgO). While only poor data is available on the toxicity of most synthesized nanoparticles, most investigations on these new materials have been done with titania and silica. At present the assessment of nanoscaled metal oxide toxicity is focused on free and primary particles. As nanoparticles are sintered and agglomerated to larger structures, they lose most of the vast toxic potential to human health that appears in primary particles.

Carbonaceous particles are generated by pyrolysis of gaseous or liquid hydrocarbons, or by spark generation between two graphite electrodes. The most common product is *carbon black*, consisting of amorphous, variably sized colloidal particles of elemental carbon (Fig. 7.2). In contrast to soot, which results from incom-

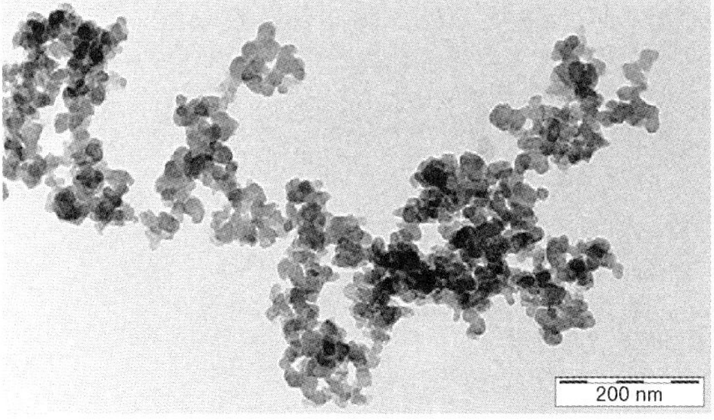

Fig. 7.2. Transmission electron micrograph of carbon black (Printex 90), showing aggregates of primary particles with an average diameter of 14 nm. (T. Detzel, ITG.)

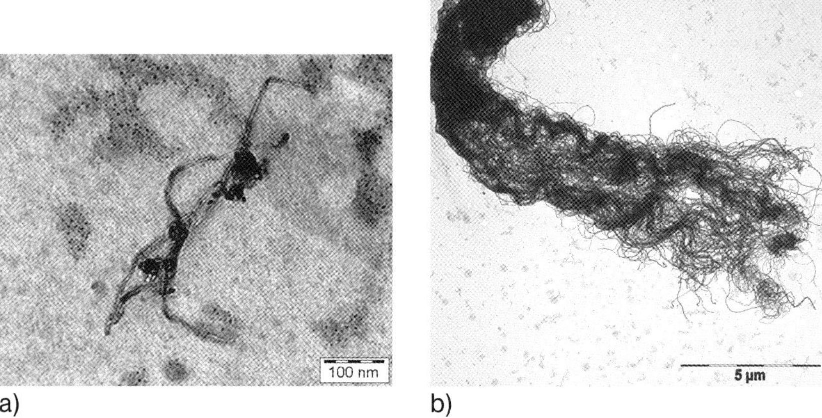

Fig. 7.3. Transmission electron micrograph of (a) single SWCNT and (b) MWCNTs forming a bundle. (T. Detzel, ITG.)

plete combustion of carbon-containing material, carbon black contains only low amounts of solvent-extractable organic matter. The material is used as an additive in the rubber of automobile tires, in inks, batteries and protective coatings.

Buckyballs (also known as fullerenes) are cage-like molecules, e.g. the spherical buckminsterfullerene, which consists of 60 carbon atoms (C_{60}) and was discovered in the 1980s [20]. They can be generated in the arc between two graphite electrodes in a helium atmosphere. The heat vaporizes the graphite, and fullerenes form as the gaseous carbon cools.

Carbon fibers consisting of amorphous carbon with a diameter of 7–10 µm have been produced by similar methods since the 1980s. The fibers are embedded in resin or plastic to produce composite materials used in aerospace, automotive, sports goods, and prosthetic industries. By modification of the production method, long tube-like *carbon nanotubes* are formed [21]. Single-walled carbon nanotubes (SWCNTs) have a diameter of 1–2 nm and are up to 100 µm long. Multi-walled carbon nanotubes (MWCNTs) consist of several layers of carbon cylinders, which increases the diameter to 10–30 nm (Fig. 7.3). This new material has high potential for new commercial products because it exhibits very interesting properties such as great tensile strength, high conductivity, or unique electronic features. Therefore, it is predicted that tons of carbon nanotubes will be produced worldwide every year in the near future [22].

Recently, carbon fibers with nanometer dimensions have been synthesized [23]. The diameters range from 60 to 200 nm and, unlike carbon nanotubes, they do not posses a helical carbon arrangement (Fig. 7.4). There is evidence that when carbon nanofibers are used as an orthopedic or dental material some of the common problems associated with implant material such as insufficient cytocompatibility may be avoided.

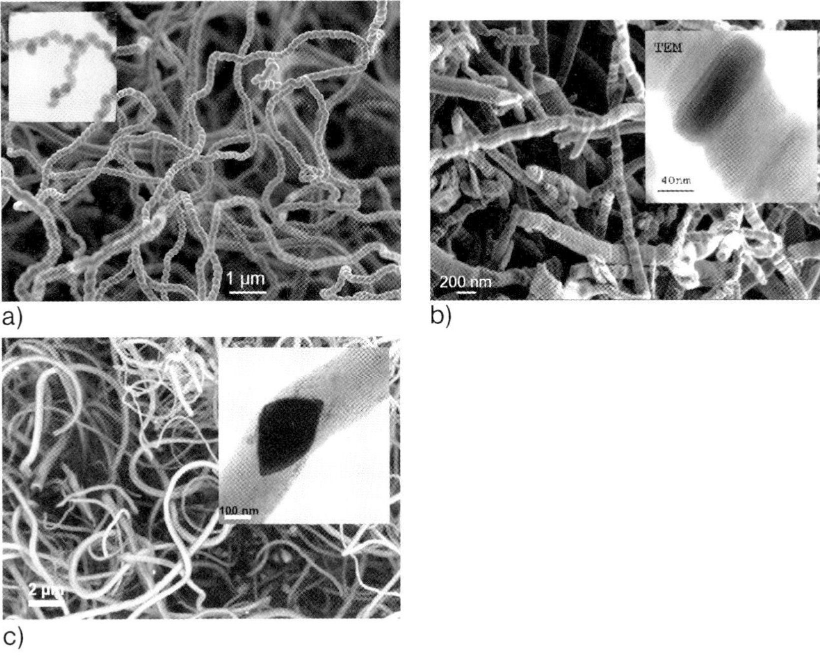

a) b)

c)

Fig. 7.4. Carbon nanofibers of different size and properties. (a) Screws (CNF-SC) are 50 to 200 nm in diameter and 1 to 10 μm long (upper left). (b) Platelets (CNF-PL) are between 150 and 250 nm in diameter and 5–50 μm long. (c) Herringbones (CNF-HB) are the thickest and shortest fibers, with a diameter of 200–600 nm and a length of 0.5 to 5 μm. These fibers are used as catalysts, part of composite materials, different polymers etc. (Reprinted with permission of FutureCarbon GmbH, Bayreuth.)

7.2.1
Health Aspects

It is of increasing concern to the public as well as toxicologists and occupational physicians that nanotechnology may create contaminants whose tiny size makes them ultra-hazardous. A further complication is that these very small materials may bind or react with other contaminants in the air or water that are harmful and facilitate their transport into living organisms, thus leading to additional adverse effects.

Most knowledge of the adverse health effects of very small particles comes from recent studies on ambient ultrafine particles (UFP) unintentionally released into the atmosphere. Epidemiological studies have shown that increased levels of UFP (<0.1 μm in aerodynamic diameter) are associated with increased respiratory and cardiovascular mortality and morbidity as well as worsening of asthma. These effects were observed in particular in susceptible persons such as the very young and old, and those with compromised respiratory and cardiovascular systems [24–

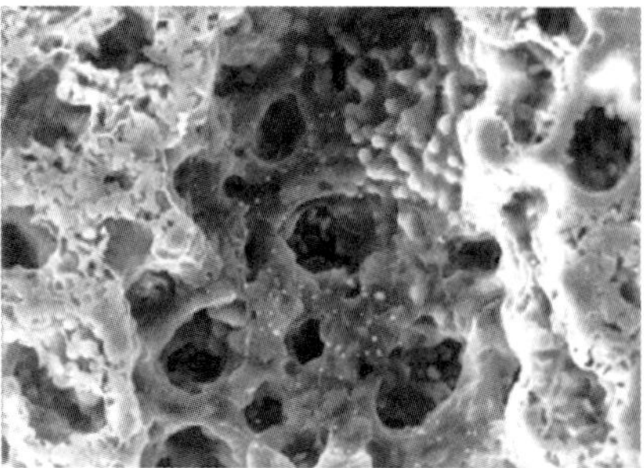

Fig. 7.5. Scanning electron micrograph demonstrating deposition of inhaled iron carbonyl particles at the alveolar duct bifurcations in the distal lung of a rat. (Reprinted from Ref. [31], with permission from Elsevier.)

26]. The composition of ambient UFP includes organic and elemental carbon, metals, chloride, nitrate and sulfate. Since the core typically consists of elemental carbon [27], it was obvious to study the biological effects using ultrafine carbonaceous particles as model particles.

7.2.2
Uptake and Possible Transport, Depots, and Accumulation in Living Organisms

Most studies on the uptake and transport of nanoparticles as well as the formation of depots and their accumulation have been performed *in vivo* (Fig. 7.5). The laboratory of Günter Oberdörster carried out fundamental studies during the last decade that delivered basic data on distribution, uptake and retention of titanium nanoparticles in lungs of rats or mice. It was demonstrated that after instillation or inhalation in rats, equivalent masses of 20 nm TiO_2 were deposited in the alveoli and accessed the pulmonary interstitium to a significantly larger extent than 250 nm TiO_2 particles [28, 29]. This resulted in a prolonged retention of the smaller particles in the lung. But these studies in rat also indicate that there is a difference between inhalation and instillation of nanoscaled TiO_2. As measured by bronchovascular lavage parameters, animals receiving particles via inhalation showed a decreased pulmonary response, in both severity and persistence, compared with animals receiving particles via instillation [30].

Dose–response curves are incompatible when the instilled mass of ultrafine and fine TiO_2 particles are used as parameters for the measured neutrophil infiltration. Alternatively, when using the particle surface area of the instilled dose, it became obvious that the inflammatory response in the lung for both ultrafine and fine TiO_2 fitted the same dose–response curve [32]. These results strongly suggest that for particles of the same surface material/chemistry, such as TiO_2, particle surface area is a more important dosimetric parameter than particle mass (or particle number). The same authors demonstrated a perfect correlation of the inflammatory response in both experimental animals by fitting the particle surface dose with the lung weights of mice and rats. This assumption to use surface area as a relevant dosimetric parameter to express toxicity data is confirmed by several studies but has to be validated in more detail (for further literature see Ref. [33]).

During the last decade several studies have shown that the response to instillation or inhalation of nanoscaled TiO_2 in various species such as hamster, rat, or mouse differs. Under conditions of equivalent lung burden with nanoscaled TiO_2, rats developed more severe inflammatory responses than mice, while the clearance of particles from the lungs was markedly impaired in both species [34]. In contrast, the clearance rate of hamsters is totally unaffected at all applied particle concentrations. The difference in clearance rates seems to be caused by a translocation of the particles into the lymph nodes of rats and mice. This was not seen in hamsters and thus the retention time in hamsters is shorter. Another variation within the three species was described by histopathological methods. After instillation or inhalation of nanoparticles an accumulation of particle-loaded macrophages surrounded by normal alveolar structures has been observed in hamster, whereas in rats the particle-loaded macrophages accumulate intra-alveolar. Moreover, interstitial fibrosis and alveolar metaplasia of lining epithelium has been demonstrated in rats and were not noted in either mice or hamsters. In contrast, no macrophage accumulation could be detected in hamsters [35]. A further study showed that, compared with mice or hamsters, rats were hypersensitive to a high lung burden of insoluble dust [34].

For ultrafine cadmium oxide particles, *in vivo* studies in rats were performed and showed that the inhaled particles cause an increasing cadmium content in lung, liver, kidney and blood. However, systemic translocation of the particles only appeared if the animals were exposed to high particle concentrations, which generated lung injury. J774 macrophages treated with CdO were swollen with poorly preserved cell membranes, cytoplasmic structures and nuclear morphology [36].

Few studies have been performed within *in vitro* systems dealing with the transport and uptake of nanoparticular metal oxides and a direct comparison of these studies is not feasible because different cell types and particles differing in size and material were used. Fundamentally, metal oxide and carbonaceous nanoparticles can be taken up by cells by very different mechanisms (Fig. 7.6). In several human cell lines nanoscaled material, such as TiO_2, SiO_2 and ZrO_2, was taken up into the cells [37]. The incorporated particles were detected in autophagic vacuoles, which also contained amorphous cellular material and membranes [38], as well as within the cytoplasm (Fig. 7.7).

Nanoscaled TiO_2 significantly impaired macrophage phagocytosis at a lower dose than its fine counterpart [39]. Thus, the slower clearance of ultrafine particles from the lung can be in part attributed to a particle-mediated impairment of macrophage phagocytosis. Nanoscaled TiO_2 caused cytoskeletal dysfunction as decreased phagosome transport and increased cytoskeletal stiffness could be observed in concentrations of 100 µg mL^{-1} per 10^6 cells and above in macrophages [40].

A more recent study demonstrated that particle size and particle composition, respectively, were responsible for the observed biological effects by using hematite

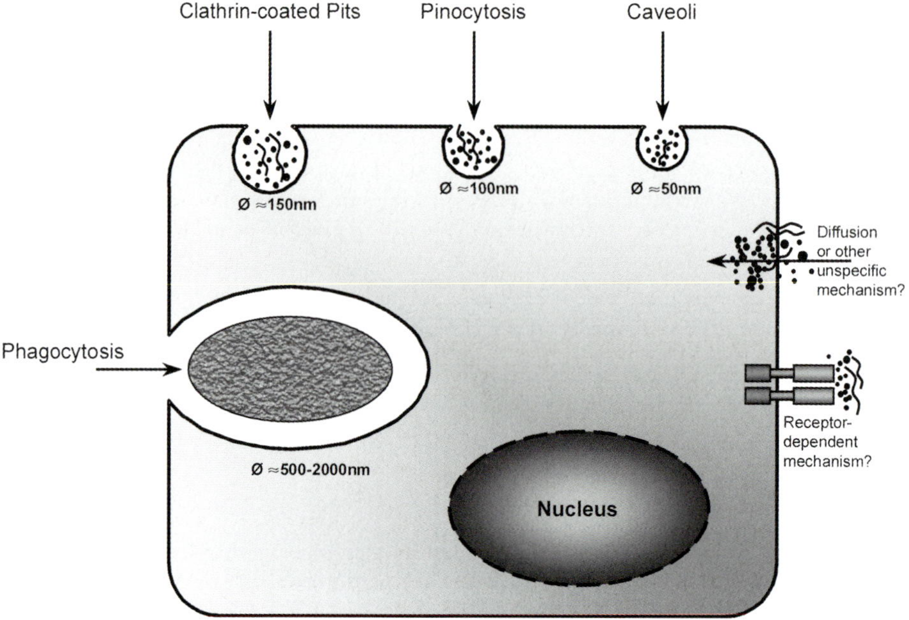

Fig. 7.6. Possible cellular uptake pathways for particles of different sizes.

(70 nm) and amorphous SiO_2 between 40 and 300 nm in diameter. Furthermore, co-culture systems of lung epithelial cells and macrophages showed an increased sensitivity to particle exposure concerning the cytokine release in comparison to the monocultures of each cell type [41].

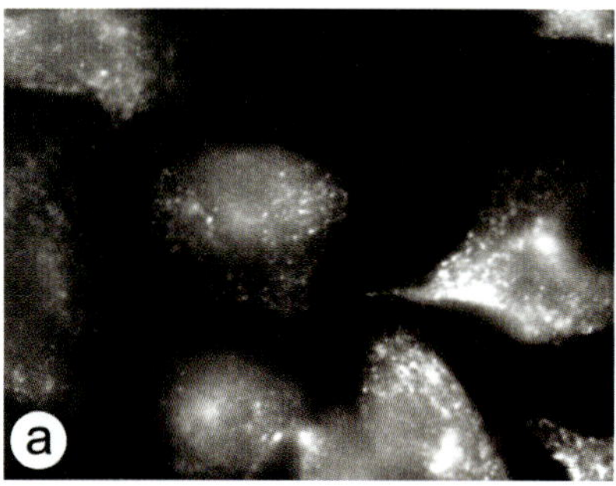

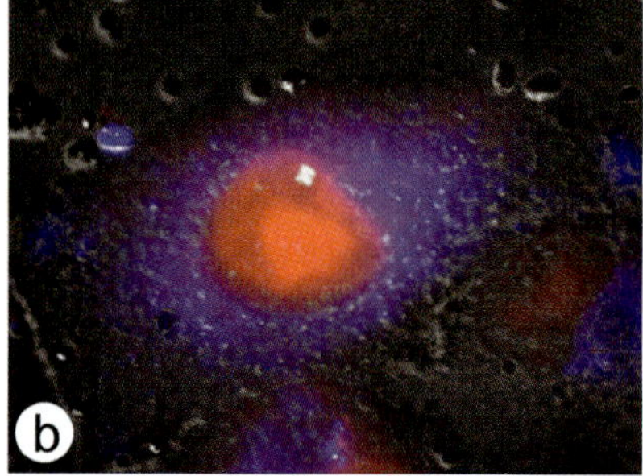

Fig. 7.7. Fluorescence micrographs of human lung epithelial cells (A549). (a) Caveolin detected with Cy3 coupled antibody, demonstrating the overall appearance of caveoli within these cells. (b) A549 cells after exposure to ZrO_2 (5 nm in diameter) coated with Coumarin 307 (blue fluorescence). The particles are equally distributed in the cytoplasm of the cells whereas none could be detected within the nuclei (counterstained with ethidium bromide – red fluorescence). Magnification: 630-fold. (K. Kern, ITG.)

The deposition and fate of inhaled ultrafine carbon particles generated by spark discharge (26 nm) was studied with the stable isotope ^{13}C [42]. In the rat model, more than 50% of the ^{13}C particles was rapidly translocated to the liver within 24 h while no significant increase in ^{13}C was detected in the other organs. In contrast, another study with insoluble ultrafine ^{192}Ir particles (15–20 nm) detected less than 1% translocation into the extracellular organs. The particles not only accumulated in the liver, but also in spleen, kidneys, brain and heart [43]. Interestingly, in this study a much lesser fraction of ultrafine particles could be removed by bronchoalveolar lavage compared with inhalation studies with larger particles. It was suggested that the ultrafine particles penetrate the epithelium or the interstitium where they are retained. A human study observed an uptake of 3–5% of radiolabeled carbonaceous ultrafine particles into the blood and translocation into the liver [44]. These results differ from those of Brown and colleagues [45] who could not find any particles outside the lungs and the cleared fractions after inhalation of an ultrafine technetium-99m-labeled carbon aerosol in their human study.

Newer studies indicate that inhaled ultrafine ^{13}C particles translocate into the brain of rats. The particles are suggested to deposit on the olfactory mucosa of the nasopharyngeal region and translocate via the olfactory nerve to the brain [46]. Despite conflicting results, it can be summarized that inhaled ultrafine particles are able to translocate to extrapulmonary organs via the blood; however, this fraction is very low.

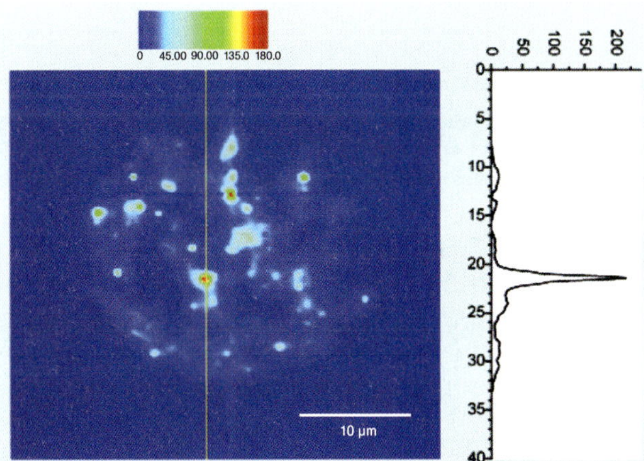

Fig. 7.8. Near-infrared fluorescence image of one macrophage-like cell incubated with SWCNTs, showing emission detected from 1125 to 1600 nm with excitation at 660 nm. Intensities are coded with false color, and the image was obtained from a z-axis series by deconvolution processing. Intensity along the yellow vertical line is plotted in the graph on the right, showing high image contrast and localized emission sources. (Adapted with permission from Ref. [50].)

Despite recent investigations in systemic transport, none of the published data strongly illustrate that a substantial amount of the inhaled dose is indeed translocated systemically [43, 47–49]. Carbonaceous particles are particularly difficult to detect in cells and tissues by electron microscopy because of low contrast and small diameters. Recently, near-IR fluorescence was applied to detect SWCNTs after phagocytosis in mouse macrophages (Fig. 7.8).

The image clearly shows the nanotubes inside the cell [50]. Because of their photostability, which is comparable to quantum dots, carbon nanotubes might be applied as fluorescent markers and contrast agents with low toxic potential for cell biology research and medical diagnosis. Diabaté et al. also demonstrated by transmission electron imaging that carbon nanotubes are taken up by macrophages and epithelial cells as bundles [51]. These large agglomerates of carbon nanotubes as well as single nanotubes separated from these bundles can be clearly detected within the cells by this method (Fig. 7.9).

There exist several studies on the uptake and distribution of nanoparticles within cells. In nearly all these experiments the investigated cells have ingested the applied nanoparticles. Only two examples demonstrate the fact that nanoscaled metal oxide particles are found in cellular systems *in vitro* (Fig. 7.10) as well as after inhalation *in vivo* (Fig. 7.11). Transmission electron microscopy is fundamentally a useful and necessary tool. It appears that nanoscaled materials can be found either enclosed within organelles like phagosomes, lysosomes, or endosomes, or are freely distributed within the cytosol [38, 41, 52].

The same is true for carbonaceous nanomaterial that has been found in various cell types after treatment *in vitro* (Figs. 7.8 and 7.9) as well as in exposed mice or rats after instillation [53, 54].

7.2.3
Biological Effects on Cellular Mechanisms

Nanoparticles produced from different materials, such as metal oxides or carbon, have enhanced properties not found in bulk materials. Unsurprisingly, therefore, the enhancement of material properties could also occur when the particles encounter biological components. With the ability to manipulate atoms and molecules, we now can create predefined nanostructures with unprecedented precision and selective affectivity. An improved understanding of the biological effects of nanoscaled materials, as described in this book, also deserves attention. Several investigations of biological interactions with nanometer-scale materials demonstrate the possible impacts on living systems (Fig. 7.12):

1. Cell membrane proteins/adhesion molecules; integrins and extracellular matrix (ECM), receptor molecules, transporters [12, 55–61];
2. phospholipid turnover and lipid mediator release [62–64];
3. ion channels [65];
4. endolysosomes [66–68];

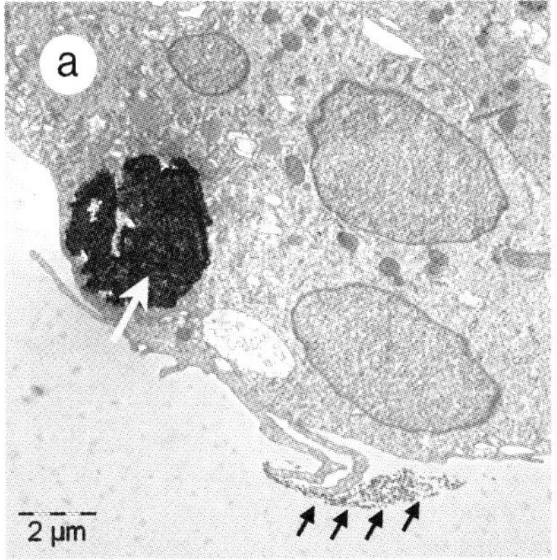
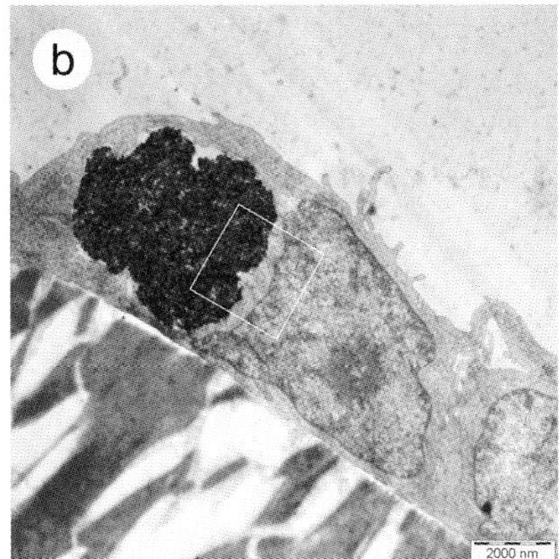
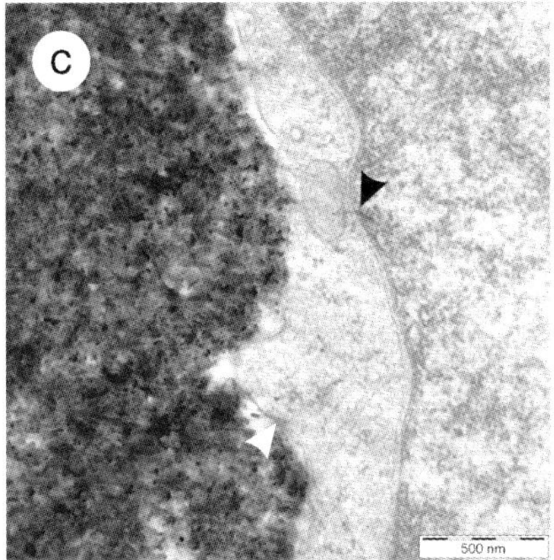

Fig. 7.9. Transmission electron micrographs of rat alveolar macrophages (NR8383) and human lung epithelial cells (BEAS2B) ingesting bundles of primary SWCNTs. (a) NR8383 cell ingested a bundle of carbon nanotubes (white arrow) and contacted another one (black arrows); (b) BEAS2B cells ingested a SWCNT- bundle of similar size, the white square is shown in higher magnification in (c), where single carbon nanotubes can be found next to the bundle in the cytosol (white arrowhead) and at the nuclear envelope (black arrowhead). (T. Detzel, ITG.)

5. mitochondria [69–71];
6. nucleus and DNA [72, 73].

7.2.3.1 Metal Oxides

Nanosized materials are easily taken up into cells and are either stored in several compartments or freely located within the cytosol (Section 7.2.2). Metal oxides interfere with membranes, proteins or other structures of the cells (see above). Incorporation takes place via caveoli, clathrin-coated pits or endocytosis (Fig. 7.6),

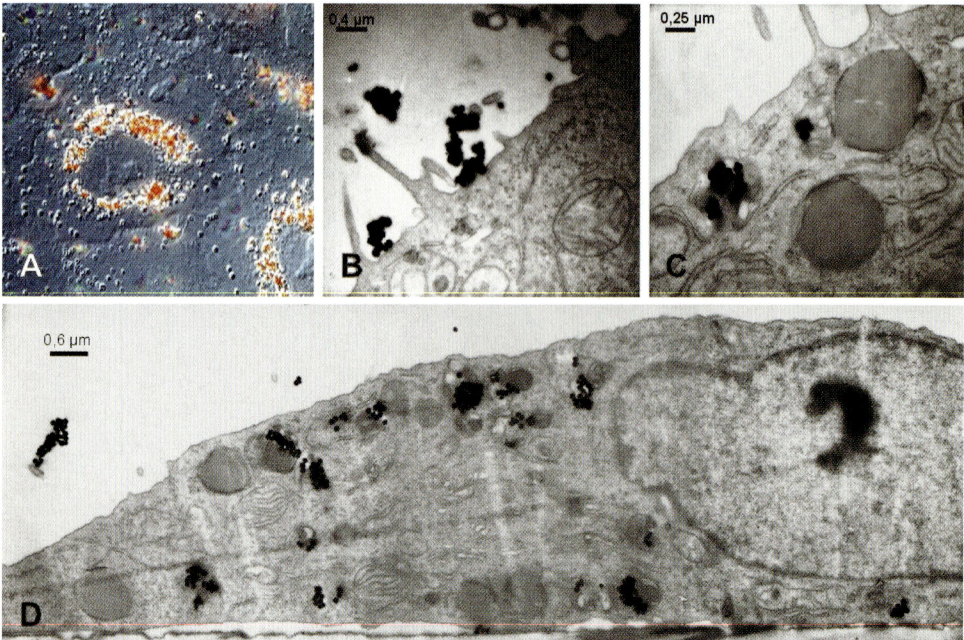

Fig. 7.10. Hematite particles are taken up into the cytosol of alveolar epithelial cells (A549). Light microscopic image of A549 cells (DIC, 630×; A) after 24 h of hematite exposure (100 µg mL^{-1} or 32 µg cm^{-2}). Transmission electron micrographs of A549 cells after 3 h (20000×; B), 6 h (30000×; C), and 16 h of exposure to hematite particles (50 µg mL^{-1} or 16 µg cm^{-2}) (12000×; D). (Reprinted from Ref. [41], with permission from Elsevier.)

leaving these nanosized metal oxides in lysosomes [38, 41]. A distinct mechanism for nanoparticle uptake has not been described yet, but it appears to be a dependency on primary particle and agglomerate size. The possible cellular mechanism for the recognition and initiation of the uptake process has recently been suggested to be a member of the Toll-like receptor family [74].

Size Dependency Recently, ultrafine (20 nm) preparations of TiO$_2$ have been shown to cause a significant loss in viability compared to fine (220 nm) particles [40], similar to results for ultrafine and fine nickel [75]. Additionally, an increase in fibrogenic mediators like procollagen can be observed that appear to be stronger for ultrafine preparations [76]. Proliferation of macrophages is impaired in these samples as well, and to a greater extent than in fine particle treated controls [40]. Inhalation studies revealed a higher pulmonary deposition in rats with ultrafine CdO (40 nm) aerosol than was measured with fine CdO particles [36]. Bermudez and his colleagues have suggested particle clearance in mice and rats is retarded

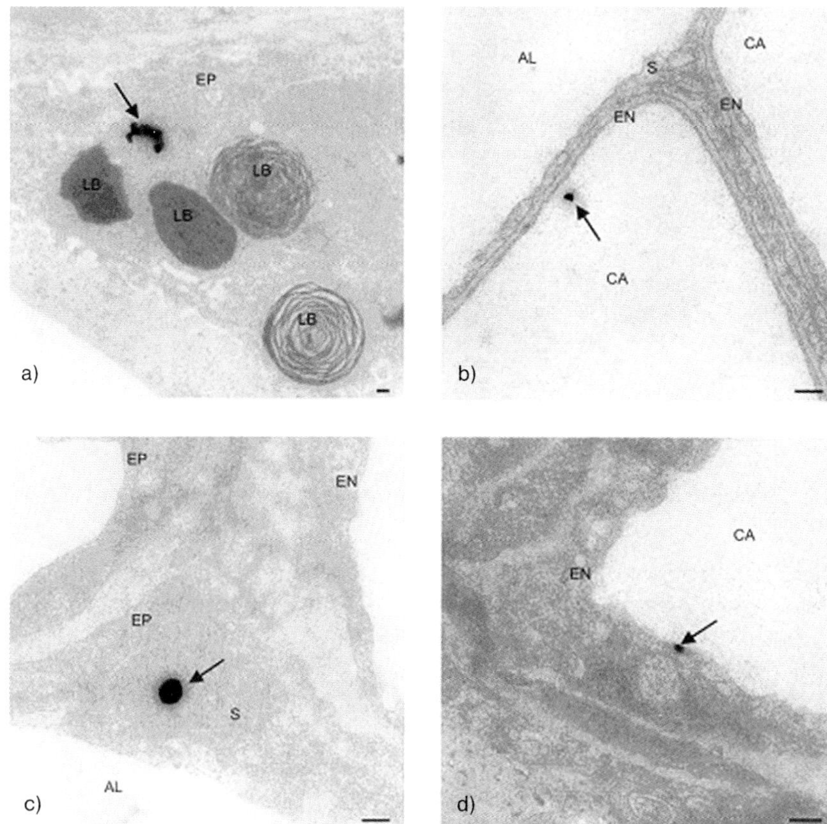

Fig. 7.11. EFTEM (energy filtering transmission electron microscopy) images taken at 0 eV of particles (arrows) on ultrathin sections of the lung parenchyma of exposed rats. (a) Type II cell (EP) close to lamellar bodies (LB). (b) A capillary (CA), near the alveolar endothelium (EN). (c) Surfactant material (S) accumulated within the surface lining layer in the corner of an alveolus, close to the epithelium (EP), alveolar lumen (AL). (d) Cytoplasm of an endothelial cell (EN). Scale bars = 100 nm. (Adapted with permission from Ref. [52].)

because of pulmonary particle overload in animals treated with high dosages of nanoscale titania [35].

Koper and his coworkers have described another size-dependent finding [77]. Nanoscale powders of MgO or CaO that tend to be nontoxic as large scale particles were doped with halogens and found to have a very strong degrading effect on certain bacteria and fungi. It was suggested that the activated nanoparticles directly interfere with proteins and nucleotides. If these formulations kill more than 90% of contacted bacteria within minutes, why should these particles (4 nm) not be

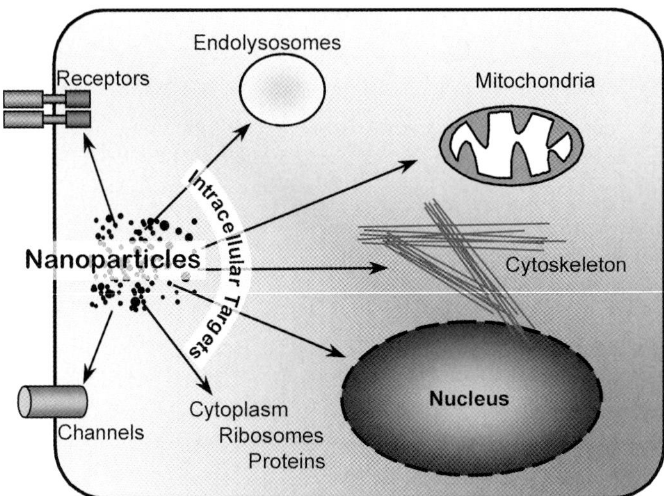

Fig. 7.12. Possible intracellular targets of nanoparticles.

harmful to human health? The key to these observations lays obviously in the large surface area of nanoparticles, which can be as high as 430 m^2 g^{-1} and therefore be very reactive. Klabunde and coworkers suggest a surface reaction of these nano-materials with P–O and P–F bonds that breaks important chemical compounds, leading to a disruption of the cellular homeostasis [78].

Inflammation Several studies have shown that incorporated metal oxide nano-particles can lead to inflammatory responses. These include the release of pro-inflammatory cytokines like IL-1, IL-6, IL-8 and TNF-α. In addition, fibrogenic fac-tors (PDGF-A and PDGF-B) can be released upon treatment with fine and ultrafine particles in rat tracheal explants [76]. Toll-like receptors (TLR) may be also involved in a nanomaterial specific manner and appeared to be induced after application of different nanoparticles, such as TiO$_2$, ZrO$_2$ and SiO$_2$, to human myelomonocytic U-937 cells [74]. These studies observed an increase in IL-1β, TNF-α and IL-1RA release in this macrophage cell line. A change in the cytological profile has been noted recently in inhalation experiments in mice, rats, and hamsters [35]. The mice had significantly elevated numbers of macrophages, lymphocytes and neutro-phils in bronchoalveolar lavage fluids even 52 weeks after the end of exposure.

Acute Toxicity Several assays have been used and developed to determine the acute toxicity of nanomaterials. By measuring the metabolic activity of mitochon-dria, many laboratories have determined the viability of various cell cultures after treatment with metal oxides or carbon nanoparticles [38, 41, 74, 79, 80]. Other studies use bacterial systems by giving the number of colony forming units as a marker for acute toxicity [77]. Consistent across different methods or biological systems, nanoscaled metal oxide particles exhibit a strong decrease in viability.

Genotoxicity To overcome severe effects on DNA such as was demonstrated for asbestos, it is very important to increase our knowledge of direct DNA damaging effects or indirect genotoxic mechanisms via oxidative stress by nanoscaled metal oxides.

TiO_2 nanoparticles caused lung tumors in rats at the end of lifetime exposure [81]. Thus, it can be speculated that metal oxide nanoparticles in general exhibit a strong reactivity to induce oxidation or electron transfer reactions because the number of metal atoms on their surface compared to those hidden within the particles is very high. Moreover, it might be assumed that these materials can induce DNA damage, enhancing the risk for tumor development (Section 7.2.3.2). Here, a broad basis of data is missing and knowledge of the retention, biological half-life and accumulation within specific target organs has to be increased.

Cytoskeletal Organization Hydroxyapatite, a major building block in bones, is used as a nanomaterial to improve adhesion of human osteoblast-like cells (HOB) to inorganic materials. Upon treatment with a high dosage of 200 particles per cell (1×10^8 per 500 000 cells) of nHA (nanosized, rod-like hydroxyapatite) HOB cells released lactate dehydrogenase into the surrounding media, indicating a loss in cell viability [79]. Surfaces coated with nHA increase the quality of focal contacts in HOB cells and support growth *in vitro* if used in lower concentrations (Fig. 7.13). However, ultrafine particles made of titania (20 nm TiO_2) cause retarded relaxation and stiffness of the cytoskeleton in macrophages that can not be observed for micronized titania (220 nm) [40]. Moreover, the alveolar ability for clearance is slowed due to reduced macrophage phagocytosis and mobility [39], as has been demonstrated by measuring the uptake of control beads. Recognition of nanoscale metal oxides may take place via a Toll-like receptor (TLR) mediated uptake process that has been described for bacteria and viruses, as suggested by Lucarelli et al. [74]. They observed changes in mRNA expression levels (TLR1-10, MD2 and CD14) upon metal oxide treatment of human differentiated myelomonocytic U-937 cells.

7.2.3.2 New Carbon Modifications

Induction of intracellular oxidative stress seems to be a key biological response to combustion generated [82] and manufactured particles, as well as organic components associated with particles. Furthermore, there is evidence for additive or synergistic interactions between ultrafine carbon black particles and soluble transition metals in causing oxidative stress and inflammation [83]. The oxidative potential of particles can be observed in cell-free systems, e.g., by electron spin resonance (ESR) spectroscopy [84].

If particles with oxidative potential find their way inside cells, that same ability may convert oxygen and other molecules into highly reactive radicals that can in-

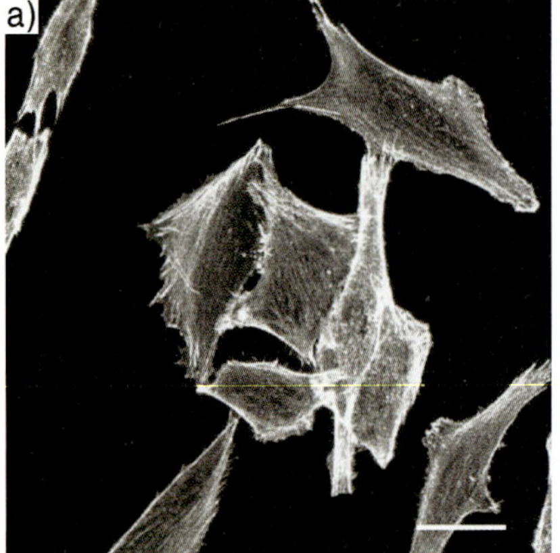
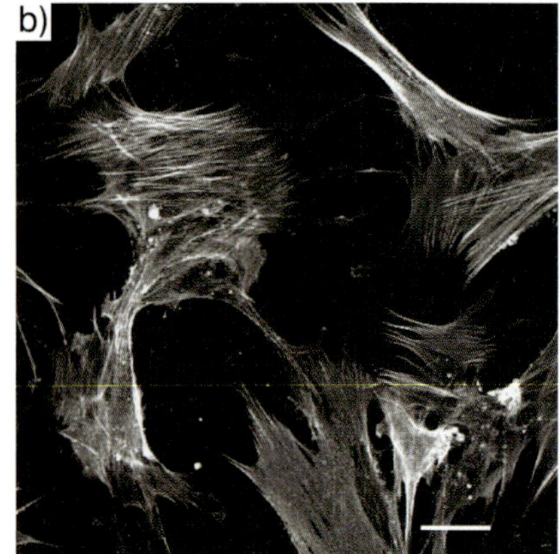
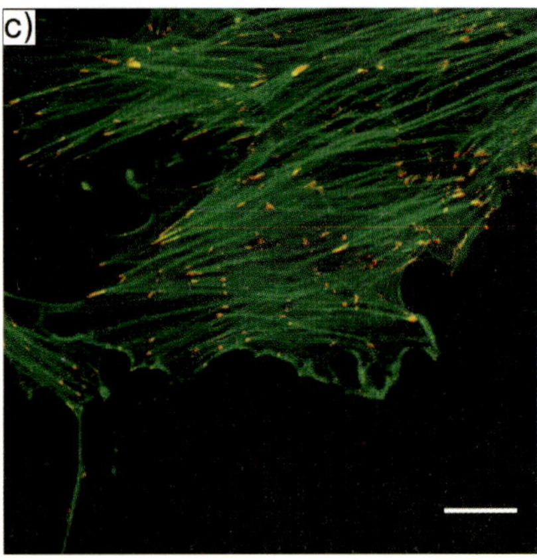

Fig. 7.13. CLSM images of actin cytoskeleton for human osteoblast-like cells on (a) control and (b) nHA-sprayed (nanosized, rod-like hydroxyapatite) substrate after 20 h of culture. Scale bar = 40 μm. (c) CLSM image of actin (green) and vinculin (red), as in (b) after two days culture. Scale bar = 20 μm. (Reprinted from Ref. [79], with permission from Kluwer Academic Publishers.)

duce intracellular signaling pathways as a defense mechanism or even damage cellular components, leading to cell death.

Figure 7.14 shows the possible sequence of events occurring after contact of particles with living cells. If particles make contact with proteins located at the outer plasma membrane, they may induce changes in the molecular conformation of these proteins. Many of these proteins are receptors that transmit external signals into the cell, and a conformational change of the receptor may activate it, leading to the onset of a cellular response. Once inside the cell, particles may induce intracellular oxidative stress by disturbing the balance between oxidant and antioxidant processes, e.g., the glutathione system. The oxidative stress may also stimulate an

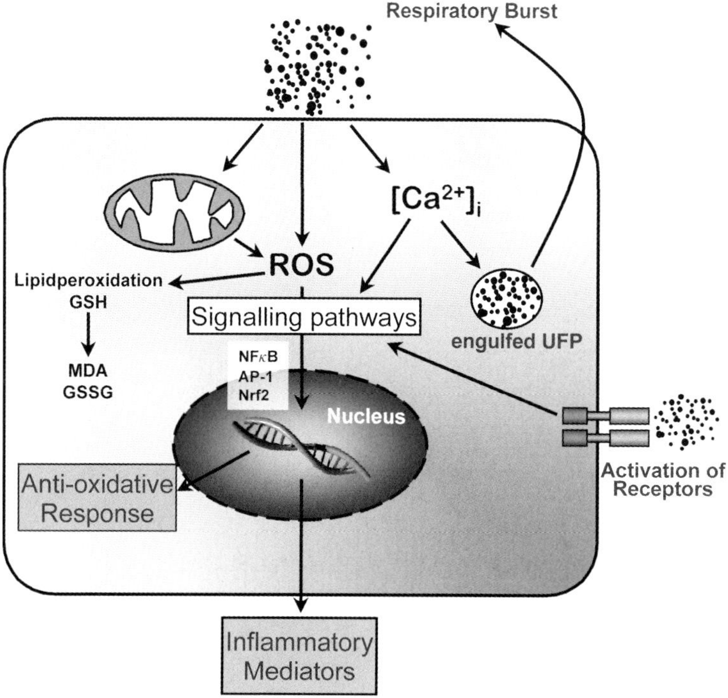

Fig. 7.14. Possible events after interaction of particles with cells. Particles may interact with receptors or may cause oxidative stress that induces an increase in intracellular calcium concentration, decrease intracellular GSH and/ or activate transcription factors via different signaling cascades. Activated transcription factors translocate into the nucleus, leading to gene activation, synthesis of antioxidant enzymes and/or inflammatory mediators.

increase of the cytosolic calcium concentration, possibly via interaction with calcium channels on the plasma membrane or on the endoplasmatic reticulum, leading to influx of Ca^{2+} from the extracellular environment or to release of Ca^{2+} from intracellular stores [85]. The intracellular calcium concentration strongly regulates signaling pathways via interaction with several proteins such as calmodulin and protein kinases. These changes cause the activation of redox-sensitive transcription factors, such as NF-κB, AP-1 or Nrf2, which translocate to the nucleus and bind to the promotor regions of the genes regulated by these transcription factors. For NF-κB these genes include TNF-α, IL-6, IL-8, ICAM-1, iNOS and others, which are highly pro-inflammatory [86]. Exceeding oxidative stress may also modify proteins, lipids (lipid peroxidation) and nucleic acids, which further stimulates the antioxidant defense system or even leads to cell death.

Experiments with rats have demonstrated that inhalation of particles consisting of elemental carbon may cause considerable injury to the lung and that the toxic potential increases with decreasing particle size and increasing particle surface area [81, 87]. Freshly prepared ultrafine carbon particles generated from pure graphite electrodes in an electric spark discharge generator [88] induce an increase in heart rate and a decrease in heart-rate variability after 24 h inhalation by healthy rats [89]. These studies indicate a systemic effect of the inhaled carbon particles (38 nm, 180 $\mu g\ m^{-3}$) without evidence of an inflammation-mediated mechanism. The authors suggest the responses may have induced an alteration of the cardiac autonomic balance mediated by a pulmonary receptor activation.

Frampton and colleagues have conducted a large clinical inhalation study of the effects of laboratory-generated ultrafine carbon particles [90]. Healthy people and people with asthma inhaled 10 or 25 $\mu g\ m^{-3}$ of ultrafine carbon particles (average diameter 25 nm) for 2 h. This dose was 10 to 100× higher than average concentrations of ultrafine particles of this size class reported in urban air. They did not detect changes in any airway inflammatory endpoint in both groups although it was calculated that about 50% more particles deposited in lungs of asthmatic people than in healthy people and over 4× as many particles were deposited in the lungs of exercising as in the lungs of resting participants. Electrocardiogram analysis after exposure showed a transient reduction of the heart rate variability and a reduced repolarization interval in healthy people and in people with asthma.

Inflammation *In vitro* studies with different cell systems also demonstrated that small carbon black particles (14 nm) induced more oxidative stress and pro-inflammatory cytokines than primary particle sizes of 250 nm [84]. These effects could not be explained by adhering transition metals since leaching of the particles with different buffers and complexing iron with desferrioxamin did not reduce the carbon-black-induced effects [91].

Freshly prepared spark discharge-generated ultrafine carbon particles have a higher oxidative potential than both aged particles and larger particles with less surface area [84]. However, the particle-induced release of arachidonic acid and derived lipid mediators in canine alveolar macrophages was independent of their oxidative potential but dependent on their ability to activate cytosolic and secretory phospholipases A_2 (PLA_2). PLA_2 hydrolyzes membrane phospholipids to release arachidonic acid, which is further metabolized to prostaglandins and leukotrienes. The particle-induced effects were also observed in human alveolar macrophages.

In normal human bronchial epithelial cells, these types of particles induced the expression of the pro-inflammatory cytokine (IL-8), possibly controlled by the p38 mitogen-activated protein kinase (MAPK) signaling pathway. Activation of the transcription factor NF-κB, however, seems not to play a role [92]. IL-8 is a potential chemoattractant for neutrophils. An elevated level of IL-8 in the lung is a characteristic of respiratory diseases such as cystic fibrosis, asthma, chronic bronchitis and acute respiratory distress symptom.

Genotoxicity Inhalation of carbon black by rats induced the formation of 8-oxo-7,8-dihydro-2′-deoxyguanosin (8-oxo-dG) in the lungs [73]. 8-Oxo-dG, a modified nucleotide, is a well-known and commonly used biomarker of free radical-induced oxidative DNA damage. This DNA modification may induce point mutations, which are widely observed in mutated oncogenes and tumor suppressor genes, and is therefore associated with many diseases such as cancer and neurodegenerative diseases. Interestingly, studies with rats showed that particles without organics (TiO_2) or with very low amounts of organics (carbon black) resulted in a similar induction of tumors compared with diesel particles, which contain considerably higher amounts of organics [81]. Therefore, tumor development is suggested to be caused by secondary genotoxic events due to particle-induced persistent inflammation and increased cell proliferation due to lung particle overload rather than by direct genotoxic effects.

While the toxic effects of carbon black or spark discharge-generated particles of elemental carbon have been studied very intensely, few studies with toxicological background exist on fullerenes and carbon nanotubes. Adelmann et al. [93] studied the effects of fullerenes in primary bovine alveolar macrophages and HL-60 macrophages using particles generated in the arc between two graphite electrodes in a helium atmosphere. The fullerenes reduced the viability of the cells and induced increased levels of the pro-inflammatory cytokines TNF-α, IL-6 und IL-8. The effects were comparable with those of graphite. Another *in vitro* study with C_{60} demonstrated low cytotoxicity compared to quartz, a moderate increase of TNF-α and IL-8 release, and no oxidative potential compared to zymosan in macrophages [94, 95].

E. Oberdörster has studied the effects of C_{60} fullerenes after exposure of fish (juvenile largemouth bass) as a model for the impact of nanoparticles produced in bulk with the potential to be released into the environment [64]. She demonstrated that a 48 h treatment with fullerenes significantly increased the lipid peroxidation in the brain and depleted the glutathione content in the gill. Both parameters are indicators of increased oxidative stress.

For carbon nanotubes, it is uncertain if there are analogous mechanisms to those of other fibrous particles such as asbestos and synthetic vitreous fibers (SVFs), which penetrate into the lung and may persist in the tissue. Large epidemiological studies of SVF manufacturing workers provided very little evidence of harmful effects in humans [96]. However, it is widely assumed that all biopersistent fibers may be harmful if inhaled in large enough doses. Long insoluble fibers are difficult to clear by phagocytic cells. The macrophages die after a long process of trying to engulf the fibers and release inflammatory cytokines into the lung. This may trigger the complex cellular response mechanisms that cause cancer after exposure to asbestos [97]. Nevertheless, a study at the University of Warsaw sought to determine if carbon nanotubes may behave like asbestos [98]. The experiments with guinea pigs revealed that carbon nanotubes do not exhibit effects similar to asbestos and it is suggested that working with soot containing carbon nanotubes is unlikely to be associated with health risks.

Two more studies from the same group in Warsaw have dealt with the dermatological and inhalation effects of fullerenes and carbon nanotubes. In the study on dermatological effects, rabbits were treated and the researchers "did not *find* any signs of health hazards related to skin irritation and allergic risks." This group recommended no special precautions with respect to both fullerenes and carbon nanotubes in the working environment; in fact, the articles were titled "Fullerenes: Null Risk of Skin Irritation and Allergy" and "Carbon Nanotubes: Null Risk of Skin Irritation and Allergy" [99, 100]. However, these studies may not have been sensitive enough because recently published experimental data suggest a variety of effects in rat, mouse or human cellular systems (see below). Moreover, one has to take into account that carbon nanotubes are always contaminated with catalytic metals used during their production.

For most production processes, predominantly Fe and Ni are used as metal catalysts. They are normally removed from the raw product, but part of the metal is encased in the tubes and cannot be removed completely. Manufacturing of carbon nanotubes leads to bundles of nanotubes forming clumps and aggregates. If they are inhaled or come into contact with the skin during handling, the potential hazard will strongly depend on the metal content and on the size of the aggregates. The hazard of exposure to carbon nanotubes at occupational settings has been investigated by only a few studies. Maynard and his coworkers measured the aerosol mass and number concentration in three laboratories where SWCNTs were generated by different processes and handled manually [17]. They observed concentrations of 0.70 to 53 μg m^{-3} of nanotubes in the atmosphere and considerable masses on gloves during handling.

Because of reports that carbon fibers induced dermal irritation such as contact dermatitis in humans after occupational exposure of the skin to carbon fibers [101], some *in vitro* studies have investigated the effects of carbon particles in immortalized keratinocyte cultures. In a companion article to Maynard et al. [17], the biological effects of SWCNTs before catalyst removal (containing 30% Fe by mass) was studied in human keratinocytes [102]. An observed dose-dependent decrease in cell viability and glutathione (GSH) levels was dramatically reversed by the metal chelator desferrioxamin. This indicates a significant role of iron in the biological effects of the SWCNTs. This study further confirmed oxidative stress in SWCNT-treated cells by the formation of free radical species, increased lipid peroxidation and decrease of the antioxidant reserve. The effects of MWCNTs in human dermal keratinocytes were also studied by Monteiro-Riviere et al. [103]. They demonstrated by transmission electron microscopy that MWCNTs were present in cytoplasmic vacuoles.

Two independent studies with rats [53] and mice [54] reported the appearance of granulomas, interstitial inflammation, and obstruction of the airways after instillation of high doses of aggregated carbon nanotubes. Granulomas are a combination of dead and live tissue surrounding the foreign material. Warheit and his coworkers [53] concluded that the acute effects are normal responses to persistent particulate material and are not specific for carbon nanotubes. Lam et al. [54] observed that the SWCNTs were more toxic than carbon black and quartz particles after instillation in mice and that nanotubes treated to remove the metals were nearly as toxic as raw nanotubes (see Chapter 6). Histological tests showed that all particles reached the alveoli and remained there even after 90 days. The biopersistence of SWCNTs and the induction of granulomatous lesions are important evidences for adverse health effects. Because of strong aggregation on the nanotubes it is necessary to study the effects either by inhalation or by *in vitro* experiments.

In vitro experiments with alveolar epithelial cells and macrophages showed that SWCNTs and MWCNTs induced oxidative stress, inflammatory responses, and cell death in a dose-dependent manner [51]. The authors found that the toxic potential of SWCNTs was similar to ultrafine carbon black (Printex 90) and higher than MWCNTs and quartz.

Recently, SWCNTs, MWCNTs and C_{60} fullerenes were tested in comparison to quartz in primary alveolar macrophages isolated from guinea pigs [104]. A sequence order of the cytotoxic potential of SWCNTs > MWCNTs > quartz > C_{60} fullerenes was found. The C_{60} fullerenes were shown to be non-toxic in the MTT test even at the highest concentration of 226 μg cm^{-2} while the SWCNTs reduced the viability by 20% at 1.4 μg cm^{-2}. These observations are compatible with results obtained with mice where carbon nanotubes are more toxic to the lung than quartz [54]. The carbon nanotubes were ~90% pure; impurities included mainly amorphous carbon and only trace amounts of the catalysts Fe, Y and Ni. The particles also reduced the phagocytic ability of the alveolar macrophages, as seen by microscopic and flow cytometry analysis. Phagocytosis of carbon nanotubes was accompanied by ultrastructural alterations, as demonstrated by transmission electron imaging, indicating the onset of apoptotic processes.

Besides the mechanical and electrical characteristics, carbon nanotubes can be functionalized with different molecules to achieve improved properties and functions such as biocompatibility and biomolecular recognition capabilities. This would enable applications in biomedical engineering and drug delivery. Amino acids and peptides coupled to SWCNTs yield SWCNT derivatives with higher water solubility that can translocate across cell membranes [105]. Furthermore, plasmid DNA associates with ammonium-functionalized CNTs, and these complexes were taken up by mammalian cells. The CNT-mediated DNA delivery to cells was very effective, resulting in a 10× higher gene expression than achieved with DNA alone [105]. These studies indicate that CNTs have a high potential in delivery systems in the molecular therapy of diseases.

7.2.4
Possible Hazards – Toxicological Impacts

We tried in the above-described toxicological issues to point to important mechanisms and studies that gave hints where the hazards of nanoparticles could be detected. Obviously, dependent on the material nanoparticles are produced from, their biological effects can be very different. Above all, metal oxides are strongly expected to be harmful because the toxicity of metals and their compounds is well described. The question is do they behave differently as nanosized particles than as dissolved ionic forms or organic compounds? Therefore, it is of interest to know if these materials are soluble in biological fluids, and how long they persist in their target tissues. Such criteria may directly influence the use of metal oxide nano-

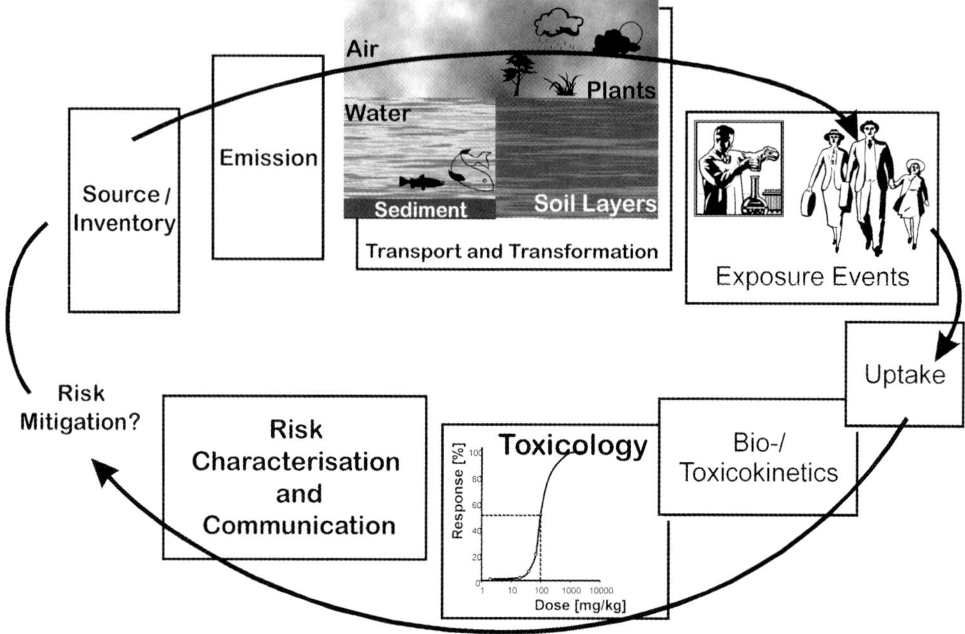

Fig. 7.15. Schematic presentation of life-cycle assessment of nanomaterials. From source to dose, an evaluation loop to the reduction of emission of nanomaterials.

particles in technical, cosmetic, and medical products with which an exposure is obvious. Carbonaceous material, though, has been less intensively investigated and long-term effects are mostly speculative. First results, from studies with animals and cellular systems, point to oxidative mechanisms that might have the potency to kill cells or could be discussed in connection to DNA damage. Regarding these effects, further investigations are needed to clarify such fundamental biological mechanisms before these materials, like fullerenes or carbon nanotubes, are produced in very high amounts and possibly released into the environment. Thus, it is important to include the entire life-cycle of nanomaterials (Fig. 7.15).

7.3
Risk Characterization – A Conclusion

As stated in Section 7.1.2 risk characterization stands at the end of the sequence of hazard and exposure identification and assessment (Fig. 7.1). To fulfill the criteria of risk assessment, besides the toxicological and pharmacological studies, exposure situations have to be recognized. Therefore, one has to keep in mind that bioaccumulation processes can lead to an enrichment in organisms or organs over several

orders of magnitude. Thus, a minimum catalogue of action has to be set up to reduce the risk at working places and within the environment.

Generally, in a recent report from the Institute of Occupational Medicine in
Great Britain [106], it has been postulated to consider:

- the existence of potential routes for human exposure;
- possible industrial sources of occupational exposure;
- the levels of exposure;
- means of, and effectiveness of control measures;
- potential numbers of humans exposed;
- trends in the (potential) use of nanotechnology;
- views as to the likely impact of the implementation of the change from research
  use to full-scale industrial use.

7.3.1
Opportunities and Risks of Nanomaterials

All the above-described mechanisms of uptake, transport and distribution of nanoparticles in cells of different species and organs are important for their implications and applications. During discussion of the adverse effects of nanoparticles it
is always important to consider the positive applications within the environment as
well as in biological systems. Only a few are listed here:

- tools in imaging and microscopy;
- diagnostics and analysis (research and therapy);
- production of bioactive compounds and materials (Lab-on-a-Chip);
- targeting and dosing of drugs;
- intervention in biological processes (cell growth);
- nutrition (bioavailability, stability, optics);
- cosmetics (UV-filter, liposomal formulations);
- sensors and detectors;
- biomolecules for information and communication technology (ICT; DNA com-
  puting).

7.3.2
New Materials without Risks?

Nanotechnological products are developed with regard to the future: for a presumed need, to solve foreseeable problems, and for a future market. Besides intended effects (the use and functions of nanomaterials), unintended effects (the
"side effects", which also include misappropriation and misuse) might occur and
might influence the overall balance of opportunities and risks. Technological risks
belong to unintended and undesirable side effects.

Not only environmental or health risks are subject to a risk assessment but also
economic risks and potential social problems like technology conflicts as well as

risks for sustainable development in a general sense. Classical types of technical or technologically-initiated risks are [107]:

- *Accidents in technical facilities*: These are disruptions of normal operation. Release of high amounts of material in such cases can enhance public distrust of new technology (e.g., poisonous gas catastrophes in Seveso and Bophal).
- *Consequences for human health*: New materials or completely new emissions can affect human health, e.g., in the production or use of technology. To these belong the known risks and side effects of medicines (e.g., diethylstilbestrol – DES), but also the dramatic history of asbestos. Entire chapters of modern regulations for handling hazardous substances are reactions to actual health problems. Special problems of risk assessment arise in the case of low dose exposures.
- *Consequences for the natural environment*: Air pollution, the ozone hole, chemical residues in ground water and in the soil are well-known, unintended consequences of the use of technology. Other than in the case of accidents in technical facilities alluded to above, these are often *gradual* processes. They are not always readily recognizable, and there is dissent on the question of tolerance limits or "cut-off-" or "threshold values", from which point on protective or remedial measures would have to be taken.
- *Social and cultural effects of technology*: Social risks connected with technology are, for instance, the loss of many jobs through rationalization and automation, especially as far as less highly qualified work is concerned. Ethical "slippery slopes" in biomedical questions are also felt by segments of the population to be "cultural" risks (e.g., positive eugenics).

Such technological risks show certain characteristics that influence the approaches to their anticipative investigation and evaluation through risk research and technology assessment. Among these, in particular, are:

- local and global effects (atmospheric emissions and the global water cycle);
- enlargement of number of people affected by hazards (even in future);
- the problem of delayed effects: Perceptible damage appears decades after its cause (e.g., ozone hole, asbestos case);
- complexity of cause-relationship connections (e.g., the mad cow disease BSE);
- inability to perceive risks (e.g., radioactivity) with human sensory organs;
- irreversibility of hazards (e.g., persistent pollutants can not be completely retrieved from the environment).

In summary, it has turned out in past decades that side effects can interfere with or even counteract the goals pursued by means of technology. This ambivalence of technology, the greater discrepancy between the intended and the (then) actually realized effects constitutes a *conditio humana* of technological civilization. Two positions are futile: the demand for absolute safe and risk-free technology ("zero risk") and the disregard or denial of the "dark side" of technology. The challenge consists much rather in addressing, analyzing, and evaluating the risks, comparing them with the expected benefits, and then taking the results of these deliberations into consideration in decision-making processes [107].

References

1 AGRAWAL, A. K., SINGHAL, A., GUPTA, C. M. Functional drug targeting to erythrocytes in vivo using antibody bearing liposomes as drug vehicles. *Biochem. Biophys. Res. Commun.* **1987**, 148, 357–361.

2 CLARK, A. P. Liposomes as drug delivery systems. *Cancer Pract.* **1998**, 6, 251–253.

3 DESMUKH, D. S., BEAR, W. D., WISNIEWSKI, H. M., BROCKERHOFF, H. Long-living liposomes as potential drug carriers. *Biochem. Biophys. Res. Commun.* **1978**, 82, 328–334.

4 FENDLER, J. H., ROMERO, A. Liposomes as drug carriers. *Life Sci.* **1977**, 20, 1109–1120.

5 GABIZON, A. Liposomes as a drug delivery system in cancer chemotherapy. *Horiz. Biochem. Biophys.* **1989**, 9, 185–211.

6 GREGORIADIS, G. Drug entrapment in liposomes. *FEBS Lett.* **1973**, 36, 292–296.

7 GREGORIADIS, G., WILLS, E. J., SWAIN, C. P., TAVILL, A. S. Drug-carrier potential of liposomes in cancer chemotherapy. *Lancet* **1974**, 1, 1313–1316.

8 KSHIRSAGAR, N. A., GOKHALE, P. C., PANDYA, S. K. Liposomes as drug delivery system in leishmaniasis. *J. Assoc. Physicians India* **1995**, 43, 46–48.

9 SPEISER, P. P. Nanoparticles and liposomes: A state of the art. *Methods Find. Exp. Clin. Pharmacol.* **1991**, 13, 337–342.

10 ALLEN, T. M., CULLIS, P. R. Drug delivery systems: Entering the mainstream. *Science* **2004**, 303, 1818–1822.

11 CHEN, Y., XUE, Z., ZHENG, D., XIA, K., ZHAO, Y., LIU, T., LONG, Z., XIA, J. Sodium chloride modified silica nanoparticles as a non-viral vector with a high efficiency of DNA transfer into cells. *Curr. Gene Ther.* **2003**, 3, 273–279.

12 GUPTA, A. K., CURTIS, A. S. Lactoferrin and ceruloplasmin derivatized superparamagnetic iron oxide nanoparticles for targeting cell surface receptors. *Biomaterials* **2004**, 25, 3029–3040.

13 JORDAN, A., WUST, P., SCHOLZ, R., TESCHE, B., FAHLING, H., MITROVICS, T., VOGL, T., CERVOS-NAVARRO, J., FELIX, R. Cellular uptake of magnetic fluid particles and their effects on human adenocarcinoma cells exposed to AC magnetic fields in vitro. *Int. J. Hyperthermia* **1996**, 12, 705–722.

14 LI, K. C., GUCCIONE, S., BEDNARSKI, M. D. Combined vascular targeted imaging and therapy: A paradigm for personalized treatment. *J. Cell Biochem.* **2002**, 39(Suppl), 65–71.

15 SHI KAM, N. W., JESSOP, T. C., WENDER, P. A., DAI, H. Nanotube molecular transporters: Internalization of carbon nanotube-protein conjugates into Mammalian cells. *J. Am. Chem. Soc.* **2004**, 126, 6850–6851.

16 OTSUKA, H., NAGASAKI, Y., KATAOKA, K. PEGylated nanoparticles for biological and pharmaceutical applications. *Adv. Drug Deliv. Rev.* **2003**, 55, 403–419.

17 MAYNARD, A. D., BARON, P. A., FOLEY, M., SHVEDOVA, A. A., KISIN, E. R., CASTRANOVA, V. Exposure to carbon nanotube material: Aerosol release during the handling of unrefined single-walled carbon nanotube material. *J. Toxicol. Environ. Health A* **2004**, 67, 87–107.

18 KRUG, H. F., KERN, K., DIABATÉ, S. Toxikologische Aspekte der Nanotechnologie. Versuch einer Abwägung. *Technikfolgenabsch.: Theorie Praxis* **2004**, 13, 58–64.

19 KRUG, H. F., KERN, K., WÖRLE-KNIRSCH J. M., DIABATÉ, S. Ultrafine

particles. Health risk and possible applications. *Internist. prax.* **2004**, 45, 443–455.

20 KROTO, H. W., HEATH, J. R., OBRIEN, S. C., CURL, R. F., SMALLEY, R. E. C-60 – Buckminsterfullerene. *Nature* **1985**, 318, 162–163.

21 IIJIMA, S., AJAYAN, P. M., ICHIHASHI, T. Growth model for carbon nanotubes. *Phys. Rev. Lett.* **1992**, 69, 3100–3103.

22 BALL, P. Roll up for the revolution. *Nature* **2001**, 414, 142–144.

23 PRICE, R. L., ELLISON, K., HABERSTROH, K. M., WEBSTER, T. J. Nanometer surface roughness increases select osteoblast adhesion on carbon nanofiber compacts. *J. Biomed. Mater. Res. A* **2004**, 70, 129–138.

24 IBALD-MULLI, A., WICHMANN, H. E., KREYLING, W. G., PETERS, A. Epidemiological evidence on health effects of ultrafine particles. *J. Aerosol Med.* **2002**, 15, 189–201.

25 WICHMANN, H. E., SPIX, C., TUCH, T., WOLKE, G., PETERS, A., HEINRICH, J., KREYLING, W. G., HEYDER, J. Daily mortality and fine and ultrafine particles in Erfurt, Germany part I: Role of particle number and particle mass. *Res. Rep. Health Eff. Inst.* **2000**, 98, 5–86.

26 SAMET, J. M., DOMINICI, F., CURRIERO, F. C., COURSAC, I., ZEGER, S. L. Fine particulate air pollution and mortality in 20 U.S. cities, 1987–1994. *N. Engl. J. Med.* **2000**, 343, 1742–1749.

27 HUGHES, L. S., CASS, G. R., GONE, J., AMES, M., OLMEZ, I. Physical and chemical characterization of atmospheric ultrafine particles in the Los Angeles area. *Environ. Sci. Technol.* **1998**, 32, 1153–1161.

28 BAGGS, R. B., FERIN, J., OBERDÖRSTER, G. Regression of pulmonary lesions produced by inhaled titanium dioxide in rats. *Vet. Pathol.* **1997**, 34, 592–597.

29 FERIN, J., OBERDÖRSTER, G., PENNEY, D. P. Pulmonary retention of ultrafine and fine particles in rats. *Am. J. Respir. Cell Mol. Biol.* **1992**, 6, 535–542.

30 OBERDÖRSTER, G., COX, C., GELEIN, R. Intratracheal instillation versus intratracheal-inhalation of tracer particles for measuring lung clearance function. *Exp. Lung Res.* **1997**, 23, 17–34.

31 WARHEIT, D. B. Nanoparticles: Health impacts? *Mater. Today* **2004**, 7, 32–35.

32 OBERDÖRSTER, G. Toxicology of ultrafine particles: In vivo studies. *Philos. Trans. R. Soc. Lond. Ser. A – Math. Phys. Eng. Sci.* **2000**, 358, 2719–2739.

33 OBERDÖRSTER, G., OBERDÖRSTER, E., OBERDÖRSTER, J. Nanotoxicology: An emerging discipline evolving from studies of ultrafine particles. *Environ. Health Perspect.* **2005**, 113, 823–839.

34 HEXT, P. M., TOMENSON, J. A., THOMPSON, P. Titanium dioxide: Inhalation toxicology and epidemiology. *Ann. Occup. Hyg.* **2005**, 49, 461–472.

35 BERMUDEZ, E., MANGUM, J. B., WONG, B. A., ASGHARIAN, B., HEXT, P. M., WARHEIT, D. B., EVERITT, J. I. Pulmonary responses of mice, rats, and hamsters to subchronic inhalation of ultrafine titanium dioxide particles. *Toxicol. Sci.* **2004**, 77, 347–357.

36 TAKENAKA, S., KARG, E., KREYLING, W. G., LENTNER, B., SCHULZ, H., ZIESENIS, A., SCHRAMEL, P., HEYDER, J. Fate and toxic effects of inhaled ultrafine cadmium oxide particles in the rat lung. *Inhal. Toxicol.* **2004**, 16(Suppl 1), 83–92.

37 KERN, K., WÖRLE-KNIRSCH, J. M., KRUG, H. F. Nanonoxes: Nanoparticle uptake, transport and toxicity. *Signal Transduct.* **2004**, 3–4, 149.

38 PETERS, K., UNGER, R. E., KIRKPATRICK, C. J., GATTI, A. M., MONARI, E. Effects of nano-scaled particles on endothelial cell function in vitro: Studies on viability, proliferation and inflammation. *J. Mater. Sci. Mater. Med.* **2004**, 15, 321–325.

39 RENWICK, L. C., DONALDSON, K., CLOUTER, A. Impairment of alveolar macrophage phagocytosis by ultrafine particles. *Toxicol. Appl. Pharmacol.* **2001**, 172, 119–127.

40 MÖLLER, W., HOFER, T., ZIESENIS, A., KARG, E., HEYDER, J. Ultrafine particles cause cytoskeletal dysfunctions in macrophages. *Toxicol. Appl. Pharmacol.* **2002**, 182, 197–207.

41 WOTTRICH, R., DIABATÉ, S., KRUG, H. F. Biological effects of ultrafine model particles in human macrophages and epithelial cells in mono- and co-culture. *Int. J. Hyg. Environ. Health* **2004**, 207, 353–361.

42 OBERDÖRSTER, G., SHARP, Z., ATUDOREI, V., ELDER, A., GELEIN, R., LUNTS, A., KREYLING, W. G., COX, C. Extrapulmonary translocation of ultrafine carbon particles following whole-body inhalation exposure of rats. *J. Toxicol. Environ. Health A* **2002**, 65, 1531–1543.

43 KREYLING, W. G., SEMMLER, M., ERBE, F., MAYER, P., TAKENAKA, S., SCHULZ, H., OBERDÖRSTER, G., ZIESENIS, A. Translocation of ultrafine insoluble iridium particles from lung epithelium to extrapulmonary organs is size dependent but very low. *J. Toxicol. Environ. Health A* **2002**, 65, 1513–1530.

44 NEMMAR, A., HOET, P. H., VANQUICKENBORNE, B., DINSDALE, D., THOMEER, M., HOYLAERTS, M. F., VANBILLOEN, H., MORTELMANS, L., NEMERY, B. Passage of inhaled particles into the blood circulation in humans. *Circulation* **2002**, 105, 411–414.

45 BROWN, J. S., ZEMAN, K. L., BENNETT, W. D. Ultrafine particle deposition and clearance in the healthy and obstructed lung. *Am. J. Respir. Crit. Care Med.* **2002**, 166, 1240–1247.

46 OBERDÖRSTER, G., SHARP, Z., ATUDOREI, V., ELDER, A., GELEIN, R., KREYLING, W. G., COX, C. Translocation of inhaled ultrafine particles to the brain. *Inhal. Toxicol.* **2004**, 16, 437–445.

47 KHANDOGA, A., STAMPFL, A., TAKENAKA, S., SCHULZ, H., RADYKEWICZ, R., KREYLING, W. G., KROMBACH, F. Ultrafine particles exert prothrombotic but not inflammatory effects on the hepatic microcirculation in healthy mice in vivo. *Circulation* **2004**, 109, 1320–1325.

48 NEMMAR, A., HOYLAERTS, M. F., HOET, P. H., NEMERY, B. Possible mechanisms of the cardiovascular effects of inhaled particles: Systemic translocation and prothrombotic effects. *Toxicol. Lett.* **2004**, 149, 243–253.

49 SEMMLER, M., SEITZ, J., ERBE, F., MAYER, P., HEYDER, J., OBERDÖRSTER, G., KREYLING, W. G. Long-term clearance kinetics of inhaled ultrafine insoluble iridium particles from the rat lung, including transient translocation into secondary organs. *Inhal. Toxicol.* **2004**, 16, 453–459.

50 CHERUKURI, P., BACHILO, S. M., LITOVSKY, S. H., WEISMAN, R. B. Near-infrared fluorescence microscopy of single-walled carbon nanotubes in phagocytic cells. *J. Am. Chem. Soc.* **2004**, 126, 15 638–15 639.

51 DIABATÉ, S., PULSKAMP, K., KRUG, H. F. Carbon nanotubes induce oxidative stress, inflammatory responses and cell death in pulmonary epithelial cells and macrophages. *Signal Transduct.* **2004**, 3–4, 116.

52 KAPP, N., KREYLING, W. G., SCHULZ, H., IM HOF, V., GEHR, P., SEMMLER, M., GEISER, M. Electron energy loss spectroscopy for analysis of inhaled ultrafine particles in rat lungs. *Microsc. Res. Technol.* **2004**, 63, 298–305.

53 WARHEIT, D. B., LAURENCE, B. R., REED, K. L., ROACH, D. H., REYNOLDS, G. A., WEBB, T. R. Comparative pulmonary toxicity assessment of single-wall carbon nanotubes in rats. *Toxicol. Sci.* **2004**, 77, 117–125.

54 LAM, C. W., JAMES, J. T., McCLUSKEY, R., HUNTER, R. L. Pulmonary toxicity of single-wall carbon nanotubes in mice 7 and 90 days after intratracheal instillation. *Toxicol. Sci.* **2004**, 77, 126–134.

55 BERRY, C. C., WELLS, S., CHARLES, S., AITCHISON, G., CURTIS, A. S. Cell response to dextran-derivatised iron oxide nanoparticles post internalisation. *Biomaterials* **2004**, 25, 5405–5413.

56 DE CAMPOS, A. M., DIEBOLD, Y., CARVALHO, E. L., SANCHEZ, A., ALONSO, M. J. Chitosan nanoparticles as new ocular drug delivery systems: In vitro stability, in vivo fate, and cellular toxicity. *Pharm. Res.* **2004**, 21, 803–810.

57 GUALBERT, J., SHAHGALDIAN, P., COLEMAN, A. W. Interactions of amphiphilic calix[4]arene-based solid lipid nanoparticles with bovine serum albumin. *Int. J. Pharm.* **2003**, 257, 69–73.

58 KRISTL, J., VOLK, B., AHLIN, P., GOMBAC, K., SENTJURC, M. Interactions of solid lipid nano-particles with model membranes and leukocytes studied by EPR. *Int. J. Pharm.* **2003**, 256, 133–140.

59 LIDKE, D. S., NAGY, P., HEINTZMANN, R., ARNDT-JOVIN, D. J., POST, J. N., GRECCO, H. E., JARES-ERIJMAN, E. A., JOVIN, T. M. Quantum dot ligands provide new insights into erbB/HER receptor-mediated signal transduction. *Nat. Biotechnol.* **2004**, 22, 198–203.

60 MAYE, I., DE FRAISSINETTE, A., CRUZ-ORIVE, L. M., VONDERSCHER, J., RICHTER, F., GEHR, P. Comparison of the rate of phagocytosis of orthorhombic cyclosporine A (CsA) and latex particles by alveolar macrophages from hamsters. *Cell Mol. Life Sci.* **1997**, 53, 689–696.

61 PRYHUBER, G. S., HUYCK, H. L., BAGGS, R., OBERDÖRSTER, G., FINKELSTEIN, J. N. Induction of chemokines by low-dose intratracheal silica is reduced in TNFR I (p55) null mice. *Toxicol. Sci.* **2003**, 72, 150–157.

62 BECK-SPEIER, I., DAYAL, N., KARG, E., MAIER, K. L., ROTH, C., ZIESENIS, A., HEYDER, J. Agglomerates of ultrafine particles of elemental carbon and TiO2 induce generation of lipid mediators in alveolar macrophages. *Environ. Health Perspect.* **2001**, 109(Suppl 4), 613–618.

63 BECK-SPEIER, I., DAYAL, N., KARG, E., MAIER, K. L., SCHULZ, H., SCHUMANN, G., ZIESENIS, A., HEYDER, J. Formation of prostaglandin E2, leukotriene B4 and 8-isoprostane in alveolar macrophages by ultrafine particles of elemental carbon. *Adv. Exp. Med. Biol.* **2003**, 525, 117–120.

64 OBERDÖRSTER, E. Manufactured nanomaterials (fullerenes, C60) induce oxidative stress in the brain of juvenile largemouth bass. *Environ. Health Perspect.* **2004**, 112, 1058–1062.

65 PARK, K. H., CHHOWALLA, M., IQBAL, Z., SESTI, F. Single-walled carbon nanotubes are a new class of ion channel blockers. *J. Biol. Chem.* **2003**, 278, 50212–50216.

66 KREYLING, W. G. Intracellular particle dissolution in alveolar macrophages. *Environ. Health Perspect.* **1992**, 97, 121–126.

67 LUNDBORG, M., JOHARD, U., JOHANSSON, A., EKLUND, A., FALK, R., KREYLING, W. G., CAMNER, P. Phagolysosomal morphology and dissolution of cobalt oxide particles by human and rabbit alveolar macrophages. *Exp. Lung Res.* **1995**, 21, 51–66.

68 PANYAM, J., ZHOU, W. Z., PRABHA, S., SAHOO, S. K., LABHASETWAR, V. Rapid endo-lysosomal escape of poly(DL-lactide-co-glycolide) nanoparticles: Implications for drug and gene delivery. *FASEB J.* **2002**, 16, 1217–1226.

69 CRUZ, T., GASPAR, R., DONATO, A., LOPES, C. Interaction between polyalkylcyanoacrylate nanoparticles and peritoneal macrophages: MTT metabolism, NBT reduction, and NO production. *Pharm. Res.* **1997**, 14, 73–79.

70 KNAAPEN, A. M., BORM, P. J., ALBRECHT, C., SCHINS, R. P. Inhaled particles and lung cancer. Part A: Mechanisms. *Int. J. Cancer* **2004**, 109, 799–809.

71 THIBODEAU, M., GIARDINA, C., HUBBARD, A. K. Silica-induced caspase activation in mouse alveolar macrophages is dependent upon mitochondrial integrity and aspartic proteolysis. *Toxicol. Sci.* **2003**, 76, 91–101.

72 CARLISLE, R. C., BETTINGER, T., OGRIS, M., HALE, S., MAUTNER, V., SEYMOUR, L. W. Adenovirus hexon protein enhances nuclear delivery and increases transgene expression of polyethylenimine/plasmid DNA vectors. *Mol. Ther.* **2001**, 4, 473–483.

73 Gallagher, J., Sams, R., Inmon, J., Gelein, R., Elder, A., Oberdörster, G., Prahalad, A. K. Formation of 8-oxo-7,8-dihydro-2′-deoxyguanosine in rat lung DNA following subchronic inhalation of carbon black. *Toxicol. Appl. Pharmacol.* **2003**, 190, 224–231.

74 Lucarelli, M., Gatti, A. M., Savarino, G., Quattroni, P., Martinelli, L., Monari, E., Boraschi, D. Innate defence functions of macrophages can be biased by nano-sized ceramic and metallic particles. *Eur. Cytokine Netw.* **2004**, 15, 339–346.

75 Zhang, Q., Kusaka, Y., Zhu, X., Sato, K., Mo, Y., Kluz, T., Donaldson, K. Comparative toxicity of standard nickel and ultrafine nickel in lung after intratracheal instillation. *J. Occup. Health* **2003**, 45, 23–30.

76 Churg, A., Gilks, B., Dai, J. Induction of fibrogenic mediators by fine and ultrafine titanium dioxide in rat tracheal explants. *Am. J. Physiol.* **1999**, 277, L975–L982.

77 Koper, O. B., Klabunde, J. S., Marchin, G. L., Klabunde, K. J., Stoimenov, P., Bohra, L. Nanoscale powders and formulations with biocidal activity toward spores and vegetative cells of bacillus species, viruses, and toxins. *Curr. Microbiol.* **2002**, 44, 49–55.

78 Rajagopalan, S., Koper, O., Decker, S., Klabunde, K. J. Nanocrystalline metal oxides as destructive adsorbents for organophosphorus compounds at ambient temperatures. *Chemistry* **2002**, 8, 2602–2607.

79 Huang, J., Best, S. M., Bonfield, W., Brooks, R. A., Rushton, N., Jayasinghe, S. N., Edirisinghe, M. J. In vitro assessment of the biological response to nano-sized hydroxyapatite. *J. Mater. Sci. Mater. Med.* **2004**, 15, 441–445.

80 Okeson, C. D., Riley, M. R., Riley-Saxton, E. In vitro alveolar cytotoxicity of soluble components of airborne particulate matter: Effects of serum on toxicity of transition metals. *Toxicol. In Vitro* **2004**, 18, 673–680.

81 Heinrich, U., Fuhst, R., Rittinghausen, S., Creutzenberg, O., Bellmann, B., Koch, W., Levsen, K. Chronic inhalation exposure of Wistar rats and 2 different strains of mice to diesel-engine exhaust, carbon-black, and titanium-dioxide. *Inhal. Toxicol.* **1995**, 7, 533–556.

82 Völkel, K., Krug, H. F., Diabaté, S. Formation of reactive oxygen species in rat epithelial cells upon stimulation with fly ash. *J. Biosci.* **2003**, 28, 51–55.

83 Wilson, M. R., Lightbody, J. H., Donaldson, K., Sales, J., Stone, V. Interactions between ultrafine particles and transition metals in vivo and in vitro. *Toxicol. Appl. Pharmacol.* **2002**, 184, 172–179.

84 Beck-Speier, I., Dayal, N., Karg, E., Maier, K. L., Schumann, G., Schulz, H., Semmler, M., Takenaka, S., Stettmaier, K., Bors, W., Ghio, A., Samet, J. M., Heyder, J. Oxidative stress and lipid mediators induced in alveolar macrophages by ultrafine particles. *Free Radic. Biol. Med.* **2005**, 38, 1080–1092.

85 Brown, D. M., Donaldson, K., Borm, P. J., Schins, R. P., Dehnhardt, M., Gilmour, P., Jimenez, L. A., Stone, V. Calcium and ROS-mediated activation of transcription factors and TNF-alpha cytokine gene expression in macrophages exposed to ultrafine particles. *Am. J. Physiol Lung Cell Mol. Physiol.* **2004**, 286, L344–L353.

86 Castranova, V. Signaling pathways controlling the production of inflammatory mediators in response to crystalline silica exposure: Role of reactive oxygen/nitrogen species. *Free Radic. Biol. Med.* **2004**, 37, 916–925.

87 Driscoll, K. E., Carter, J. M., Howard, B. W., Hassenbein, D. G., Pepelko, W., Baggs, R. B., Oberdörster, G. Pulmonary inflammatory, chemokine, and mutagenic responses in rats after subchronic inhalation of carbon black. *Toxicol. Appl. Pharmacol.* **1996**, 136, 372–380.

88 Roth, C., Ferron, G. A., Karg, E., Lentner, B., Schumann, G., Takenaka, S., Heyder, J. Generation of ultrafine particles by spark discharging. *Aerosol Sci. Technol.* **2004**, 38, 228–235.

89 HARDER, V., GILMOUR, P., LENTNER, B., KARG, E., TAKENAKA, S., ZIESENIS, A., STAMPFL, A., KODAVANTI, U., HEYDER, J., SCHULZ, H. Cardiovascular responses in unrestrained WKY rats to inhaled ultrafine carbon particles. *Inhal. Toxicol.* **2005**, 17, 29–42.

90 FRAMPTON, M. W., UTELL, M. J., ZAREBA, W., OBERDORSTER, G., COX, C., HUANG, L. S., MORROW, P. E., LEE, F. E., CHALUPA, D., FRASIER, L. M., SPEERS, D. M., STEWART, J. Effects of exposure to ultrafine carbon particles in healthy subjects and subjects with asthma. *Res. Rep. Health Eff. Inst.* **2004**, 1–47.

91 BROWN, D. M., STONE, V., FINDLAY, P., MacNEE, W., DONALDSON, K. Increased inflammation and intracellular calcium caused by ultrafine carbon black is independent of transition metals or other soluble components. *Occup. Environ. Med.* **2000**, 57, 685–691.

92 KIM, Y. M., REED, W., LENZ, A. G., JASPERS, I., SILBAJORIS, R., NICK, H. S., SAMET, J. M. Ultrafine carbon particles induce interleukin-8 gene transcription and p38 MAPK activation in normal human bronchial epithelial cells. *Am. J. Physiol Lung Cell Mol. Physiol.* **2005**, 288, L432–L441.

93 ADELMANN, P., BAIERL, T., DROSSELMEYER, E., POLITIS, C., POLZER, G., SEIDEL, A., SCHWEGLER-BERRY, D., STEINLEITNER, C. Effects of fullerenes on alveolar macrophages in vitro. In: *Toxic and Carcinogenic effects of Solid Particles in the Respiratory Tract* (ed. MOHR, U., DUNGWORTH, D. L., MAULDERLY, J., OBERDÖRSTER, G.), ILSI Press, Washington DC, **1994**.

94 BAIERL, T., DROSSELMEYER, E., SEIDEL, A., HIPPELI, S. Comparison of immunological effects of fullerene C60 and raw soot from fullerene production on alveolar macrophages and macrophage like cells in vitro. *Exp. Toxicol. Pathol.* **1996**, 48, 508–511.

95 BAIERL, T., SEIDEL, A. In vitro effects of fullerene C-60 and fullerene black

on immunofunctions of macrophages. *Fullerene Sci. Technol.* **1996**, 4, 1073–1085.

96 HESTERBERG, T. W., HART, G. A. Synthetic vitreous fibers: A review of toxicology research and its impact on hazard classification. *Crit. Rev. Toxicol.* **2001**, 31, 1–53.

97 GODLESKI, J. J. Role of asbestos in etiology of malignant pleural mesothelioma. *Thorac. Surg. Clin.* **2004**, 14, 479–487.

98 HUCZKO, A., LANGE, H., CALKO, E., GRUBEK-JAWORSKA, H., DROSZCZ, P. Physiological testing of carbon nanotubes: Are they asbestos-like? *Fullerene Sci. Technol.* **2001**, 9, 251–254.

99 HUCZKO, A., LANGE, H., CALKO, E. Fullerenes: Experimental evidence for a null risk of skin irritation and allergy. *Fullerene Sci. Technol.* **1999**, 7, 935–939.

100 HUCZKO, A., LANGE, H. Carbon nanotubes: Experimental evidence for a null risk of skin irritation and allergy. *Fullerene Sci. Technol.* **2001**, 9, 247–250.

101 EEDY, D. J. Carbon-fibre-induced airborne irritant contact dermatitis. *Contact Dermatitis* **1996**, 35, 362–363.

102 SHVEDOVA, A. A., CASTRANOVA, V., KISIN, E. R., SCHWEGLER-BERRY, D., MURRAY, A. R., GANDELSMAN, V. Z., MAYNARD, A., BARON, P. Exposure to carbon nanotube material: Assessment of nanotube cytotoxicity using human keratinocyte cells. *J. Toxicol. Environ. Health A* **2003**, 66, 1909–1926.

103 MONTEIRO-RIVIERE, N. A., NEMANICH, R. J., INMAN, A. O., WANG, Y. Y., RIVIERE, J. E. Multi-walled carbon nanotube interactions with human epidermal keratinocytes. *Toxicol. Lett.* **2005**, 155, 377–384.

104 JIA, G., WANG, H., YAN, L., WANG, X., PEI, R., YAN, T., ZHAO, Y., GUO, X. Cytotoxicity of carbon nanomaterials: Single-wall nanotube, multi-wall nanotube, and fullerene. *Environ. Sci. Technol.* **2005**, 39, 1378–1383.

105 PANTAROTTO, D., BRIAND, J. P., PRATO, M., BIANCO, A. Translocation of bioactive peptides across cell membranes by carbon nanotubes. *Chem. Commun.* **2004**, 16–17.

106 AITKEN, R. J., CREELY, K. S., TRAN, C. L. Nanoparticles: An occupational hygiene review. HSE Books, Inst. of Occupational Medicine, Edinburgh, UK, **2004**, Vol. 274.

107 KRUG, H. F., GRUNWALD, A. Risk assessment and risk management. In: *Assessment and Perspectives of Nanotechnology* (ed. BRUNE, H., ERNST, H., GRUNWALD, A., GRÜNWALD, W., HOFMANN, H., JANICH, P., KRUG, H. F., MAYOR, M., SCHMID, G., SIMON, U., VOGEL, V., GETHMANN, C. F.), Springer, Berlin, **2005**.

III
Environment

8
Nanomaterials for Environmental Remediation

Glen E. Fryxell and Shas V. Mattigod

8.1
Introduction

Over the last 10–15 years there has been an explosion of activity in the design and synthesis of nanomaterials built around a wide variety of basic architectures. More recently, a portion of this effort has focused on the environmental impacts and environmental applications of these nanomaterials. Why all this interest in nanomaterials? What advantages might these tiny structures provide to environmental remediation efforts? This chapter overviews research in this area, and outlines some of the advantages that these materials provide to environmental clean-up efforts.

The most obvious advantage that nanostructured materials provide for environmental remediation is that they offer very high specific surface areas (measured in square meters per gram). Thus, for a base material of a given density, nanostructured materials can concentrate large amounts of surface area into a very small volume. When the goal is to selectively remove a toxic contaminant from a large volume feed stream (industrial effluent, contaminated groundwater, polluted river, etc.), the ability to selectively treat (sorb or react) a specific contaminant with a small amount of material has clear advantages, both in terms of efficacy and cost. Nanomaterials can do exactly this.

For certain remediation applications, especially those dealing with dilute or trace level contamination, mass-transfer issues can dominate the kinetics of the treatment process. An *in situ* treatment process can address some of this by sending the treatment out after the contaminant in a hunter–seeker sense. The facile dispersion of nanomaterials, especially nanoparticulate materials, facilitates the efficiency of this treatment strategy, particularly in highly channeled flow systems with high tortuosity (e.g., soil matrices).

Another advantage provided by nanomaterials is that the bulk of the reagent is not buried deep beneath the surface, inaccessible to solution-borne contaminants. Whether the intent is sorption or chemical modification of the contaminant, for the treatment method to be effective the contaminant species must be able to access an active reaction/binding site. Anything buried deep beneath the surface is wasted. Because of the high surface area to mass ratios of nanomaterials, most of the material is at, or adjacent to, an accessible surface.

Nanotechnologies for the Life Sciences Vol. 5
Nanomaterials – Toxicity, Health and Environmental Issues. Edited by Challa S. S. R. Kumar
Copyright © 2006 WILEY-VCH Verlag GmbH & Co. KGaA, Weinheim
ISBN: 3-527-31385-0

For macroparticulate porous materials, portions of the material in the core of the particle may be kinetically inaccessible if the diffusion path-length from free solution is too long. By tailoring the particle size (or macroporosity), to provide adequate access to the nanoporosity, this limitation can be overcome. Synthetic methods have been actively investigated to develop hierarchical pore structures to address these issues.

The revolution in nanomaterials synthesis started with researchers exploring what shapes and structural motifs could be made (spherical nanoparticles, hexagonal pores, cylindrical nanorods, etc.). This was followed by an exploration of chemical compositions (SiO_2, ZrO_2, CdS, etc.). More recently there has been a great deal of interest in making *functional* nanomaterials, either using self-assembly, surface modification chemistry, or by tuning the chemical composition of the material itself for the job at hand. The results of these efforts are a wide variety of functional nanomaterials that have been tailored to address the environmental remediation of several chemical contaminants, including dense non-aqueous phase liquids (DNAPLs), organophosphonate pesticides, polycyclic aromatic hydrocarbons (PAHs), heavy metals, radionuclides, oxometallate anions, CO_2 management, and more. These functional nanomaterials are aimed at ensuring that we have clean air to breathe and clean water to drink. Clean water, and access to clean water, is emerging as one of the key global political/economic issues of the 21st century. Nanomaterials are being designed and synthesized to address these needs. This chapter summarizes nanoparticle-based remediation technologies that use acid–base chemistry, redox chemistry and absorption to remove specifically targeted contaminants. Also included are hybrid nanoporous materials that contain chemically selective ligand fields, based on metal phosphonate chemistry and organosilane self-assembly. Examples of field tests on actual waste streams for reactive nanoparticle and hybrid nanoporous sorbents are also summarized.

8.2
Nanoparticle-based Remediation Materials

The simplest, geometrically, entry into the class of nanomaterials is the spherical nanoparticle. These have been made by imposing either kinetic or thermodynamic controls on the production processes, and by confining chemical reactions and/or nucleation and growth in confined spaces [1]. Nanoparticle synthetic methods result in nanoparticles that range in size from a just a few nanometers in diameter (e.g., the reverse micelle templated synthesis of gold nanoparticles), to methods that produce nanoparticles a couple of hundred nanometers in diameter (e.g., Stöber sphere synthesis).

Nanoparticles tend to be more reactive than the corresponding bulk material because of the increased chemical potential resulting from the high degree of curvature of the interface [2]. This property can make nanoparticle-based remediation methods particularly effective since they are easily dispersed and undergo the targeted chemical reaction more readily.

8.2.1
Acid–Base Chemistry

An example of the unusual reactivity of nanomaterials and how it has been exploited for the benefit of the environment is found in the work of Klabunde's group [3–6]. For the last 10 years, this group has systematically studied the destruction of halogenated hydrocarbons by nanoparticulate metal oxide aerogels (e.g., CaO). In this work nanoparticulate CaO aerogels were prepared using the "autoclave method" (hydrolysis of calcium methoxide, followed by heating under vacuum to 500 °C for 6 hours). These aerogels typically have surface areas of the order of 120 m^2 g^{-1}, and are composed of aggregates of spherical 25 nm nanoparticles. They were found to effectively destroy various chlorocarbon solvents, including CCl_4, $CHCl_3$, trichloroethylene and tetrachloroethylene. Generally, the reaction products are $CaCl_2$ and CO_2 (or CO) [Eq. (1)]. With the less reactive tetrachloroethylene, higher temperatures were required, and this led to the partial formation of $CaCO_3$ by reaction of the CO_2 product with CaO starting material.

$$2CaO + CCl_4 \rightarrow 2CaCl_2 + CO_2 \tag{1}$$

Similar studies were carried out with aerogel MgO [7]. In this case the autoclave method resulted in material with a surface area of 364 m^2 g^{-1}, a crystallite size of 4.7 nm and average pore diameter of 98.7 Å. Compressing these materials with loads of up to 20 000 lbs did not significantly change the surface area of the MgO aerogel; however, the pore volume and average pore diameter could be systematically reduced with increasing compression, introducing an interesting method of nanostructural control. A slight reduction in crystallite size was also noted. These MgO aerogels, of varying pore diameters, were evaluated for their abilities to sorb alcohols of different chain lengths; some size discrimination was noted.

Nanoparticulate aerogel MgO also reacts with 1-chlorobutane at elevated temperatures [8]. Here, the products are a mixture of butanes and $MgCl_2$. At 200 °C the reaction stops when a monolayer of $MgCl_2$ is formed, but at higher temperatures the rate and degree of conversion are enhanced considerably. Coordination of organic molecules onto the surface of these materials has been modeled to gain a better understanding of their reactivity [9].

The "autoclave method" was improved to include a hypercritical drying procedure [3]. This resulted in dry $Mg(OH)_2$ powders with surface areas as high as 1100 m^2 g^{-1}, more than twice those observed previously. Solvent effects in this hydrolysis and crystallization were carried out, and increasing the amount of toluene in the alkoxide hydrolysis reaction mixture resulted in faster gelation and higher surface areas in the final product [10]. Subsequent mechanistic studies revealed that solvation of the alkoxide/alcohol mixtures is important to the gelation process and the structure of the subsequently formed dry gel [11]. This was rationalized with a partial charge model, and was found to be purely a solvent effect, and not the result of the high-temperature hypercritical procedure.

An interesting manifestation of the unusual reactivity of nanoparticle interfaces is the unusual biocidal activity of aerogel MgO [12]. While aerogel MgO nanoparticles exhibited several properties that made them desirable as a potential disinfectant, when doped with a small amount of Cl_2 (or Br_2) they displayed effective biocidal action against Gram positive bacteria, Gram negative bacteria and spore cells. This was explained by the fact that many of these small particles could very effectively coat the bacterium, and deliver a localized high concentration of active halogen to the cell membrane. ζ-Potential measurements have shown that the aerogel MgO nanoparticles have a positively charged surface, and thus will experience a Coulombic attraction with the negatively charged cell membrane, helping to drive this targeted delivery process. These chlorinated nanoparticulate metal oxides are also selective catalysts for the chlorination of alkanes [13].

Aerogel MgO nanoparticles have also been coated with various surfactants to improve their dispersability in non-polar media [14]. This is important for the destruction of pesticides (which tend to be applied in non-polar solvents) or chemical warfare agents in the event of a leak or spill. In all cases, the surfactant-coated aerogel MgO nanoparticles dispersed more readily in organic solvents after they were treated with surfactants. The surfactant-coated MgO nanoparticles effectively destroyed Paraoxon (a pesticide); however, there was some variation in efficacy from one surfactant to another. A similar trend was observed for reaction with 2-chloroethyl ethyl sulfide (a "mustard" derivative). The surfactant coating decreased the reactivity of the MgO nanomaterials in all cases, presumably by sterically blocking surface reaction sites.

Similarly, these aerogel MgO nanomaterials have been subjected to CVD carbon treatment to increase their hydrophobicity [15]. These materials have surface areas in the range of 409 to 467 $m^2\ g^{-1}$, and pore volumes of 0.39 to 0.48 $cm^3\ g^{-1}$. The carbon formed "nanoislands" on the particles, which were estimated to be 1 or 2 graphite layers thick. Partial carbon coating of these nanoparticles has a beneficial effect on their ability to destroy hazardous materials [16]. Detailed characterization of these materials has shown that carbon is first deposited inside the pores of the aerogel aggregates, and the outer surfaces are covered with carbon only after the pores are filled.

One of the many benefits provided by alkaline earth metal oxide nanomaterials is the ability to treat a wide variety of hazardous materials. In addition to those already touched upon (chlorinated solvents, microbes, pesticides and "mustards"), these materials are effective at removing hydrogen sulfide, which is very toxic, corrosive and odiferous, from gas streams. At lower temperatures (e.g., ≤100 °C), ZnO nanoparticles destroyed H_2S more effectively than CaO or MgO nanoparticles [Eq. (2)], which is consistent with the superior thermodynamic driving force for the Zn sulfidation [17]. However, at higher temperatures (e.g., >250 °C), the CaO nanoparticles were the better choice, due to sintering of the ZnO nanocrystals. ZnO nanocrystals also effectively destroy chlorinated solvents, SO_2, and Paraoxon [18].

$$ZnO + H_2S \rightarrow ZnS + H_2O \tag{2}$$

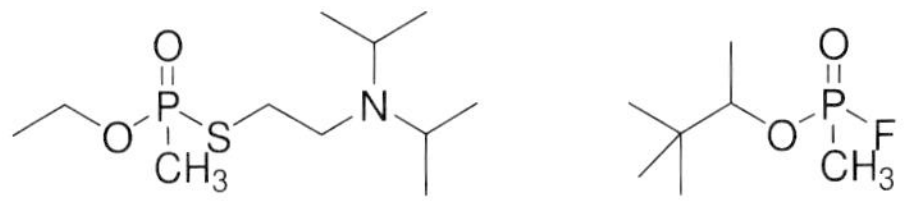

VX   GD

Fig. 8.1. Structures of the chemical warfare agents VX and GD.

These materials have also shown clear efficacy in the destruction of chemical warfare agents. For example, reaction of aerogel MgO nanoparticles with 2-chloroethyl ethyl sulfide [a mimic for bis(2-chloroethyl)sulfide, a.k.a. "HD", "distilled mustard" or "mustard gas"] in pentane solution destroyed between 25% and 65% of the mustard derivative in 4 h [19]. Addition of trace amounts of water to the mixture slightly enhanced the reaction rate, while larger amounts of water reduced both reaction rate and degree of conversion. When the reaction was performed in tetrahydrofuran (THF), the rate was slower than in pentane. Reaction in methanol resulted in solvolysis. These materials have also been applied against organophosphonate chemical warfare agents, like VX and GD (a.k.a. "Soman") (Fig. 8.1) [20]. The aerogel MgO nanomaterials were highly reactive towards GD, VX and HD. The rate was a function of surface tension and vapor pressure (these studies were carried out neat, in the absence of carrier or solvent). Similar results were obtained with CaO nanomaterials [21]. In this case, trace levels of water appear to induce an autocatalytic reaction. These chemical warfare agents were also destroyed at room temperature using nanosized Al_2O_3 [22]. Here, the reactions proceed to the particle core, resulting in extremely large reaction capacities for these nerve agents.

Similar enhanced reactivities towards halogenated solvents, SO_2 and Paraoxon were seen for nanocrystalline CuO and NiO [23].

Bimetallic nanocrystalline analogs to these materials have also been made and studied. For example, a mixed Al_2O_3/MgO phase was prepared and found to have a high surface area (559–834 m^2 g^{-1}), to display remarkably high thermal stability (minimal sintering at 700 °C), and to be effective in the destruction of CCl_4, Paraoxon and SO_2 [24]. In a variation on this theme, the alumina may be replaced with ferric oxide; the alkaline earth oxide serves as the support for the second metal oxide. A small amount of Fe_2O_3 was layered on top of the CaO support, resulting in sub-monolayer coverage, and no visible island formation [25]. This layered mixed metal oxide system effectively destroyed chlorinated solvents, organophosphonates, carbon disulfide and carbonyl sulfide. The small overlayer of Fe_2O_3 induced a remarkable enhancement in reactivity.

A similar layered bimetallic nanomaterial was prepared using an Fe_2O_3 overlayer atop of SrO nanoparticles [26]. This system was chosen since its K-edge energy absorption was a better fit for EXAFS analysis. These studies revealed that the Fe_2O_3 overlayer enhanced the reactivity of the SrO inner layers, indicating that this is not a surface-limited reaction in these bimetallic nanomaterials.

8.2.2
Redox Chemistry

Environmental remediation reactions are not limited to the acid–base reactions described above, in which electrophilic hazardous materials are converted into benign species by reaction with a nucleophilic oxide. Another area that has received much attention in recent years is the use of zero-valent metal nanoparticles to reduce certain highly oxidized species that are of environmental concern. Common targets in this area are the DNAPLs that contaminate certain groundwater supplies. As an early example, high surface area ("cryo prepared") Zn and Sn particles were found to be more effective at reducing chlorinated solvents (e.g., CCl_4, $CHCl_3$, etc.) than their bulk counterparts [27]. Magnesium is a more electropositive metal than either Zn or Sn, and hence might be expected to more efficiently reduce the chlorinated hydrocarbons. However, Mg reacted preferentially with water, emphasizing the need to balance the reactivity of these species for successful *in situ* remediation.

Zhang and coworkers extended this work to use zero-valent iron (ZVI) nanoparticles, and demonstrated their use for *in situ* remediation of chlorocarbon contaminated groundwaters [28]. In this work, the ZVI nanoparticles (made by $NaBH_4$ reduction of $FeCl_3$ in water) were more effective than either commercial iron powders or palladized iron powders (Pd enhances reactivity in these reductions) for the dechlorination of trichloroethylene (TCE) [Eq. (3)]. The dechlorination process was complete in approximately 15 min (initial concentration was 20 ppm). Palladized nanoparticle ZVI was even more rapid and more effective than bare nanoparticle ZVI. PCBs were completely dechlorinated in 17 h at ambient temperature with the Fe/Pd nanoparticles, while bare Fe nanoparticles induced less than 25% conversion under the same conditions. These Fe nanoparticles were 1–200 nm in diameter (with most between 100 and 200 nm) and had a bulk surface area of $33.5 \ m^2 \ g^{-1}$.

$$Fe(0) \ (\text{in excess}) + CHCl=CCl_2 \rightarrow FeCl_2 + CH_2=CH_2 \tag{3}$$

The mechanistic role of the Pd islands has been studied in detail [29]. Chemisorption of tetrachloroethylene onto the Pd surface was studied by high-resolution XPS, revealing that dissociation of C_2Cl_4 was complete at temperatures above 291 K. Systematic laboratory testing on all the chlorinated ethenes has shown that these ZVI nanoparticles are tens to hundreds of times faster than commercially available iron powders [30].

Similar reductions were performed using Fe/Ag nanoparticles in the dechlorination of chlorinated benzenes [31]. These materials were noticeably slower than the Fe/Pd nanoparticles reported earlier. Other bimetals (e.g., Cu/Al) have also shown promise for dehalogenating chloromethanes [32].

ZVI has also shown promise for the immobilization/fixation of As(III) in groundwater, both *in situ* and *ex situ* [33].

Mallouk and coworkers have actively studied the synthesis and application of bimetallic nanoparticles for environmental remediation of DNAPLs and toxic heavy metals. For example, they have shown that ZVI nanoparticles (10–30 nm in diameter) supported on a PolyFlo resin (the authors refer to this adduct as a "Ferragel") is a very effective reductive sorbent for Cr(vi) or Pb(ii) contamination, and suggest that it could be useful as an *in situ* remediation strategy [34]. The surface chemistry and electrochemistry of these Ferragels have been studied in detail [35]. In addition, the Mallouk group has also looked at the synthesis and chemistry of zero-valent Ni-Fe nanoparticles [36]. Previous work had shown the value of incorporating catalytic islands onto the ZVI nanoparticle surface, and in this work they replaced the Pd with the more affordable Ni. These materials were found to have crystallite sizes of 3–5 nm, particle diameters of 10–30 nm and surface areas of 59 m^2 g^{-1}. The $NaBH_4$ reduction resulted in residual boron content within these nanoparticles (~5%). These Ni/Fe nanoparticles dehalogenated all of the TCE in approximately 2 h (initial concentration 23.4 ppm). This is considerably faster than commercial Fe powders, or Fe/Ni powders, but not quite as fast as the Fe/Pd nanoparticles; this difference was attributed to the better ability of Pd to catalyze the hydrogenation reaction. These Fe and Ni/Fe nanoparticles have also been coated with "hydrophilic carbon" (i.e., carbon rendered hydrophilic by reaction with the diazonium salt of sulfanilic acid). Similar materials were coated with poly(acrylic acid). These coatings tend to lower the aggregation tendencies and sticking coefficients of these ZVI nanoparticles, thereby enhancing their transport and delivery to the contamination site. This strategy was effective in some soil types, but not all. These coated materials were also effective at dehalogenating TCE.

8.2.3
Field Deployments of ZVI

An actual field assessment of the Fe/Pd nanoparticle remediation technology was undertaken to evaluate how effectively these materials could treat a known industrial contamination site [37]. Gravity injection of nanoparticle suspensions into well holes was used to deliver the nanoparticles to the contaminated plume. TCE concentrations declined rapidly after nanoparticle injection (as much as 96.5%), with significant variability from monitoring site to monitoring site (this pattern was consistent with known colloid transport and chemistry in porous media). Details of how these materials react with contaminants in soil and water over extended periods, and *in situ* reactions of the nanoparticles in sub-surface environments, have also been summarized [38].

Quinn and coworkers have reported detailed study of an actual field deployment of *emulsified* ZVI [39]. The thinking behind this strategy is that emulsification of the nanoparticles enhances dechlorination by increasing the contact between the DNAPL and the ZVI, as well as providing vegetable oil, which is hypothesized to increase biological activity (thought to be important to certain stages of the overall reduction process). Significant reductions in TCE levels were observed at nearly all monitoring sites (commonly > 80%).

8.2.4
Absorption Chemistry

Nanoparticles do not have to chemically alter the target species to effectively remove it from the environment. Lion and coworkers have devised an interesting strategy for removing polynuclear aromatic hydrocarbons (PAHs) from contaminated soils using amphiphilic polyurethane (APU) nanoparticles [40]. These are made by emulsifying and crosslinking certain precursor polymer chains in water, and result in APU particles 17–97 nm in diameter [41]. Several variations on the basic formulation were evaluated for sorption and transport, and it was found that by increasing the size of the hydrophobic backbone it was possible to enhance the APU particle's affinity for the PAH. Increasing the number of ionic groups reduced the APU particle aggregation, and replacing carboxylates with PEG [poly(ethylene glycol)] chains prevented particle aggregation while greatly enhancing particle stability and mobility in the soils.

Similar PEG-modified urethane acrylate (PMUA) nanoparticles enhance bioremediation of PAH contaminants by increasing their bioavailability [42]. PAHs commonly sorb to soil particles and non-aqueous phase liquids (NAPLs), which limits their bioavailability. The PMUA nanoparticles released the sorbed PAHs, thereby increasing their availability to bacteria, suggesting that this might be an effective strategy for *in situ* bioremediation.

8.3
Hybrid Nanostructured Remediation Materials

Hybrid organic/inorganic materials allow for the incorporation of complex ligands into the nanomaterial structure, thereby empowering a high degree of chemical selectivity or molecular recognition. Work in this area is typified by two slightly different strategies: The first incorporates the organic ligand into the fundamental building block of the nanomaterial before construction of the scaffold, and the second entails construction of the scaffold first, followed by decoration with the organic ligand. Each approach has its advantages, and both allow for the incorporation of complex ligands and various structural backbones.

8.3.1
Nanostructured Metal Phosphonates

Abe Clearfield and coworkers have studied in great depth the synthesis and chemistry of nanostructured transition metal phosphonates. Portions of this elegant body of work have been reviewed [43, 44]. Part of the motivation behind this work is to use these nanostructured hybrid materials as ion exchangers [45]. The basic strategy is to build a scaffold based on the strong metal–phosphonate interactions to form the backbone of the material, and to tether an organic ligand to the phosphonic acid. This ligand may participate in dictating the final structure of the ma-

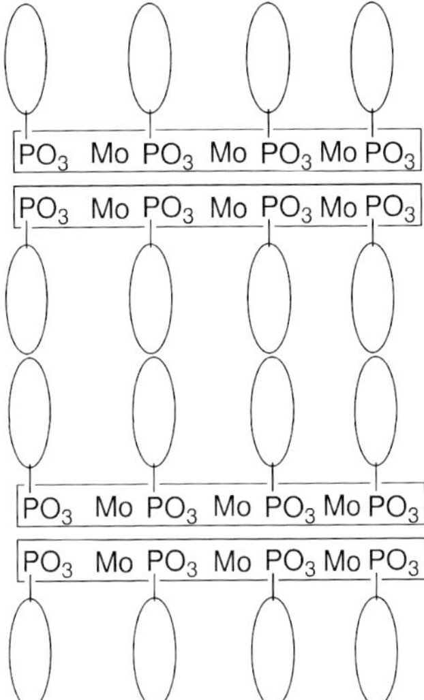

Fig. 8.2. Simplified schematic showing the structure of $MoO_2(O_3PC_6H_5)\,H_2O$.

terial, or it may simply be a spectator. The advantages of this synthetic strategy are that, in general, it is procedurally simple and the products are generally robust and not subject to hydrolysis or other modes of degradation.

It is useful to introduce this class of hybrid materials with the phenylphosphonic acid derivatives to showcase some basic structural features. For example, molybdenyl phenylphosphonate forms double-stranded chains, with the molybdenyl chains held together with hydrogen bonds [46]. This leaves the phenyl rings oriented roughly perpendicular to the double stranded chain, creating a hydrophobic pocket between the polar molybdenyl backbones (Fig. 8.2). Molybdenyl phenylphosphonate was found to intercalate short-chain alkyl amines, but not short-chain alcohols. Zinc phenylphosphonate, however, forms a pleated sheet, in which the zinc–phosphonate interactions form the backbone of the sheet, and the pendant phenyl rings form hydrophobic pockets between the sheets [47]. The zinc materials were also intercalated alkylamines, in this case in a 1:1 stoichiometry. Detailed characterization of the intercalation process revealed that the N atom is coordinated to the Zn center, disrupting a portion of the structure of the zinc phosphonate sheet (opening holes), but the layered sheet structure remained intact. This intercalation can be reversed by washing the adduct with dilute HCl.

A unique structural motif may be created by tying the metal-phosphonate layers together with a diphosphonic acid to create a pillared layer nanostructure. For example, Cu and Zn complexes of short-chain (C_2 and C_3) diphosphonic acids form such pillared layered structures [48]. In general, these complexes appear to be densely packed, with no open spaces for molecular intercalation. An exception is the Cu(II) complex of propylenebis(phosphonate), in which the organic groups are about 6.8 Å apart. This is due to the presence of water molecules in the Cu-phosphonate lattice. Longer tethers might allow for increased interlayer spacing, providing more room for molecular intercalation.

Zinc biphenylenebis(phosphonate) is an interesting material [49]. When produced at pH 1.6, it forms a linear chain compound since only two of the three phosphonate oxygen atoms are ionized. When made at pH 4.5, all of the oxygen atoms are ionized and the pillared layer structure is once again obtained. To increase the microporosity (and ion-exchange capacity) of these materials, analogs were made in which a portion of the bis(phosphonate) pillars was replaced with phosphate groups. These mixed systems have notably higher surface areas than the pure pillared structures (35–136, vs. 20–28 m^2 g^{-1}), and these phosphate groups were chemically accessible, as shown by acid–base titration.

The structurally similar Zr(IV) biphenylenebis(phosphonate), and the terphenylene analog, were also prepared under acidic conditions [50]. If the HF/Zr ratio was 20 or less, these materials have remarkable surface areas of ∼400 m^2 g^{-1} and pore diameters on the order of 10–20 Å. These pillared aromatics were also able to be sulfonated by either fuming sulfuric acid or gaseous SO_3. Acid–base titration revealed an impressive acid functional density of 3.2 mmol g^{-1}. These sulfonic acids proved to be effective acid catalysts for several reactions, and would clearly also make fine ion-exchange materials.

8.3.1.1 Iminodiacetic Acids and Related Chelating Ligands

Layered metal phosphonate structures have also been prepared using tethered iminodiacetic acid (IDAA) moieties (the functional subunit of EDTA) [51, 52]. These materials have alternating Zr-phosphonate layers, and iminodiacetic acid layers. The Zr-phosphonate layers provide the structural backbone, while the pendant iminodiacetic acids can serve for molecular recognition and ion-exchange materials (Fig. 8.3). These compounds were found to be effective intercalation hosts for various alkylamines, which packed into the structures in well-ordered bilayers [52]. The complex in which the Zr center was fully substituted (i.e., there were four IDAA phosphonate ligands per metal center, resulting in only IDAA ligands in the interstitial layers) exhibited a surprisingly low affinity for transition metals and lanthanide cations at pH 2 (K_ds < 100 mL g^{-1} in all cases), presumably due to steric congestion. However, in compounds where a fraction of the IDAA-phosphonates were substituted with phosphate (resulting in vacancies in the interstitial layer), much higher affinities for transition metals resulted, particularly for the lanthanides. In some cases, K_ds as high as 25 000 were measured. Presumably, this is due to a combination of steric relief in the interstitial layer, as well as the ion-exchange capacity of the exposed phosphate hydroxyls in these vacancies.

Analogously, nitrilotris(methylenephosphonic acid) (NTP) has also been used to create functional nanostructured host materials [53]. These ligands undergo a unique self-assembly process when allowed to react with various aromatic amines (e.g., 1,10-phenanthroline, quinoline, acridine, etc.). The products are three-dimensional networks that are extensively stabilized by short, symmetrical hydro-

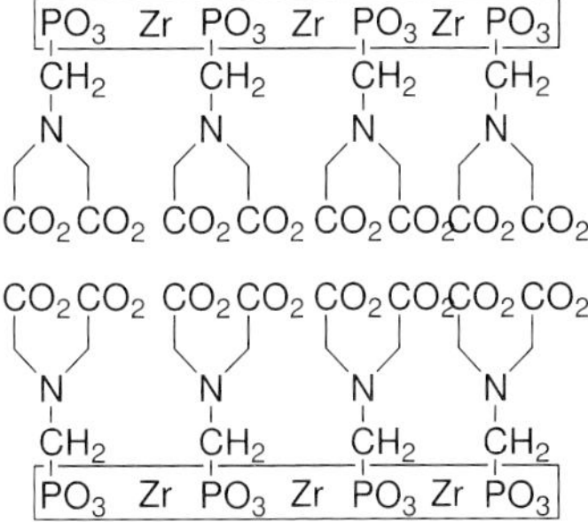

Fig. 8.3. Simplified schematic showing the structure of Zr N-(phosphonomethyl)iminodiacetic acid.

gen bonds, forming complex open porous structures. Some of these supramolecular structures (e.g., that formed between NTP and 1,7-phenanthroline) have unique chromophoric properties, forming charge-transfer complexes when protonated, suggesting that perhaps these materials might be useful in sensing/detection applications. Metal complexes of NTP have also been prepared [54]. Once again, open, highly hydrogen-bound structures were obtained. These materials have the potential to be excellent ion-exchange materials, as well as molecular hosts.

8.3.1.2 Macrocycle Metal Phosphonates

The Clearfield group has achieved an elegant level of molecular sophistication with their design and synthesis of layered metal phosphonates containing pendant macrocyclic ligands, such as crown ethers and azamacrocycles [55]. Once again, in these compounds the metal phosphonate/phosphate forms the structural backbone of the material, while the phosphate hydroxyls and macrocyclic ligands fill the interstitial layers (Fig. 8.4). The presence of an acidic, exchangeable phosphate proton in the immediate proximity of a chemoselective macrocyclic ligand is of obvious value for ion-exchange applications. Bridging diphosphonic acid crown ether complexes have also been prepared, and shown to have anion exchange capability (these azacrowns were protonated and could serve as anion hosts) [56]. Changing the metal from Zr to Cu(ii) has a profound impact on the structure of these complexes [57], with the Cu(ii) ion chelated within the macrocycle, and wrapped up by the phosphonate "arms". The units were then arrayed in a linear fashion, held together by hydrogen bonds between the phosphonate units. Cadmium(ii) was also chelated, in this case to form a dinuclear complex (one in the ring and one between the phosphonates); the bent geometry of this dinuclear complex ultimately gave rise to a convoluted cyclic arrangement of hydrogen-bonded complexes. For optimal exploitation of chemical selectivity of the macrocyclic ligand, it appears that Zr is the preferred metal for making these nanostructured crown ether complexes.

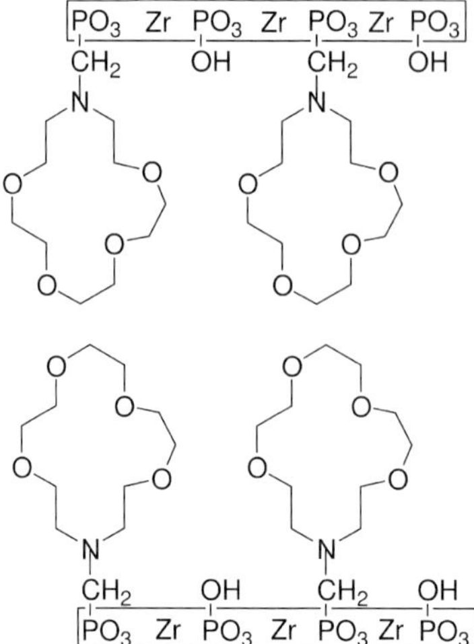

Fig. 8.4. Schematic showing the structure of Zr 1-aza-15-crown-5 phosphonate/phosphate.

8.3.2
Self-assembled Monolayers on Mesoporous Supports (SAMMS)

The surfactant-templated synthesis of mesoporous ceramic oxides has provided an excellent foundation for environmental sorbent and sensor materials [58, 59]. This foundation provides for the ability to install a wide variety of chemically selective ligand fields with which to bind the target analyte. The first work in this area addressed the need to remove mercury from groundwater [60–62]. This was accomplished by anchoring an organosilane monolayer terminated with a thiol group inside the pore surfaces of these mesoporous materials. The resulting product is called self-assembled monolayers on mesoporous supports (SAMMS) (Fig. 8.5). These hierarchical materials have unprecedented capacity and kinetics for sequestering mercury [63–65]. In addition, the mercury laden sorbent passed the EPA TCLP leachate test, revealing that the Hg is indeed very tightly bound within the mesoporous matrix. By trapping the Hg inside the nanoporous matrix, it is inaccessible to microbial attack and subsequent methylation and release (a limitation of polymer-based ion-exchange resins). Other "soft" heavy metals, like Cd, Au and Ag can also be bound by thiol-terminated SAMMS [63]. Other groups have also attached thiol-terminated monolayers inside similar mesoporous materials [66–68]. Jaroniec has tackled Hg by attaching some related sulfur-based ligand systems inside mesoporous silica phases [69]. Jaroniec's materials have excellent Hg binding capacity, but can only be partially regenerated.

8.3.2.1 Thiol SAMMS Performance with Actual Waste
Bench-scale treatability tests were carried out at Pacific Northwest National Laboratory (PNNL) on actual waste to evaluate the mercury adsorption performance of

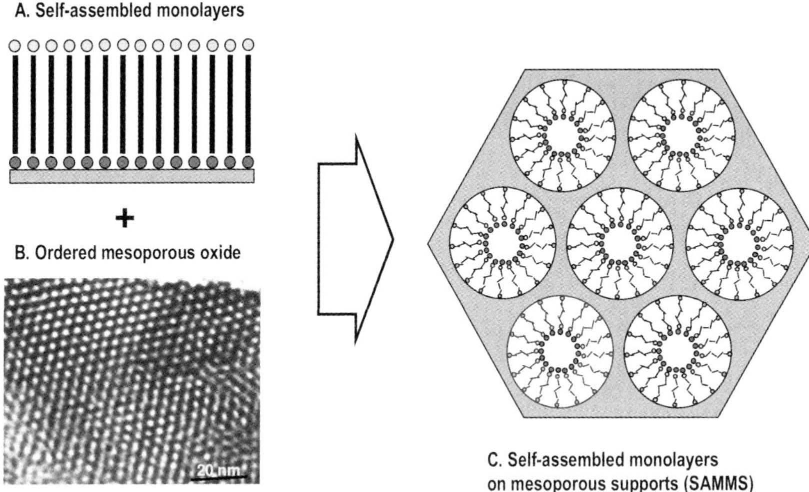

Fig. 8.5. Schematic showing the structure of self-assembled monolayers on mesoporous supports (SAMMS).

thiol-SAMMS from a real-world waste stream, originating from a pilot-scale waste glass melter operation. The principal dissolved components in this alkaline waste stream (pH 8.5) consisted mainly of sodium borate ($\sim$30 mM) and sodium fluoride ($\sim$9 mM), with minor concentrations of sodium chloride ($\sim$3 mM), sodium nitrite ($\sim$0.9 mM), sodium sulfate ($\sim$0.8 mM), sodium nitrate ($\sim$0.6 mM) and sodium iodide ($\sim$0.2 mM). The mercury concentration in solution was measured to be 4.64 mg L^{-1}. Iodide ion is a strong mercury complexing ligand and the speciation calculations for this mixture indicated that all the soluble mercury existed as iodide complexes (HgI_2^{0}, $\sim$52%; HgI_3^{-}, $\sim$47%; and HgI_4^{2-}, $\sim$1%). Other dissolved components such as Al, Ba, Ca, Cd, Co, Cr, Cu, Fe, Mo, Ni, PO_4, Pb and Zn were present in trace concentrations ($<$2 mg L^{-1}). To test the effectiveness of thiol-SAMMS in adsorbing mercury from this complexing matrix, a 50 mL aliquot of the filtered solution was treated with 40 mg of thiol-SAMMS (solution to solid ratio 1250 mL g^{-1}) and after 8 h of equilibration the residual mercury concentration was determined.

The results (Table 8.1) indicate that thiol-SAMMS effectively removes mercury to meet the U.S. Environmental Protection Agency's (EPA) regulatory level of 0.15 mg L^{-1} from both filtered and unfiltered samples of waste solution. However, filtering prior to thiol-SAMMS treatment was more effective than treating unfiltered waste solution. By filtering the solution prior to thiol-SAMMS treatment, we can remove $\sim$99% of mercury in the waste stream, leaving a residual concentration of only 0.05 mg L^{-1}. Notably, the residual mercury concentration resulting from the first experiment is about an order of magnitude less than the EPA regulatory limit [70]. These results confirmed previous observations that thiol-SAMMS can effectively remove iodide-complexed mercury from solutions [65].

Tab. 8.1. Thiol-SAMMS mercury removal data for melter condensate waste solution.

Tr #	Waste solution	Conc (mg L^{-1})	Removal (%)	pH
0	Untreated	4.640	–	8.5
1	Filtered – treated[a]	0.052	98.9	8.0
1A	Unfiltered – treated[b]	0.179	96.1	8.0
2A	Unfiltered – treated[c]	0.148	96.8	8.0

EPA discharge limit for Hg in treated effluent is 0.15 mg L^{-1}.
[a] 50 mL of waste solution filtered (0.45 μm) and treated with 40 mg of 35/thiol-SAMMS.
[b] 50 mL unfiltered waste solution treated with 60 mg of 35/thiol-SAMMS.
[c] 50 mL unfiltered waste solution treated with 40 mg of 65/thiol-SAMMS.

8.3.2.2 Thiol SAMMS Performance on Contaminated Oil

Not all chemical separation needs to involve aqueous media, and sorbent materials that are effective in aqueous systems are not always effective in hydrophobic media. Previous testing at the Oak Ridge National Laboratory demonstrated that thiol-SAMMS is effective at removing Hg from contaminated vacuum pump oil [71]. These batch tests consisted of equilibrating 50 mL aliquots of oil containing about 650 ppb mercury with thiol-SAMMS (w/v 0.1–1%) for 24 h. Following equilibration, the residual mercury concentration in the treated oil was measured. The data (Fig. 8.6) showed that the residual concentration of mercury decreased logarithmically as a function of thiol-SAMMS added to the solution (w/v). Thus, these

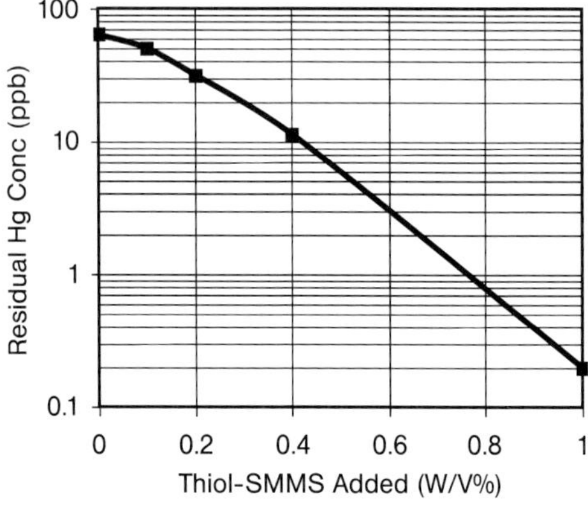

Fig. 8.6. Mercury adsorption by thiol-SAMMS from vacuum pump oil.

tests demonstrated that thiol-SAMMS could also effectively remove mercury from viscous, hydrophobic media, such as contaminated vacuum pump oil.

8.3.2.3 **Anion SAMMS**

Anion exchange technologies are not as mature as cation exchange methods, and anion exchange materials tend to be built around quaternary ammonium salt functionalities. While this approach works, it doesn't generally provide a direct anion–cation interaction, and therefore selectivity is dictated by indirect properties, such as anion hydration energies, adduct solubilities, etc. SAMMS have been made in which the pore surfaces have been lined with cationic transition metal complexes, thereby allowing a cationic "docking station" for the target anion [72–75]. With a metal-tris(ethylenediamine) complex (Cu-EDA), the complex contains a trifurcate cleft on the upper surface, with three exofacial protons, creating a stereospecific binding site that is ideally structured for binding tetrahedral oxometallate anions, like arsenate and chromate [72]. By selecting the metal and oxidation state appropriately, it is possible to install a metal cation such that, while forced into an octahedral coordination geometry by the chelating diamines, the cation is electronically predisposed to undergo Jahn–Teller distortion to alleviate orbital degeneracy. This provides a driving force for departing from an octahedral coordination environment and facilitates a direct bond between the anion and the metal center. Indeed, detailed EXAFS studies have revealed that Cu-EDA SAMMS loses two of the primary amine nitrogen ligands and adopts a trigonal bipyramidal coordination environment when binding oxometallate anions (Fig. 8.7) [73]. Sulfate is a ubiquitous tetrahedral anion, commonly encountered in many groundwaters, and while it is indeed bound by Cu-EDA SAMMS, the binding is reversible and it can be displaced by more problematic anions like arsenate and chromate [72]. Binding kinetics are fast, with equilibrium achieved in minutes.

This sort of anion binding can also be used as a synthetic strategy to install more complex interfacial functionality. For example, Cu-EDA SAMMS treated with ferrocyanide anion affords the corresponding Cu-EDA ferrocyanide adduct (Fig. 8.8), which has excellent affinity for binding cesium [76]. Cesium binding capacity is

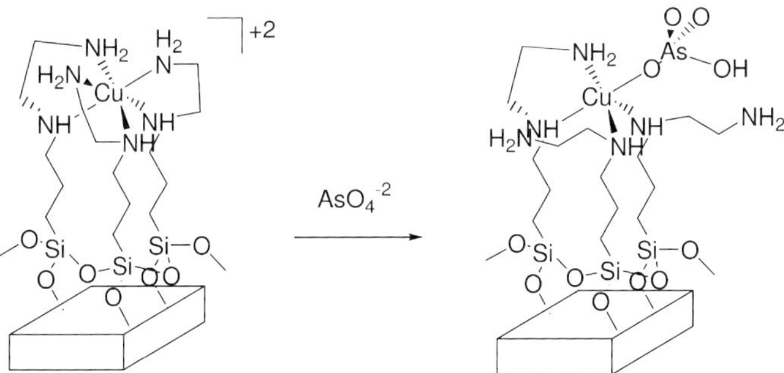

Fig. 8.7. Schematic showing Cu-EDA SAMMS binding an arsenate anion.

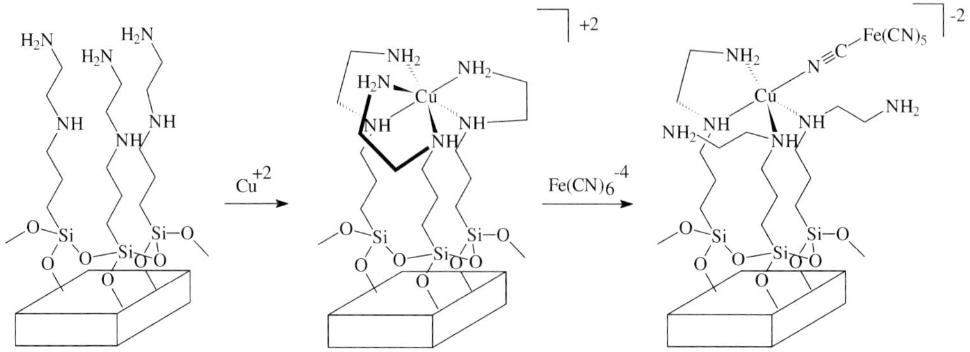

Fig. 8.8. Schematic showing the synthesis of ferrocyanide Cu-EDA SAMMS.

good and binding kinetics are moderately fast, with equilibrium achieved in less than half an hour.

Metallation can also be used as a synthetic strategy to create new classes of SAMMS. For example, thiol-SAMMS used to bind Hg or Ag have an excellent affinity for radioiodide [77].

8.3.2.4 Actinide SAMMS

More sophisticated ligand fields have also been incorporated into mesoporous materials to create highly selective sorbent materials for radioactive actinides [78–81]. In this case, CMPO-inspired [78, 79], or HOPO classes [80, 81] of ligand were installed inside a mesoporous silica matrix and found to have exceptional selectivity and affinity for binding lanthanide and actinide species from complex mixtures. For example, the acetamide-phosphonate (a.k.a. "Ac-Phos") ligand can be employed in either the ester or acid form, making it well-suited for either hydrophobic (e.g., contaminated cutting oils) or hydrophilic waste streams, respectively. The phosphonate-based ligands exhibited very good affinities for all actinides studied (except NpO_2^+), down to pHs as low as 0.5. The actinide-binding affinities of similar carboxylate-based ligand systems drop off at lower pHs [78, 79]. The HOPO class of ligands is the result of 20+ years of ligand design and development by Ken Raymond's group at Berkeley, and these ligands are superb actinide binders [82–84]. Incorporation of the HOPO ligands into the SAMMS superstructure has resulted in a superior class of actinide sorbent material [80, 81]. In this case the 3,2-HOPO ligand was the most effective actinide ligand, capable of even binding the hard-to-grab NpO_2^+ cation very effectively, with K_ds in excess of 150 000 (Fig. 8.9).

8.3.3
Functional CNTs

Self-assembly is not limited to organosilanes on metal oxide surfaces. Chen et al. have reported an interesting variation on this theme [85]. Using this approach, Fif-

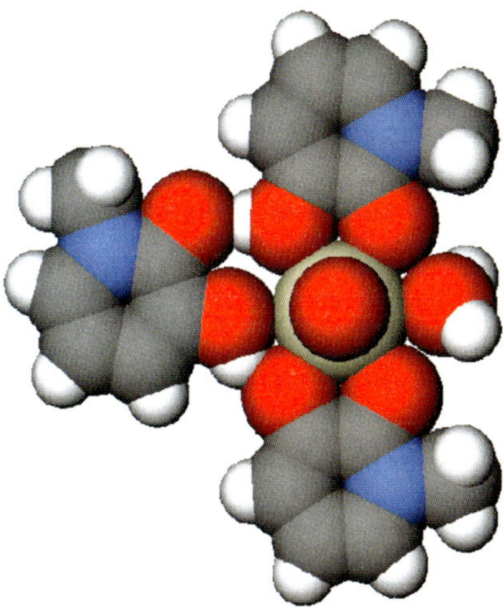

Fig. 8.9. Hexagonal bipyramidal geometry of the complex formed between 3,2-HOPO and the NpO$_2$ cation (looking down the O=Np=O bond). The brown atom in the center is Np, and the red atoms surrounding it are O atoms. The pair of bidentate 3,2-HOPO ligands (top and bottom) are strongly hydrogen bound to the monodentate 3,2-HOPO ligand (left). The remaining equatorial coordination site is filled by a water molecule (on the right).

ield and coworkers have successfully modified carbon nanotubes for selective binding of catalysts, radionuclides and CO$_2$ [86]. Here, supercritical CO$_2$ was an advantageous reaction medium for the self-assembly process, as it can dissolve the modified pyrenes used as anchors, but provides poor solvent shielding, so that the pyrene/CNT π-stacking interaction is allowed to predominate, thereby facilitating self-assembly. These selectively functionalized CNT's are envisioned to be very useful in environmental sensing and detection applications.

8.4
Conclusions

Nanomaterials offer many useful properties for environmental remediation: high surface area, enhanced interfacial reactivity, easy dispersability, and facile sorption kinetics. Nanoparticle-based strategies have been built around alkaline earth oxide materials, zero-valent metals and crosslinked polymers. These different classes of materials offer widely differing chemistries that can be tailored to address differing remediation needs, from DNAPLs to chemical warfare agents to PAHs. Hybrid nanostructured materials are also finding application in environmental chemistry. The elegant work of Clearfield and coworkers has shown that layered metal phosphonates can be made in which the interstitial layers are composed of a wide variety of ligand structures, including crown ethers, amino acids and arylsulfonates. Surfactant templated synthesis of mesoporous ceramic oxides has provided a versatile foundation upon which a wide range of environmental sorbents can be made, tailored to sequester "soft" heavy metals, oxometallate anions, cesium and the actinides. Similar self-assembly routes are currently being explored as ways to chemically modify carbon nanotubes. Functional nanomaterials constitute a versatile, and powerful, toolbox for environmental remediation. The future is indeed bright for environmental nanomaterials.

References

1 Cao, G. *Nanostructures and Nanomaterials: Synthesis Properties and Applications*, Imperial College Press, London, **2004**, pp. 51–104.

2 Cao, G. *Nanostructures and Nanomaterials: Synthesis Properties and Applications*, Imperial College Press, London, **2004**, pp. 26–31.

3 Koper, O. B., Lagadic, I., Volodin, A., Klabunde, K. J., Alkaline-earth oxide nanoparticles obtained by aerogel methods. Characterization and rational for unexpectedly high surface chemical reactivities, *Chem. Mater.* **1997**, 9, 2468–2480.

4 Koper, O. B., Wovchko, E. A., Glass, J. A. Jr., Yates, J. T. Jr., Klabunde, K. J., Decomposition of CCl$_4$ on CaO, *Langmuir* **1995**, 11, 2054–2059.

5 Koper, O. B., Lagadic, I., Klabunde, K. J., Destructive adsorption of chlorinated hydrocarbons on ultrafine (nanoscale) particles of calcium oxide. 2, *Chem. Mater.* **1997**, 9, 838–848.

6 Koper, O. B., Klabunde, K. J., Destructive adsorption of chlorinated hydrocarbons on ultrafine (nanoscale) particles of calcium oxide. 3. Chloroform, trichloroethene, and tetrachloroethene, *Chem. Mater.* **1997**, 9, 2481–2485.

7 Richards, R., Li, W., Decker, S., Davidson, C., Koper, O., Zaikovski, V., Volodin, A., Rieker, T., Klabunde, K. J., Consolidation of metal oxide nanocrystals. Reactive pellets with controllable pore structure that represent a new family of porous inorganic materials, *J. Am. Chem. Soc.* **2000**, 122, 4921–4925.

8 Fenelonov, V. B., Mel'gunov, M. S., Mishakov, I. V., Richards, R. M., Chesnokov, V. V., Volodin, A. M., Klabunde, K. J., Changes in texture and catalytic activity of nanocrystalline MgO during its transformation to MgCl$_2$ in the reaction with 1-chlorobutane, *J. Phys. Chem. B* **2001**, 105, 3937–3941.

9 Kakkar, R., Kapoor, P. N., Klabunde, K. J., Theoretical study of the adsorption of formaldehyde on magnesium oxide nanosurfaces: Size effects and the role of low-coordinated and defect sites, *J. Phys. Chem. B* **2004**, 108, 18 140–18 148.

10 Diao, Y., Walawender, W. P., Sorensen, C. M., Klabunde, K. J., Ricker, T., Hydrolysis of magnesium methoxide. Effects of toluene on gel structure and gel chemistry, *Chem. Mater.* **2002**, 14, 362–368.

11 Ranjit, K. T., Klabunde, K. J., Solvent effects in the hydrolysis of magnesium methoxide, and the production of nanocrystalline magnesium hydroxide. An aid in understanding the formation of porous inorganic materials, *Chem. Mater.* **2005**, 17, 65–73.

12 Stoimenov, P. K., Klinger, R. L., Marchin, G. L., Klabunde, K. J., Metal oxide nanoparticles as bactericidal agents, *Langmuir* **2002**, 18, 6679–6686.

13 Sun, N., Klabunde, K. J., Nanocrystal metal oxide-chlorine adducts: Selective catalysts for chlorination of alkanes, *J. Am. Chem. Soc.* **1999**, 121, 5587–5588.

14 Jeevanandam, P., Klabunde, K. J., Redispersion and reactivity studies on surfactant-coated magnesium oxide nanoparticles, *Langmuir* **2003**, 19, 5491–5495.

15 Mel'gunov, M. S., Mel'gunov, E. A., Zaikovski, V. I., Fenelonov, V. B., Bedilo, A. F., Klabunde, K. J.,

Carbon dispersion and morphology in carbon-coated nanocrystalline MgO, *Langmuir* **2003**, 19, 10 426–10 433.

16 BEDILO, A. F., SIGEL, M. J., KOPER, O. B., MEL'GUNOV, M. S., KLABUNDE, K. J., Synthesis of carbon-coated MgO nanoparticles, *J. Mater. Chem.* **2002**, 12, 3599–3604.

17 CARNES, C. L., KLABUNDE, K. J., Unique chemical reactivities of nanocrystalline metal oxides towards hydrogen sulfide, *Chem. Mater.* **2002**, 14, 1806–1811.

18 CARNES, C. L., KLABUNDE, K. J., Synthesis, isolation, and chemical reactivity studies of nanocrystalline zinc oxide, *Langmuir* **2000**, 16, 3764–3772.

19 NARSKE, R. M., KLABUNDE, K. J., FULTZ, S., Solvent effects on the heterogeneous adsorption and reactions of (2-chloroethyl)ethyl sulfide on nanocrystalline magnesium oxide, *Langmuir* **2002**, 18, 4819–4825.

20 WAGNER, G. W., BARTRAM, P. W., KOPER, O., KLABUNDE, K. J., Reactions of VX, GD, and HD with nanosize MgO, *J. Phys. Chem. B* **1999**, 103, 3225–3228.

21 WAGNER, G. W., KOPER, O., LUCAS, E., DECKER, S., KLABUNDE, K. J., Reactions of VX, GD, and HD with nanosize CaO: Autocatalytic dehydrohalogenation of HD, *J. Phys. Chem. B* **2000**, 104, 5118–5123.

22 WAGNER, G. W., PROCELL, L. R., O'CONNOR, R. J., MUNAVALLI, S., CARNES, C. L., KAPOOR, P. N., KLABUNDE, K. J., Reactions of VX, GD, and HD with nanosize Al_2O_3: Formation of aluminophosphates, *J. Am. Chem. Soc.* **2001**, 123, 1636–1644.

23 CARNES, C. L., STIPP, J., KLABUNDE, K. J., Synthesis, characterization, and adsorption studies of nanocrystalline copper oxide and nickel oxide, *Langmuir* **2002**, 18, 1352–1359.

24 CARNES, C. L., KAPOOR, P. N., KLABUNDE, K. J., Synthesis, characterization, and adsorption studies of nanocrystalline aluminum oxide and a bimetallic nanocrystalline aluminum oxide/magnesium oxide, *Chem. Mater.* **2002**, 14, 2922–2929.

25 DECKER, S. P., KLABUNDE, J. S., KHALEEL, A., KLABUNDE, K. J., Catalyzed destructive adsorption of environmental toxins with nanocrystalline metal oxides. Fluoro-, chloro-, bromocarbons, sulfur, and organophosphorous compounds, *Environ. Sci. Technol.* **2002**, 36, 762–768.

26 DECKER, S., LAGADIC, I., KLABUNDE, K. J., EXAFS observation of the Sr and Fe site structural environment in SrO and Fe_2O_3-coated SrO nanoparticles used as carbon tetrachloride destructive adsorbents, *Chem. Mater.* **1998**, 10, 674–678.

27 BORONINA, T., KLABUNDE, K. J., Destruction of organohalides in water using metal particles: Carbon tetrachloride/water reactions with magnesium, tin, and zinc, *Environ. Sci. Technol.* **1995**, 29, 1511–1517.

28 WANG, C. B., ZHANG, W. X., Synthesizing nanoscale iron particles for rapid and complete dechlorination of TCE and PCBs, *Environ. Sci. Technol.* **1997**, 31, 2154–2156.

29 PARK, K. T., KLIER, K., WANG, C. B., ZHANG, W. X., Interaction of tetrachloroethylene with Pd(100) studied by high-resolution x-ray photoemission spectroscopy, *J. Phys. Chem. B* **1997**, 101, 5420–5428.

30 LIEN, H. L., ZHANG, W. X., Nanoscale iron particles for complete reduction of chlorinated ethenes, *Colloids Surf. A: Physicochem. Eng. Aspects* **2001**, 191, 97–105.

31 XU, Y., ZHANG, W. X., Subcolloidal Fe/Ag particles for reductive dehalogenation of chlorinated benzenes., *Ind. Eng. Chem. Res.* **2000**, 39, 2238–2244.

32 LIEN, H. L., ZHANG, W. X., Enhnaced dehalogenation of halogenated methanes by bimetallic Cu/Al, *Chemosphere* **2002**, 49, 371–378.

33 KANEL, S. R., MANNING, B., CHARLET, L., CHOI, H., Removal of arsenic(III) from groundwater by nanoscale zero-valent iron, *Environ. Sci. Technol.* **2005**, 39, 1291–1298.

34 PONDER, S. M., DARAB, J. G., MALLOUK, T. E., Remediation of Cr(VI) and Pb(II) aqueous solutions using supported, nanoscale zero-valent

iron, *Environ. Sci. Technol.* **2000**, 34, 2564–2569.

35 PONDER, S. M., DARAB, J. G., BUCHER, J., CAULDER, D., CRAIG, I., DAVIS, L., EDELSTEIN, N., LUKENS, W., NITSCHE, H., RAO, L., SHUH, D. K., MALLOUK, T. E., Surface chemistry and electrochemistry of supported zerovalent iron nanoparticles in the remediation of aqueous metal contaminants, *Chem. Mater.* **2001**, 13, 479–486.

36 SCHRICK, B., BLOUGH, J. L., JONES, A. D., MALLOUK, T. E., Hydrodechlorination of trichloro-ethylene to hydrocarbons using bimetallic nickel-iron nanoparticles, *Chem. Mater.* **2002**, 14, 5140–5147.

37 ELLIOTT, D. W., ZHANG, W. X., Field assessment of nanoscale bimetallic particles for groundwater treatment, *Environ. Sci. Technol.* **2001**, 35, 4922–4926.

38 ZHANG, W. X., Nanoscale iron particles for environmental remediation: An overview, *J. Nanoparticle Res.* **2003**, 5, 323–332.

39 QUINN, J., GEIGER, C., CLAUSEN, C., BROOKS, K., COON, C., O'HARA, S., KRUG, T., MAJOR, D., YOON, W. S., GAVASCAR, A., HOLDSWORTH, T., Field demonstration of DNAPL dehalogenation using emulsified zero-valent iron, *Environ. Sci. Technol.* **2005**, 39, 1309–1318.

40 KIM, J. Y., COHEN, C., SCHULER, M. L., LION, L. W., Use of amphiphilic polymer particles for in situ extraction of sorbed phenanthrene from a con-taminated aquifer material, *Environ. Sci. Technol.* **2000**, 34, 4133–4139.

41 TUNGITTIPLAKORN, W., LION, L. W., COHEN, C., KIM, J. Y., Engineered polymeric nanoparticles for soil remediation, *Environ. Sci. Technol.* **2004**, 38, 1605–1610.

42 TUNGITTIPLAKORN, W., COHEN, C., LION, L. W., Engineered polymeric nanoparticles for bioremediation of hydrophobic contaminants, *Environ. Sci. Technol.* **2005**, 39, 1354–1358.

43 CLEARFIELD, A., KRISHNAMOHAN SHARMA, C. V., ZHANG, B. P., Crystal engineered supramolecular metal phosphonates: Crown ethers and iminodiacetates, *Chem. Mater.* **2001**, 13, 3099–3112.

44 CLEARFIELD, A., Organically pillared micro- and mesoporous materials, *Chem. Mater.* **1998**, 10, 2801–2810.

45 CLEARFIELD, A., Inorganic ion exchangers: A technology ripe for development, *Ind. Eng. Chem. Res.* **1995**, 34, 2865–2872.

46 POOJARY, D. M., ZHANG, Y., ZHANG, B., CLEARFIELD, A., Synthesis, x-ray powder structure, and intercalation behavior of molybdenyl phenyl-phosphonate, $MoO_2(O_3PC_6H_5)$ H_2O, *Chem. Mater.* **1995**, 7, 822–827.

47 POOJARY, D. M., CLEARFIELD, A., Coordinative intercalation of alkylamines into layered zinc phenylphosphonate. Crystal structures from x-ray powder diffraction data, *J. Am. Chem. Soc.* **1995**, 117, 11 278–11 284.

48 POOJARY, D. M., ZHANG, B., CLEARFIELD, A., Pillared layered metal phosphonates. Synthesis and x-ray powder structures of copper and zinc alklyenebis(phosphonates), *J. Am. Chem. Soc.* **1997**, 119, 12 550–12 559.

49 ZHANG, B., POOJARY, D. M., CLEARFIELD, A., Synthesis and characterization of layered zinc biphenylenebis(phosphonate) and three mixed-component arylenebisphosphonate/phosphates, *Inorg. Chem.* **1998**, 37, 1844–1852.

50 WANG, Z., HEISING, J. M., CLEARFIELD, A., Sulfonated microporous organic-inorganic hybrids as strong Bronsted acids, *J. Am. Chem. Soc.* **2003**, 125, 10 375–10 383.

51 ZHANG, B., POOJARY, D. M., CLEARFIELD, A., Synthesis, charac-terization, and amine intercalation behavior of zirconium N-(phosphono-methyl)iminodiacetic acid layered compounds, *Chem. Mater.* **1996**, 8, 1333–1340.

52 ZHANG, B., POOJARY, D. M., CLEARFIELD, A., Synthesis and crystal structure of the linear chain zirconium organophosphonate $(NH_4)Zr[F_2][H_3\{O_3PCH_2 NH(CH_2CO_2)_2\}_2]\cdot3H_2O\cdot NH_4Cl$, *Inorg. Chem.* **1998**, 37, 249–254.

53 KRISHNAMOHAN SHARMA, C. V.,

CLEARFIELD, A., Three-dimensional hexagonal structures from a novel self-complementary molecular building block, *J. Am. Chem. Soc.* **2000**, 122, 4394–4402.

54 CABEZA, A., OUYANG, X., KRISHNAMOHAN SHARMA, C. V., ARANDA, M. A. G., BRUQUE, S., CLEARFIELD, A., Complexes formed between nitrilotris(methylenephosphonic acid) and M^{2+} transition metals: Isostructural organic-inorganic hybrids, *Inorg. Chem.* **2002**, 41, 2325–2333.

55 ZHANG, B., CLEARFIELD, A., Crown ether pillared and functionalized layered zirconium phosphonates: A selective new strategy to synthesize novel ion selective materials, *J. Am. Chem. Soc.* **1997**, 119, 2751–2752.

56 CLEARFIELD, A., POOJARY, D. M., ZHANG, B., ZHAO, B., DERECSKEI-KOVACS, A., Azacrown ether pillared layered zirconium phosphonates and the crystal structure of N,N′-bis(phosphonomethyl)-1,10-diaza-18-crown-6, *Chem. Mater.* **2000**, 12, 2745–2752.

57 MAO, J. G., WANG, Z., CLEARFIELD, A., Synthesis, characterization, and crystal structures of two new divalent metal complexes of N,N′-bis(phosphonomethyl)-1,10-diaza-18-crown-6: A hydrogen bonded 1D array and a 3D network with a large channel, *Inorg. Chem.* **2002**, 41, 3713–3720.

58 FRYXELL, G. E., ADDLEMAN, R. S., MATTIGOD, S. V., LIN, Y., ZEMANIAN, T. S., WU, H., BIRNBAUM, J. C., LIU, J., FENG, X., Environmental and sensing applications of molecular self-assembly, in *Encyclopedia of Nanoscience and Nanotechnology*, Marcel-Dekker, **2004**, pp. 1135–1145.

59 LIN, Y., YANTASEE, W., FRYXELL, G. E., Electrochemical sensors based on functionalized nanoporous silica, in *Encyclopedia of Nanoscience and Nanotechnology*, Marcel-Dekker, **2004**, pp. 1051–1061.

60 FENG, X., FRYXELL, G. E., WANG, L. Q., KIM, A. Y., LIU, J., Organic monolayers on ordered mesoporous supports, *Science* **1997**, 276, 923–926.

61 LIU, J., FENG, X., FRYXELL, G. E., WANG, L. Q., KIM, A. Y., GONG, M., Hybrid mesoporous materials with functionalized monolayers, *Adv. Mater.* **1998**, 10, 161–165.

62 MERCIER, L., PINNAVAIA, T. J., Heavy metal ion adsorbents formed by the grafting of a thiol functionality to mesoporous silica molecular sieves: Factors affecting Hg(II) uptake, *Chem. Mater.* **1997**, 9, 2491–2498.

63 MATTIGOD, S. V., FRYXELL, G. E., FENG, X., LIU, J., Self-assembled monolayers on mesoporous supports for metal separation, in *Metal Separation Technologies Beyond 2000: Integrating Novel Chemistry with Processing*, ed. K. C. LIDDELL, D. J. CHAIKO, The Minerals, Metals and Materials Society, Warrendale, PA, USA, **1999**, pp. 71–79.

64 KEMNER, K. M., FENG, X., LIU, J., FRYXELL, G. E., WANG, L.-Q., KIM, A. Y., GONG, M., MATTIGOD, S. V., Investigation of the local chemical interactions between Hg and self assembled monolayers on mesoporous supports, *J. Synchotron Rad.* **1999**, 6, 633–635.

65 MATTIGOD, S. V., FENG, X., FRYXELL, G. E., LIU, J., GONG, M., Separation of complexed mercury from aqueous wastes using self-assembled mercaptan on mesoporous silica, *Sep. Sci. Technol.* **1999**, 34, 2329–2345.

66 KIM, Y., LEE, B., YI, J., Effect of framework and textural porosities of functionalized mesoporous silica on metal ion adsorption capacities, *Sep. Sci. Technol.* **2004**, 39, 1427–1442.

67 WALCARIUS, A., ETIENNE, M., LEBEAU, B., Rate of access to the binding sites in organically modified silicates. 2. Ordered mesoporous silicas grafted with amine or thiol groups, *Chem. Mater.* **2003**, 15, 2161–2173.

68 KANG, T., PARK, Y., YI, J., Highly selective adsorption of Pt^{2+} and Pd^{2+} using thiol-functionalized mesoporous silica, *Ind. Eng. Chem. Res.* **2004**, 43, 1478–1484.

69 ANTOCHSHUK, V., OLKHOVYK, O., JARONIEC, M., PARK, I.-S., RYOO, R., Benzoylthiourea-modified mesoporous silica for mercury(II) removal, *Langmuir* **2003**, 19, 3031–3034.

70 *Universal Treatment Standards 40CFR §
268.38*, United States Environmental
Protection Agency, Washington, D.C.,
USA.

71 KLASSON, K. T., *Treatment of Mercury
Contaminated Oil from the Mound site.
Topical Report.* Oak Ridge National
Laboratory, Oak Ridge, TN 37831, **2000**.

72 FRYXELL, G. E., LIU, J., GONG, M.,
HAUSER, T. A., NIE, Z., HALLEN, R. T.,
QIAN, M., FERRIS, K. F., Design and
synthesis of selective mesoporous
anion traps, *Chem. Mater.* **1999**, 11,
2148–2154.

73 KELLY, S., KEMNER, K. M., FRYXELL,
G. E., LIU, J., MATTIGOD, S. V.,
FERRIS, K. F., An X-ray absorption fine
structure spectroscopy study of the
interactions between contaminant
tetrahedral anions to self-assembled
monolayers on mesoporous supports,
J. Phys. Chem. B **2001**, 105, 6337–6346.

74 YOSHITAKE, H., YOKOI, T., TATSUMI,
T., Adsorption of chromate and
arsenate by amino-functionalized
MCM-41 and SBA-1, *Chem. Mater.*
2002, 14, 4603–4610.

75 YOSHITAKE, H., YOKOI, T., TATSUMI,
T., Adsorption behavior of arsenate at
transition metal cations captured by
amino-functionalized mesoporous
silicas, *Chem. Mater.* **2003**, 15, 1713–
1721.

76 LIN, Y., FRYXELL, G. E., WU, H.,
ENGLEHARD, M., Selective sorption of
cesium using self-assembled
monolayers on mesoporous supports
(SAMMS), *Environ. Sci. Technol.* **2001**,
35, 3962–3966.

77 MATTIGOD, S. V., FRYXELL, G. E.,
SERNE, R. J., PARKER, K. E., MANN,
F. M., Evaluation of novel getters for
adsorption of radioiodine from
groundwater and waste glass leachates,
Radiochim. Acta **2003**, 91, 539–545.

78 FRYXELL, G. E., WU, H., LIN, Y.,
SHAW, W. J., BIRNBAUM, J. C.,
LINEHAN, J. C., NIE, Z., KEMNER,
K. M., KELLY, S., Lanthanide selective
sorbents: Self-assembled monolayers
on mesoporous supports (SAMMS),
J. Mater. Chem. **2004**, 14, 3356–3363.

79 FRYXELL, G. E., LIN, Y., FISKUM, S.,
BIRNBAUM, J. C., WU, H., KEMNER,
K. M., KELLY, S., Actinide

sequestration using self-assembled
monolayers on mesoporous supports
(SAMMS), *Environ. Sci. Technol.* **2005**,
39, 1324–1331.

80 LIN, Y., FISKUM, S. K., YANTASEE, W.,
WU, H., MATTIGOD, S. V., FRYXELL,
G. E., RAYMOND, K. N., XU, J.,
Incorporation of hydroxypyridinone
(HOPO) ligands into self-assembled
monolayers on mesoporous supports
(SAMMS) for selective actinide
sequestration., *Environ. Sci. Technol.*
2005, 39, 1332–1337.

81 YANTASEE, W., FRYXELL, G. E., LIN, Y.,
WU, H., RAYMOND, K. N., XU, J.,
Hydroxypyridinone (HOPO)
functionalized self-assembled
monolayers on mesoporous supports
(SAMMS) for sequestering rare earth
cations, *J. Nanosci. Nanotechnol.* **2005**,
5(9), 1537–1540.

82 GORDEN, A. E. V., XU, J., RAYMOND,
K. N., DURBIN, P. W., Rational design
of sequestering agents for plutonium
and other actinides, *Chem. Rev.* **2003**,
103, 4207–4282.

83 XU, J., WHISENHUNT, D. W. JR.,
VEECK, A. C., UHLIR, L. C., RAYMOND,
K, N., Thorium(IV) complexes of
bidentate hydroxypyridinonates, *Inorg.
Chem.* **2003**, 42, 2665–2674.

84 XU, J., DURBIN, P. W., KULLGREN, B.,
EBBE, S. N., UHLIR L. C., RAYMOND,
K. N., Synthesis and initial evaluation
for in vivo chelation of Pu(IV) of a
mixed octadentate spermine-based
ligand containing 4-carbamoyl-3-
hydroxy-1-methyl-2(1H)-pyridinone
and 6-carbamoyl-1-hydroxy-2(1H)-
pyridinone, *J. Med. Chem.* **2002**, 45,
3963–3971.

85 CHEN, R. J., ZHANG, Y., WANG, D.,
DAI, H., Noncovalent sidewall
functionalization of single-walled
carbon nanotubes for protein
immobilization, *J. Am. Chem. Soc.*
2001, 123, 3838.

86 FIFIELD, L. S., DALTON, L. R.,
ADDLEMAN, R. S., GALHOTRA, R. A.,
ENGELHARD, M. H., FRYXELL, G. E.,
AARDAHL, C. L., Noncovalent
functionalization of carbon nanotubes
with molecular anchors using
supercritical fluids, *J. Phys. Chem. B*
2004, 108, 8737–8741.

9
Nanomaterials for Water Treatment

Peter Majewski

9.1
Introduction

Water is one of the most valuable substances in the world and its availability in the form of potable and drinking water is of great importance for any society. However, the conversion of water from various sources into potable and good quality drinking water can be very demanding.

The recent report entitled "Water for People – Water for Life" of the World Water Assessment Program of UNESCO [1] emphasizes that the availability of potable and drinking water is of major socio-economic importance worldwide. It also clearly states the importance of foreseeing water economy in terms of use, availability, quality, and technologies. The report also emphasizes that organic pollutants in water are the main reason for water-related diseases, such as diarrhea, worm infections, and infectious diseases, and more than 6000 people die every day due to water-related diseases.

Water use is increasing everywhere. The world's six billion inhabitants are already appropriating 54% of all the accessible freshwater contained in rivers, lakes and underground aquifers. By 2025 humankind's share will be 70%. This estimate reflects the impact of population growth alone. If per capita consumption of water resources continues to rise at its current rate, humankind could be using over 90% of all available freshwater within 25 years, leaving just 10% for all other living beings. Currently, on a global basis, 69% of all water withdrawn for human use on an annual basis is soaked up by agriculture (mostly as irrigation); industry accounts for 23% and domestic use (household, drinking water, sanitation) accounts for about 8%. These global averages vary a great deal between regions. In Africa, for instance, agriculture consumes 88% of all water withdrawn for human use, while domestic use accounts for 7% and industry for 5%. In Europe, most water is used in industry (54%), while agriculture and domestic use take 33% and 13% respectively [1].

Shrinking fresh water resources, increasing salinity of bore water, especially in arid areas of the world, and increasing demand for potable, drinking, and irriga-

Nanotechnologies for the Life Sciences Vol. 5
Nanomaterials – Toxicity, Health and Environmental Issues. Edited by Challa S. S. R. Kumar
Copyright © 2006 WILEY-VCH Verlag GmbH & Co. KGaA, Weinheim
ISBN: 3-527-31385-0

tion water requires the utilization of sea, brackish, and saline bore water for fresh water supply. The conventional industrial way to desalinate water is reverse osmosis (RO), which is widely used, especially in arid costal areas, such as Arabia, the USA, especially Florida, Australia, and on ships. Although RO has proven to be a robust desalination method, its major drawback is its high demand in electric energy and the related high costs as well as biofouling of the membranes [2]. This fact requires either a relevant infrastructure of electric energy or additional equipment for local supply of electric energy, both of which are not always provided, especially in re-mote and/or underdeveloped areas as well as disaster and combat areas. Therefore, a simple "coffee filter"-like desalination method without the need of electric energy is highly desired and would significantly facilitate the utilization of sea, brackish, and saline bore water for the supply of potable, drinking, and irrigation water. An-other important aspect, especially for underdeveloped and disaster areas, of such a "low-tech" method would be that the actual desalination treatment could be per-formed by a non-trained person.

Natural organic matter (NOM) is one of the key water quality parameters that affects treatment processes. NOM reduces the effectiveness of water treatment by interfering with the flocculation process, makes treatment with activated carbon and membrane filtration less efficient and is a precursor to the formation of disin-fectant by-products (DBP). Furthermore, NOM acts as a food source for microor-ganisms, resulting in bacterial regrowth in distribution systems. These concerns have resulted in removal of NOM from raw water being of prime concern for water authorities. This has been acknowledged by the United States in the Disinfectant and Disinfection By-product Rule. This rule requires removal of NOM, as mea-sured by total organic carbon (TOC), to minimize the formation of DBP. Enhanced coagulation has been designated as the best available technology for TOC reduc-tion, with the required removal determined by influent TOC concentration and al-kalinity. TOC of less than 2 mg L^{-1} prior to disinfection requires no treatment while TOC levels above this require removal of between 20 to 50%, with the higher removal required at higher TOC and lower alkalinity.

NOM is also well-known in drinking water. It is a complex matrix of heteroge-neous organic material, consisting of particulate and dissolved fractions, harmless and harmful pollutants like pyrogens, such as endotoxins, proteins, and bacteria, as well as colored and non-colored components [3, 4]. Therefore, the ability to re-move NOM is a key factor in determining the efficiency of novel methods for water treatment.

DBPs consist of various organic compounds, the best known of which are triha-lomethanes (THMs). The presence of chloroform and other THMs in finished drinking water was first associated with the chlorination of drinking water in 1974. It was discovered that, in addition to killing microorganisms, disinfectants react with organic and inorganic substances naturally present in the water to pro-duce various DBPs, which include THMs. DBPs associated with chlorination are THMs, haloacetic acids, haloacetonitriles and halopicrins. Chlorite and chlorate are by-products of chlorine dioxide disinfection, while ozonation may result in bromate formation. Nitrosodimethylamine (NDMA) is a by-product of chlorination and chloramination.

Certain DBPs have been shown to be detrimental to health in laboratory animal studies. As a result the Environmental Protection Agency (EPA) regulated the most prevalent DBPs, the THMs, in 1979, setting the limit at 100 mg L^{-1}. With the promulgation of the Stage 1 Disinfectants and Disinfection By-product Rule in December 1998, the trihalomethane limit has since been reduced to 80 µg L^{-1} and limits of 60 µg L^{-1} for haloacetic acid, 1.0 mg L^{-1} for chlorite and 10 µg L^{-1} for bromate have been introduced. While haloacetonitriles, halopicrins, chlorate and NDMA have been identified as health hazards, further research needs to be conducted before identifying appropriate regulatory levels. It is important to achieve a balance between reducing exposure to DBPs and maintaining control of water-borne diseases through regulatory efforts (see also: U.S. Environmental Protection Agency, http://yosemite.epa.gov/water/).

Conventional treatment process, i.e., coagulation/flocculation – sedimentation – filtration, is one of the most widely used treatment methods in drinking water treatment. However, especially organic matter with high molecular weight ($\gg 10$ kDa), such as pyrogens, are often not or not completely removed by coagulation and flocculation, with major consequences for peoples' health. Consequently, water has to be disinfected, mainly by chlorination and chloramination. However, as mentioned above, DBPs are a health risk and some are known carcinogens. Therefore, a reduction of disinfection of water is highly desired.

Even artificial organic matter such as pesticides, herbicides, and fungicides, drugs like antibiotics, detergents like alkyl aryl sulfonates and laureth sulfonates, organic solvents, such as trichloroethylene (TCE), and also organics from the food processing industry are often not, or not completely, removable from potable water because of their chemical stability against disinfectants. In addition, organic residues are the most frequent pollutants in wastewater of major industry, such as pulp and paper, textiles and leather, iron and steel, and petrochemicals and refineries.

Novel nanomaterials for water treatment and remediation, which in some cases are close to industrial applications, could provide some new innovative concepts to match the above outlined demands. In some cases they also represent more efficient alternatives to conventional treatment technologies, such as coagulation and flocculation, and provide novel concepts for desalination. However, in most cases there is still a long way to go before large-scale applications, i.e., the treatment of giga liters per year by nanomaterials, are commercially and technically feasible. However, the first important steps have already been carried out.

The present chapter provides a comprehensive overview of the most promising approaches to applying nanomaterials for water treatment and remediation. The nanomaterials will be introduced and their synthesis, physical properties, and operating issues reviewed. In addition, as far as the existing literature provides sufficient data about operational issues of the material, their possible application will be assessed.

Tab. 9.1. Common environmental contaminants that can be
transformed by nanoscale iron particles [5].

Chlorinated methanes	Trihalomethanes
Carbon tetrachloride (CCl_4)	Bromoform ($CHBr_3$)
Chloroform ($CHCl_3$)	Dibromochloromethane ($CHBr_2Cl$)
Dichloromethane (CH_2C_2)	Dichlorobromomethane ($CHBrCl_2$)
Chloromethane (CH_3Cl)	*Chlorinated ethenes*
Chlorinated benzenes	Tetrachloroethene (C_2Cl_4)
Hexachlorobenzene (C_6Cl_6)	Trichloroethene (C_2HCl_3)
Pentachlorobenzene (C_6HCl_5)	*cis*-Dichloroethene ($C_2H_2Cl_2$)
Tetrachlorobenzenes ($C_6H_2Cl_4$)	*trans*-Dichloroethene ($C_2H_2Cl_2$)
Trichlorobenzenes ($C_6H_3Cl_3$)	1,1-Dichloroethene ($C_2H_2Cl_2$)
Dichlorobenzenes (C_6HCl_2)	Vinyl chloride (C_2H_3Cl)
Chlorobenzene (C_6H_5Cl)	*Other polychlorinated hydrocarbons*
Pesticides	PCBs
DDT ($C_{14}H_9Cl_5$)	Dioxins
Lindane ($C_6H_6Cl_6$)	Pentachlorophenol (C_6HCl_5O)
Organic dyes	*Other organic contaminants*
Orange II ($C_{16}H_{11}N_2NaO_4S$)	*N*-Nitrosodimethylamine ($C_4H_{10}N_2O$)
Chrysoidine ($C_{12}H_{13}N_4Cl$)	TNT ($C_7H_5N_3O_6$)
Tropaeolin O ($C_{12}H_9N_2NaO_5S$)	*Inorganic anions*
Acid Orange	Dichromate ($Cr_2O_7{}^{2-}$)
Acid Red	Arsenic ($AsO_4{}^{3-}$)
Heavy metal ions	Perchlorate ($ClO_4{}^-$)
Mercury (Hg^{2+})	Nitrate ($NO_3{}^-$)
Nickel (Ni^{2+})	
Silver (Ag^+)	
Cadmium (Cd^{2+})	

Table 9.1 lists several pollutants in water that have been successfully degraded by
iron nanoparticles [5]. Figure 9.1 shows the effect of the addition of iron nanoparticles on water contaminated with TCE. After only a few days the concentration of
TCE has been reduced significantly at the well, but also at some distance from the
well.

Equations (1) and (2) clearly show that adding iron nanoparticles to water should
produce a characteristic increase in pH and decline in the redox potential (E_H) of
the solution. A highly reducing environment ($E_H \ll 0$) is created due to the rapid
consumption of oxygen and other potential oxidants as well as production of hydrogen. Figure 9.2 shows that the oxidation–reduction potential (ORP) decreases
after the addition of the nanoscale iron particles into a solution. Typically, in a
closed batch reactor, a pH increase of 2–3 units was reported while ORP reduction
was in the range 500–900 mV [5, 31].

In poorly buffered water, the increase in pH can be significant. However, most

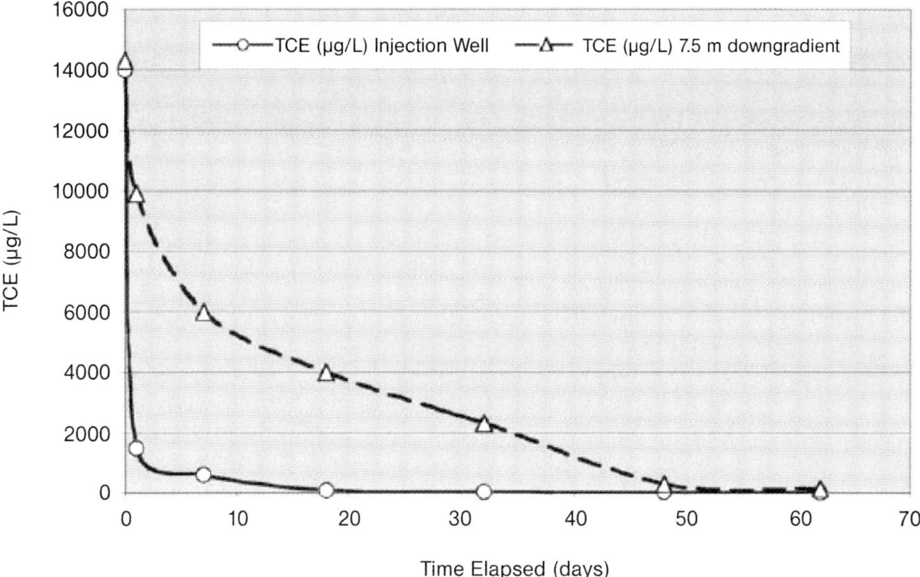

Fig. 9.1. Effect of the addition of iron nanoparticles on water contaminated with TCE. After only a few days the TCE concentration has been reduced significantly at the well, and also at some distance from the well [5].

often, natural water reservoirs contain some buffering capacity and, therefore, the anticipated increase in pH would be less pronounced.

Nanoparticles of iron and FeO surely provide an innovative, simple and cheap alternative to conventional methods. However, crucially, wide range applications, especially for drinking and potable water, require removal of the particles or the resulting iron cations from the water, which will be one of the major challenges facing a future application of the method.

9.3
Inorganic Photocatalysts

Increased public concern over environmental pollutants as well as waterborne pyrogens has prompted the need to develop novel treatment methods, with photocatalysis attracting a lot of attention concerning the degradation of pollutants [37, 38]. Much of the natural purification of aqueous system lagoons, ponds, streams, rivers and lakes is caused by sunlight initiating the breakdown of organic molecules into simpler molecules and ultimately into carbon dioxide and other mineral products [39].

Photocatalytic detoxification of wastewater combines heterogeneous catalysis with solar technologies [40]. Semiconductor photocatalysis, with a primary focus on TiO_2, has been applied to various problems of environmental interest in addition

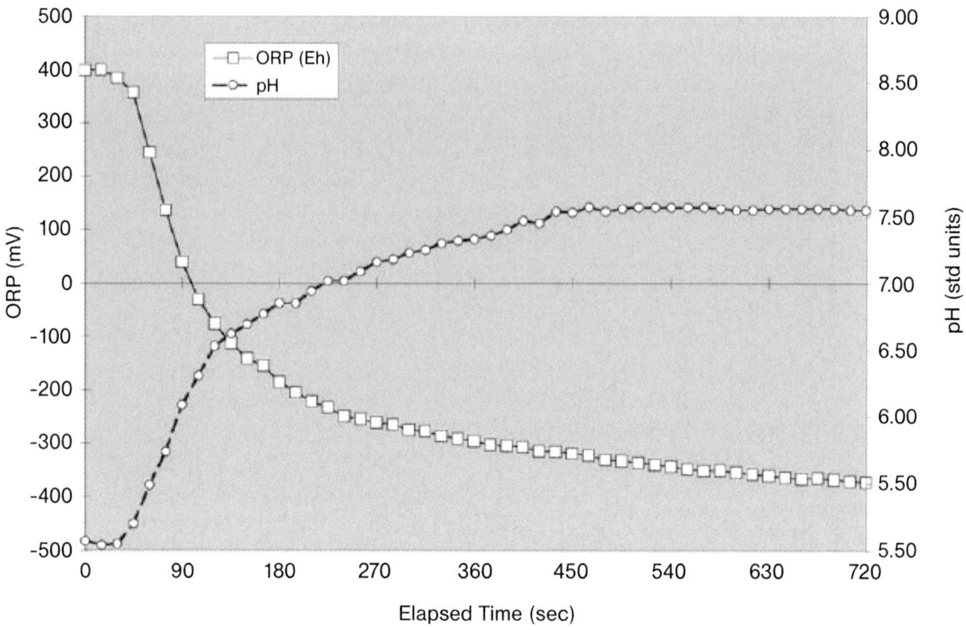

Fig. 9.2. Water chemistry of nanoscale iron particles. Rapid pH and oxidation–reduction potential (ORP) changes were observed after the addition of 0.11 g L^{-1} of Pd/Fe nanoparticles in water. A very low ORP can be established in 200 s [5].

to water and air purification. The application of illuminated semiconductors for degrading undesirable organics dissolved in air or water is well documented and has been successful for a wide variety of compounds [41].

Organic compounds such as alcohols, carboxylic acids, amines, herbicides and aldehydes have been degraded via photocatalysis in both laboratory and field studies. The photocatalytic process can mineralize the hazardous organic molecules into carbon dioxide, water and simple mineral acids [42]. Thus, a major advantage of the photocatalytic process over existing technologies is that it provides an almost complete degradation of the organic toxins, whereas other methods, such as conventional coagulation and flocculation, most often only concentrates the toxins, which then have to be separated from the water by filtering or sedimentation of the flocculant. Compared to other advanced oxidation technologies, especially those using oxidants such as hydrogen peroxide and ozone, additional oxidizing chemicals are not required when applying photocatalysis, because ambient oxygen is the oxidant [38].

The photocatalytic process can also be applied to destroy nuisance odors, taste and odor compounds, and naturally occurring organic matter, which contains the precursors to trihalomethanes formed during the chlorine disinfection step in drinking water treatment [43].

Tab. 9.2. Band gap energy of various photocatalysts [44].

Photocatalyst	Band gap energy (eV)	Photocatalyst	Band gap energy (eV)
Si	1.1	ZnO	3.2
TiO_2 (rutile)	3.0	TiO_2 (anatase)	3.2
WO_3	2.7	CdS	2.4
ZnS	3.7	$SrTiO_3$	3.4
SnO_2	35	WSe_2	1.2
Fe_2O_3	2.2	$X\text{-}Fe_2O_3$	3.1

During photocatalysis, illumination of a semiconductor photocatalyst with ultraviolet (UV) radiation activates the catalyst, establishing a redox environment in the aqueous solution [40]. The physical and chemical fundamentals underlying this phenomenon are comprehensively described by Bhatkhande et al. [38] and Beydoun et al. [44].

An ideal photocatalyst should be stable, inexpensive, non-toxic and, of course, highly photoactive. Another primary criteria for the degradation of organic compounds is that the redox potential of the $H_2O/{}^{\cdot}OH$ couple ($OH^- + {}^{\cdot}OH + e^-$; $E^0 = -2.8$ V) lies within the bandgap of the semiconductor [41]. Several semiconductors have bandgap energies sufficient for catalyzing a wide range of chemical reactions. These include TiO_2, WO_3, $SrTiO_2$, α-Fe_2O_3, ZnO and ZnS (Table 9.2). Of these materials, TiO_2 seems to be the most promising for photocatalytic destruction of organic pollutants [45–63]. This semiconductor provides the best compromise between catalytic performance and stability in aqueous media [64]. The anatase phase of titanium dioxide is the material with the highest photocatalytic detoxification properties [65]. Binary metal sulfide semiconductors such as CdS, CdSe or PbS are insufficiently stable for catalysis, at least in aqueous media, because they undergo photoanodic corrosion [45, 66], and, more important with respect to drinking water treatment, these materials are also toxic. The iron oxides are not suitable semiconductors as they are quickly subjected to photocathodic corrosion to FeOOH and other iron hydroxide species. The band gap of ZnO (3.2 eV) is equal to that of anatase. However, ZnO is also not stable in water and forms $Zn(OH)_2$ on the surface of the ZnO particles, resulting in deactivation of the catalyst [45].

Nanocrystalline photocatalysts are of increasing interest due to their unique photophysical and photocatalytic properties [44, 65, 67]. Several review articles deal with the photophysical properties of nanocrystalline semiconductors and various studies have demonstrated that some properties of nanocrystalline semiconductor particles differ from those of bulk materials [37, 65, 68–71]; for a comprehensive review see Ref. [44].

Nanosized particles, with diameters ranging between 1 and 10 nm, possess properties that represent a transition between the molecular and bulk phases [65]. In bulk material, the electron excited by light absorption finds a high density of states in the conduction band, where it can exist with different kinetic energies [67]. However, the size of a nanoparticle can be the same as or even smaller than the size of the first excited state. Thus, the electron and hole generated by illumination cannot fit into such a particle unless a state of higher kinetic energy is created [72].

Hence, as the size of the semiconductor particle is reduced below a critical diameter, the spatial confinement of the charge carriers within a potential well causes them to behave quantum mechanically [41, 73, 74]. In solid-state terminology this means that the bands split into discrete electronic states representing quantized levels in the valence and conduction bands [72] and the nanoparticle behaves like a giant atom [67]. Nanosized semiconductor particles that exhibit size-dependent optical and electronic properties are called quantized particles or quantum dots [75]. Quantum-size effects occur when the semiconductor particles become smaller than the Bohr radius of the first excitation state [74]. This has also been stated by other authors when the particle size of a colloidal particle becomes comparable to the DeBroglie wavelength of the charge carriers [72, 76].

The excitation radius of TiO_2 nanoparticles has been reported to be between 0.3 and 2 nm [73, 77, 78]. Quantization effects of TiO_2 nanoparticles have been observed at particles sizes at and below about 10 nm [45, 72, 77–81].

One important advantage of the application of quantum-sized particles is the increase in bandgap energy with decreasing particle size. At the critical radius and below, the charge carriers begin to behave quantum mechanically and the confinement of the charge results in a series of discrete electronic states, prompting an increase in the effective band gap and a shift of the band edges. Thus, by varying the size of the semiconductor particles, it is possible to enhance the redox potential of the valence-band holes and the conduction band electrons [82], which results in an increase in the rate constants for charge transfer at the surface [41, 81]. Consequently, the use of quantum-sized semiconductor particles may result in increased photoactivity for systems in which the rate-limiting step is interfacial charge transfer [81]. Hence, semiconductor nanoparticles of sufficient size can possess enhanced photoredox chemistry that might not otherwise exist in bulk materials [73, 79, 81, 83–87].

Another factor that could be advantageous is the fact that a very large fraction of atoms are located at the surface of a nanoparticle [67]. Quantum-sized particles also have high surface area to volume ratios, which further enhances their catalytic activity [82]. One disadvantage of nanosized particles is the need for light with a shorter wavelength for photocatalyst activation – a smaller percentage of a polychromatic light source will be useful for photocatalysis.

The maximum efficiency of photocatalysis of TiO_2 has been observed at a particles size of about 10 nm. However, even a slight decrease or increase of particle size results in a decrease in efficiency [77, 79, 86, 88, 89].

Nanoparticles of titania can be synthesized by flame synthesis [90], chemical vapor deposition [91], wet chemical methods, such as chloride method using $TiCl_4$

solutions [92], hydrothermal processing [93], and the alkoxide method [94, 95]. The alkoxide method, better known as the sol–gel method, which is widely used in the synthesis of ceramic materials, has the advantage that it is capable of producing photocatalysts of better homogeneity as well as high purity, involves lower processing temperatures, and can be tailored to control the materials properties [96].

Although photocatalysis exhibits various significant advantages over conventional methods such as flocculation and coagulation, its major disadvantage is that it needs illumination, preferably by UV light. However, with respect to operational aspects of the treatment of drinking, potable, waste, and process water, solar illumination is not always possible or, if provided by UV lamps, causes considerable costs. Thus, even if photocatalysts represent a very efficient, reliable, and innovative alternative to conventional treatment or other oxidizing treatments, such as ozone or H_2O_2 treatment, operational issues and often the costs will be the main criteria when considering their application.

9.4
Functionalized Self-assembled Monolayers

Another nanomaterial for water treatment consists of a particular silica (quartz sand) or silicate material with a nanometer-sized coating, a so-called functionalized self-assembled monolayer (SAM). SAMs for water treatment have been deposited onto solid μm-sized silicate particles as well as on so-called mesoporous silicates. Mesoporous silicates are silicate particles that contain a nanometer-sized open porosity. This porosity provides an enormous increase of the effective surface area available for interactions between the SAM and pollutants. A few grams of mesoporous silicate provide a surface area the size of three football fields [97].

Although applications of silane-based self-assembled monolayers have been elaborated for surface modification in terms of controlling its hydrophobicity, introducing and controlling the surface chemistry, and the synthesis of crystalline oxide thin films via self-assembled monolayers, the removal of pollutants from water via SAMs has not been considered widely.

A SAM is a close packed, highly ordered array of chained hydrocarbon molecules containing 3–17 CH_2 units (Fig. 9.3). Especially, long-chain molecules (17 CH_2-units, e.g., octadecyltrichlorosilane) form a highly ordered array of surfactant molecules due to stabilizing van der Waals forces between the molecules, whereas short-chain molecules (three CH_2-units, e.g., aminopropyltrimethoxysilane) are less perfectly arrayed. Hence, the length of the hydrocarbon molecules and the related thickness of the SAM varies between about 0.6 nm (three CH_2-units) and about 2.5 nm (17 CH_2-units) [98].

The SAM is simply described as a hydrocarbon with the general formula X-$(CH_2)_n$-Y. Y represents the "bonding group", such as trichlorosilyl ($-SiCl_3$), trimethoxysilane [$Si(OCH_3)_3$] etc., forming tightly covalent Si–O–Si bonds to the surface atoms of silicon and silica [99, 100]. X denotes the "head group", chosen from

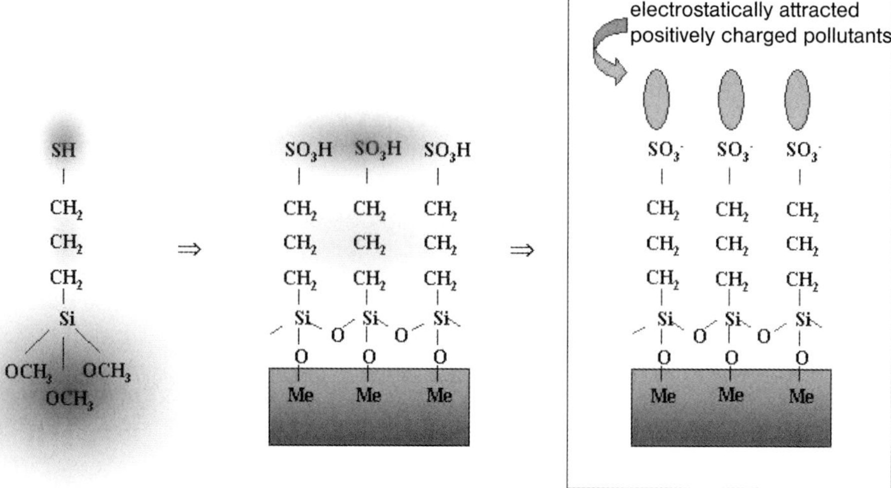

Fig. 9.3. The surfactant 3-mercaptopropyltrimethoxysilane
(left) and the resulting sulfonate SAM on a substrate (middle)
and immersed into contaminated water (right).

among several possible species, such as NH_2, $NHNH_2$, NH_3Cl, SH, SO_3H, COOH, PO_4H_2 etc. When immersed into aqueous solutions, the head groups deprotonate and form negatively surfaces or capture protons at low pH and form positively charged surfaces. Therefore, by carefully choosing the self-assembled monolayer and pH of the aqueous solution in which the SAM is immersed, negatively as well as positively charged surfaces can be obtained [101, 102]. Removal of charged pollutants is due either to electrostatic attraction between the SAM and the counter-charged pollutant or to chemisorption of the pollutants by the functional group of the SAM.

Surfaces are coated with SAMs by simply immersing the substrate material into an organic solution, such as toluene, that contains surfactant molecules, such as 3-mercaptopropyltrimethoxysilane [101–105]. After thoroughly washing with ethanol and water, the SAMs-coated powder is ready for application.

SAMs for water treatment are mainly employed on silica and silicate particles, which are dispersed in the water during treatment, and subsequently filtered. Mesoporous silicates coated with SAMs containing a thiol-functionality (SH functionality), so-called thiol-SAMMS (self-assembled monolayers on mesoporous supports), have been developed specifically for the removal of mercury from liquid media [106–118]. They have the unique ability to bind cationic, organic, metallic, and complexed forms of mercury. Because of the high surface area, extensive binding sites, and tailored functional group, results of tests with mercury demonstrate the high loadings (up to 635 mg Hg per gram of SAMMS), high affinity (distribution coefficient, or K_d, $\sim 1 \times 10^8$) and rapid kinetics (minutes) possible through the

Tab. 9.3. Binding affinity of thiol-SAMMS for selected metal species [97].

Metal	Initial concentration (μg L^{-1})	Final concentration (μg L^{-1})	Metal loading (mg g^{-1})	Distribution coefficient (K_d) (mL g^{-1})
Ag(I)	90	1	0.0089	8900
Ca(II)	2070	2070	0	0
Cd(II)	4670	32	0.4638	14 467
Co(II)	2810	2670	0.0140	5
Cu(II)	2240	<5	>0.2235	>44 700
Eu(III)	9010	1220	0.7790	639
Hg(II)	487	0	1.0146	1×10^8
Mg(II)	1580	1580	0	0
Pb(II)	3040	300	0.2740	913
Zn(II)	2790	2410	0.0380	16

use of these molecularly-engineered materials. The efficiency of SAMMS in a non-aqueous system (oil) has also been demonstrated with mercury with excellent results.

Selectivity, isotherm, kinetic, stability, and regeneration data on thiol-SAMMS are available for multiple metals. Table 9.3 summarizes the binding affinity of thiol-SAMMS for selected metal species. Results show that thiol-SAMMS can selectively adsorb, in addition to mercury, other soft acid cations: silver, cadmium, copper, and lead. In all cases, thiol-SAMMS show minimal interference from alkali and alkali earth metals, such as Na$^+$, Mg^{2+}, and Ca^{2+}. Kinetic experiments exploring the adsorption of mercury in 10 and 500 ppm mercury solutions have demonstrated the rapid binding kinetics of thiol-SAMMS. The SAMMS rapidly reduced the mercury concentration from 500 to 0.5 ppb within 5 min. At a mercury concentration of 10 ppm, a reduction down to 3.1 ppb was observed within 5 min, which is significantly faster than the kinetics of commercial mercury absorbers.

In addition, the use of metal-chelated ligands immobilized on mesoporous silica as novel anion-binding materials for toxic anions such as chromate, arsenate, pertechnetate, and selenite has also been demonstrated. Novel chemical interfaces with chelate-SAMMS have shown selective removal of toxic metal oxoanions by a ligand exchange mechanism. This approach allows the construction of binding sites that satisfy the stereoelectronic requirements of tetrahedral anions. This SAMMS form can remove chromate and arsenate to low levels, even with competing sulfate ion present.

Nearly complete removal of arsenate and chromate has been reported in the presence of interfering anions for solutions containing up to 100 ppm toxic metal anions under various conditions. The material remains effective for even higher concentration solutions (in excess of 1000 ppm anions). Anion loading of more than 130 mg g^{-1} (1.12 mmol g^{-1}) of SAMMS and distribution coefficients of >100 000 have been observed.

Anion removal tests have been performed in water containing 1, 10, and 100 ppm arsenate and chromate with a solution-to-silica (SAMMS) ratio of 100. In all tests, essentially 100% of the chromate was removed in a single treatment. The addition of 150-ppm-sulfate-competing anions had little effect on the adsorption behavior. At the same solution-to-silica ratio (100 mL g^{-1}), chromate concentrations > 1000 ppm began to produce saturation of the binding sites. The maximum adsorbing capacity is about 130 mg g^{-1} or 1.12 mmol g^{-1}. For a much higher solution-to-silica ratio (500 mL g^{-1}), almost 100% removal of the chromate is observed for chromate concentrations up to 100 ppm. Higher concentrations of chromate under these conditions result in saturation of the binding sites.

Similar results were also obtained for arsenate removal. The maximum loading capacity is 140 mg g^{-1} or 1.0 mmol g^{-1}. Under the same conditions, the residual concentrations of arsenate are all slightly higher than chromate at low anion concentrations. This suggests the binding chemistry has higher affinity for chromate than arsenate.

Bulk chemical analysis of treated water samples of various sources clearly indicates the capability of SAMs to remove metal cations from water. The amount of powder per liter of water and the treatment time have been identified as important factors in increasing the efficiency of the method. However, the concentration of the contaminant appears to be less crucial, as metal cations of very high as well as very low concentrations could be removed with almost the same efficiency. Recent experiments with seawater show that this method has considerable potential for desalination [119, 120]. The concentration of all analyzed metals could be significantly reduced by the treatment, whereas that of chloride was only reduced by about 20%. However, it may only be a matter of further optimization before the desalination of water by SAMs is technically possible.

These recent studies have also shown that organic matter can be removed by functionalized SAMs [119, 120]. Figure 9.4 shows the absorbance of organic matter of the Myponga Valley reservoir near Adelaide, Australia, at a wavelength of 260 nm versus the molecular weight of the molecules. Clearly, after only a few hours of treatment with a few grams of SAMs-coated silica particles per liter, organic matter with a molecular weight of more than 1 kDa has been almost completely removed. Organic matter with a molecular weight between 300 and 1000 Da has been removed very efficiently (>90%), whereas compounds with a molecular weight below 300 Da are almost untouched. The fact that especially organic matter with high molecular weight, such as pyrogens, could significantly be reduced indicates that the amount of disinfection of the water, such as by chlorination and chloramination, could be decreased significantly when SAMs are applied to water treatment.

Besides natural organic matter, contaminants like sodium alkyl aryl sulfonate and sodium laureth sulfonate based detergents and drugs could also be removed from water. In all cases, the contaminants were removed almost completely after a treatment of about 30 min, as indicated by measuring the UV absorbance of the water

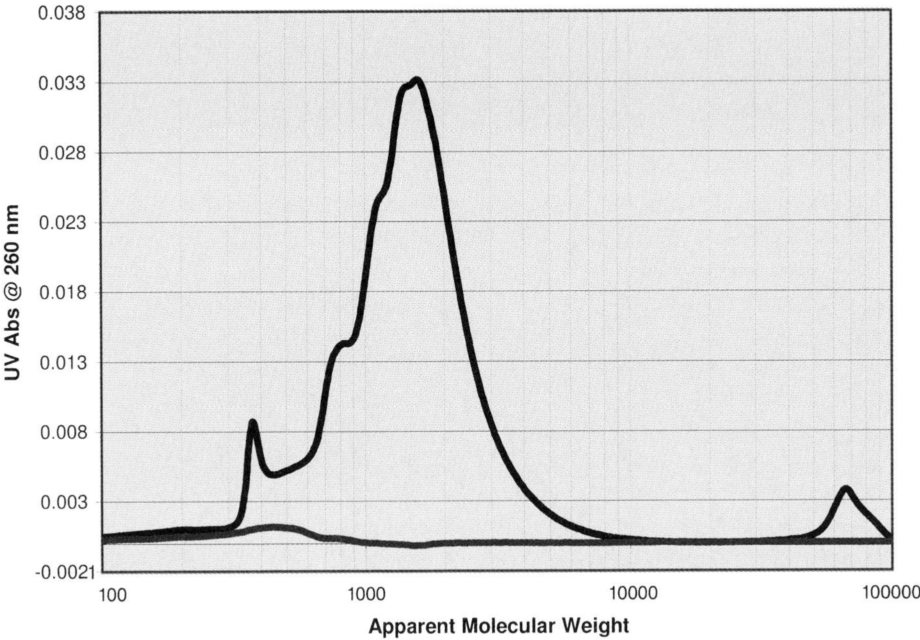

Fig. 9.4. UV absorbance at 260 nm of Myponga raw water (black) and treated water (lighter, almost horizontal line) [120].

samples [119, 120]. Although these results are very promising, further detailed studies are necessary to optimize the efficiency of the removal of organic detergents.

Although SAMs for water treatment have great potential, the technology is still at the beginning. For a technical application, SAMs will have to show that they can be recycled, are stable under physical conditions during treatment, and do not contaminate the water due to release of the SAM layers from the substrate during treatment. However, the main innovative aspect of the method is the fact that the actual treatment can be performed without the need of electric or thermal power as well as illumination, which may be a significant advantage over conventional methods such as reverse osmosis for desalination and microfiltration for the removal of pathogens [121].

9.5
Other Materials

Some other nanomaterials as well as nanocomposites have been studied for water treatment and remediation [122, 123]. Because these studies are very recent, it is yet not possible to assess whether these approaches offer significant advantages over conventional methods or more established nanomaterials as described in Chapters 2–4. However, some of the approaches are very promising and further research will clarify their potential.

Paknikar et al. [124] have studied iron sulfide nanoparticles for the degradation of lindane. Lindane is an organochlorine pesticide and a persistent organic pollutant. Lindane residues have been detected in drinking water sources as well as beverages and the development of viable methods for their removal is highly desirable. Iron nanoparticles have already been shown to transform effectively chlorinated organic compounds. However, their use in the treatment of drinking water and beverages has toxicity concerns. This study employed FeS nanoparticles, which were synthesized by a standard wet chemical method and were stabilized by novel polymers of microbial origin. The authors could show that the stabilized FeS nanoparticles dechlorinate lindane rapidly with very high efficiency. Dechlorinated compounds and the stabilized polymers could be completely degraded in a subsequent microbiological treatment, which facilitated precipitation of bulk iron. This novel integrated nano-biotechnological method may offer a safe, viable, and cost-effective solution to the removal of pollutants from various drinking water sources.

Peng et al. [125] have developed a novel adsorbent, consisting of nanoparticles of ceria supported on carbon nanotubes (CeO_2-CNTs), for the removal of arsenate from water. Their experiments showed that CeO_2-CNTs are an effective adsorbent for arsenate, and that the adsorption depends on the pH of the water. Cations of Ca and Mg significantly enhance the adsorption capacity, suggesting that this material may be a promising adsorbent for drinking water purification. The loaded adsorbent could be efficiently regenerated by dilute NaOH; a regeneration efficiency of some 90% was achieved. However, the mechanism of adsorption is not yet clear.

9.6
Magnetic Iron Exchange Resin (MIEX)

MIEX is, so far, the only approach employing a nanomaterial that is already commercially applied in water treatment [126–130]. MIEX, which is licensed by Orica Australia Pty. Ltd, is used as an alternative material for flocculation and coagulation. It has a very rapid reactivity and high capacity for the removal of NOM and forms the basis of a novel process for NOM removal. MIEX consists of nanoparticles (10–100 nm) of Fe_2O_3, which has weak permanent magnetic properties, bounded in a granular resin. Individual particles of the resin have average grain sizes of about 150 µm and, under moderate agitation, will completely separate from one another, giving a high surface area and rapid reaction rates. As soon as agitation is reduced, the particles behave like individual magnets and clump together in an open structure, which has high settling rate. The resin can easily be regenerated and recycled. To provide optimum magnetic properties of the individual granules it is crucial to have the nanoparticular Fe_2O_3 homogeneously distributed within the MIEX granules, requiring some expertise in the organic synthesis of the material [130].

References

1 *Water for People – Water for Life*, World Water Assessment Program, UNESCO, www.unesco.org/water/wwap.

2 Summary Report *Introduction to Desalination Technologies in Australia*, National Dryland Salinity Program, Department of Agriculture, Fisheries and Forestry – Australia, July **2002**.

3 D. M. OWEN, G. L. AMY, Z. K. CHOWDHURY, R. PAODE, G. McCOY, K. VISCOSIL, NOM characterisation and treatability, *J. Australian Water & Wastewater Assoc.* **1995**, 87, 46–63.

4 C. W. K. CHOW, J. A. VAN LEEUWEN, M. DRIKAS, R. FABRIS, K. M. SPARK, D. W. PAGE, The impact of the character of natural organic matter in conventional treatment, *Water Sci. Technol.* **1999**, 40, 97–104.

5 W.-X. ZHANG, Nanoscale iron particles for environmental remediation: An overview, *J. Nanoparticle Res.*, **2003**, 5, 323–332.

6 C. R. O'MELIA, *Aquatic Chemistry: Interfacial and Interspecies Processes*, ed. C. HUANG, C. R. O'MELIA, J. MORGAN, American Chemical Society: Washington, DC, **1995**, ACS Advances in Chemistry Series 244, pp. 315–337.

7 W. STUMM, J. J. MORGAN, *Aquatic Chemistry*, 3rd edn., John Wiley & Sons, New York, **1996**, Ch. 14.

8 J. CLEASBY, *Water Quality and Treatment, a Handbook of Community Water Supply*, 4th edn., American Water Works Association, McGraw-Hill, New York, **1990**, Ch. 8.

9 C. R. O'MELIA, C. L. TILLER, Natural organic matter and colloidal stability: Models and measurements, in *Environmental Particles*, ed. J. BUFFLE, H. VAN LEEUWEN, Lewis Publishers, Chelsea, MI, **1993**, Ch. 8, pp. 353–386.

10 M. ELIMELECH, J. GREGORY, X. JIA, R. A. WILLIAMS, *Particle Deposition and Aggregation: Measurement, Modeling, and Simulation*, Butterworth-Heinemann, Oxford, **1995**.

11 K.-M. YAO, M. T. HABIBIAN, C. R. O'MELIA, *Environ. Sci. Technol.* **1970**, 5, 1105–1112.

12 R. RAJAGOPALAN, C. TIEN, *Am. Inst. Chem. Eng. J.* **1976**, 22, 523–533.

13 J. E. TOBIASON, *Physical Aspects of Particle Deposition in Porous Media*, Doctoral Dissertation, The Johns Hopkins University, Baltimore, MD, **1987**, p. 280.

14 R. W. GILLHAM, S. F. O'HANNESIN, Enhanced degradation of halogenated aliphatics by zero-valent iron, *Ground Water* **1994**, 32, 958–967.

15 P. G. TRATNYEK, Correlation analysis of the environmental reactivity of organic substances, in *Perspectives in Environmental Chemistry*, ed. D. L. MACALADY, Oxford University Press, New York, **1998**, pp. 167–194.

16 W. S. ORTH, R. W. GILLHAM, Dechlorination of trichloroethene in aqueous solution using Fe(0), *Environ. Sci. Technol.* **1996**, 30, 66–71.

17 R. W. PULS, C. J. PAUL, R. W. POWELL, The application of in situ permeable reactive (zero-valent iron) barrier technology for the remediation of chromate-contaminated groundwater: A field test, *Appl. Geochem.* **1999**, 14, 989–1000.

18 S. NAM, P. G. TRATNYEK, Reduction of azo dyes with zero-valent iron, *Water Res.*, **2000**, 34, 1837–1845.

19 D. P. SIANTAR, C. G. SCHREIER, C.-S. CHOU, M. REINHARDT, Treatment of 1,2-dibromo-3-chloropropane and nitrate-contaminated water with zero-valent iron or hydrogen/palladium catalysts, *Water Res.* **1996**, 30, 215–232.

20 S. F. O'HANNESIN, R. W. GILLHAM, Long-term performance of an in situ "ironwall" for remediation of VOCs, *Ground Water* **1998**, 36, 164–170.

21 EPA (US Environmental Protection Agency), **2003**c. Databases of innovative technologies. http://www.epa.gov/tio/databases/.

22 C. WANG, W. ZHANG, Nanoscale metal particles for dechlorination of PCE and PCBs, *Environ. Sci. Technol.* **1997**, 31(7), 2154–2156.

23 W. ZHANG, C. WANG, H. LIEN, Catalytic reduction of chlorinated hydrocarbons by bimetallic particles, *Catal. Today* **1998**, 40, 387–395.

24 H. LIEN, W. J. ZHANG, Reactions of chlorinated methanes with nanoscale metal particles, *Environ. Eng.* **1999**, 125, 1042–1047.

25 Y. XU, W. ZHANG, Subcolloidal Fe/Ag particles for reductive dehalogenation of chlorinated benzenes, *Ind. Eng. Chem. Res.* **2000**, 39, 2238–2244.

26 H. LIEN, W. ZHANG, Enhanced dehalogenation by bimetallic Cu/Al, *Chemosphere* **2002**, 49, 371–378.

27 C. GRITTINI, M. MALCOMSON, Q. FERNANDO, N. KORTE, Rapid dechlorination of polychlorinated biphenyls on the surface of a Pd/Fe bimetallic system, *Environ. Sci. Technol.* **1995**, 29, 2898–2900.

28 R. MUFTIKIAN, Q. FERNANDO, N. KORTE, A method for the rapid dechlorination of low molecular weight chlorinated hydrocarbons in water, *Water Res.* **1995**, 29, 2434–2439.

29 Z. KARPINSKI, K. EARLY, J. J. D'ITRI, Catalytic hydrodechlorination of 1,1-dichlorotetrafluoroethane by Pd/Al$_2$O$_3$, *J. Catal.* **1996**, 164, 378–386.

30 K. PARK, K. KLIER, C. WANG, W. J. ZHANG, Interaction of tetrachloroethylene (PCE) with Pd (100) Studied by high resolution X-ray emission spectroscopy (HRXPS), *J. Phys. Chem. B* **1997**, 101, 5420–5428.

31 D. W. ELLIOTT, W.-X. ZHANG, Field assessment of nanoscale bimetallic particles for groundwater treatment, *Environ. Sci. Technol.* **2001**, 35, 4922–4926.

32 T. MASCIANGIOLI, W. ZHANG, Environmental technology at the nanoscale, *Environ. Sci. Technol.* **2003**, 37, 102A–108A.

33 R. GLAZIER, R. VENKATAKRISHNAN, F. GHEORGHIU, L. WALATA, R. NASH, W. ZHANG, Nanotechnology takes root. *Civil Eng.* **2003**, 73, 64–69.

34 H. LIEN, W. ZHANG, Complete dechlorination of chlorinated ethenes with nanoparticles, *Colloids Surf. A* **2001**, 191, 97–105.

35 W. ZHANG, Iron awe – nanoparticles for mine site remediation, *Mater. World* **2004**, 33–34.

36 National Research Council, *Natural Attenuation for Groundwater Remediation*, National Academy Press, Washington, DC, **2000**.

37 W. A. ZELTNER, M. A. ANDERSON, The use of nanoparticles in environmental applications, in *Fine Particles Science and Technology*, ed. E. PELIZZETTI, Kluwer Academic Publishers, Boston, MA, USA, **1996**, pp. 643–656.

38 D. S. BHATKHANDE, V. G. PANGARKAR, A. A. C. M. BEENACKERS, Photocatalytic degradation for environmental applications—a review, *J. Chem. Technol. Biotechnol.* **2001**, 77, 102–116.

39 R. MATTHEWS, Photocatalysis in water purification: Possibilities, problems and prospects, in *Photocatalytic Purification and Treatment of Water and Air*, ed. D. F. OLLIS, H. AL-EKABI, Elsevier Science Publishers, Oxford, UK, **1993**, 121–139.

40 Y. ZHANG, J. C. CRITTENDEN, D. W. HAND, D. L. PERRAM, Fixed-bed photocatalysis for solar decontamination of water, *Environ. Sci. Technol.* **1994**, 28, 435–442.

41 M. R. HOFFMANN, S. T. MARTIN, W. CHOI, D. BAHNEMANN, Environmental applications of semiconductor photocatalysis, *Chem. Rev.* **1995**, 95, 69–96.

42 S. AHMED, D. F. OLLIS, Solar assisted catalytic decomposition of the chlorinated hydrocarbons trichloroethylene and trichloromethane, *Solar Energy* **1984**, 32(5), 597–601.

43 Y. ZHANG, J. C. CRITTENDEN, D. W. HAND, The solar photocatalytic decontamination of water, *Chem. Ind.* **1994**, 714–717.

44 D. BEYDOUN, R. AMAL, G. LOW, S. McEVOY, Role of nanoparticles in photocatalysis, *J. Nanoparticle Res.* **1999**, 1, 439–458.

45 R. F. HOWE, Recent developments in photocatalysis, *Dev. Chem. Eng. Mineral Process* **1998**, 6(1), 55–84.

46 W. A. ZELTNER, C. G. HILL, JR., M. A. ANDERSON, Supported titania for photodegradation, *Chemtech* **1993**, 23, 21–28.

47 K. Hofstadler, R. Bauer, S. Novalic, G. Heiser, New reactor design for photocatalytic wastewater treatment with TiO2 immobilized on fused-silica glass fibers: Photomineralization of 4-chlorophenol, *Environ. Sci. Technol.* **1994**, 28, 670–674.

48 M. Tomkiewicz, G. Dagan, Z. Zhu, Morphology and photocatalytic activity of TiO2 aerogels, *Res. Chem. Intermed.* **1994**, 20, 701–710.

49 N. Xu, S. Zaifeng, Y. Fan, J. Dong, J. Shi, M. Z. U. Hu, Effects of particle size of TiO2 on photocatalytic degradation of methylene blue in aqueous suspensions, *Ind. Eng. Chem. Res.* **1999**, 38, 373–379.

50 H. Hidaka, Photodegradation of surfactants with TiO2 semiconductor for the environmental wastewater treatment, *Proc. Indian Acad. Sci.* **1998**, 110, 215–228.

51 I. Mazzarino, P. Piccinini, Photocatalytic oxidation of organic acids in aqueous media by supported catalyst, *Chem. Eng. Sci.* **1999**, 54, 3107–3111.

52 J. F. Tanguay, S. L. Suib, R. W. Coughlin, Dichloromethane photodegradation using titanium dioxide, *J. Catal.* **1989**, 117, 335–347.

53 D. F. Ollis, C. Y. Hsiao, L. Budiman, L. L. Chung, Heterogeneous photoassisted catalysis: conversions of perchloroethylene, dichloroethylene, chloroacetic acids and chlorobenzene, *J. Catal.* **1984**, 88, 89–96.

54 R. D. Barreto, K. A. Gray, K. Anders, Photocatalytic degradation of methyl-tert-butyl ether in TiO2 slurries: A proposed reaction scheme, *Water Res.* **1995**, 29, 1243–1248.

55 R. W. Matthews, S. R. McEvoy, Destruction of phenol in water with sun sand and photocatalysis, *Solar Energy* **1992**, 49, 507–513.

56 V. Augugliaro, F. Inglese, L. Palmisano, M. Schiavello, Annular flow photoreactor for phenol degradation in aqueous titanium dioxide dispersion. *Chem. Biochem. Eng. Q.* **1992**, 6, 63–70.

57 W. F. Jardim, S. G. Moraes, M. M. K. Takiyama, Photocatalytic degradation of aromatic chlorinated compounds using TiO2: Toxicity of intermediates, *Water Res.* **1997**, 31, 1728–1732.

58 C. Minero, V. Maurino, E. Pelizzetti, Photocatalytic degradation of free and chemically bound silicones on irradiated titanium dioxide, *Langmuir* **1995**, 11, 4440–4444.

59 R. W. Matthews, Photooxidative degradation of coloured organics in water using supported catalysts. TiO2 on sand, *Water Res.* **1991**, 25, 1169–1176.

60 A. Topalov, D. M. Gabor, J. Csanad, Photocatalytic oxidation of the fungicide metalaxyl dissolved in water over TiO2, *Water Res.* **1999**, 33, 1371–1376.

61 J. Chen, D. F. Ollis, W. H. Rulkens, H. Bruning, Photocatalyzed oxidation of alcohols and organochlorides in the presence of native TiO2 and metallized TiO2 suspensions. Part (I): Photocatalytic activity and pH influence, *Water Res.* **1999**, 33, 661–668.

62 M. M. Kondo, W. F. Jardim, Photodegradation of chloroform and urea using Ag-loaded titanium dioxide as catalyst, *Water Res.* **1991**, 25, 823–827.

63 R. M. Alberici, W. F. Jardim, Photocatalytic degradation of phenol and chlorophenols using Ag-TiO2 in a slurry reactor, *Water Res.* **1994**, 28, 1845–1849.

64 S. T. Aruna, K. C. Patil, Synthesis and properties of nanosized titania, *J. Mater. Synthesis Process.* **1996**, 4(3), 175–179.

65 D. W. Bahnemann, Ultrasmall metal oxide particles. Preparation, photophysical characterisation, and photocatalytic properties, *Isr. J. Chem.* **1993**, 33, 115–136.

66 Ch.-H. Fischer, J. Lillie, H. Weller, L. Katsikas, A. Henglein, Photo-chemistry of colloidal semiconductors. 29. Fractionation of CdS sols of small particles by exclusion chromatography, *Ber. Bunsenges, Phys. Chem.* **1989**, 93, 61–64.

67 A. Henglein, Nanoclusters of

semiconductors and metals – colloidal nanoparticles of semiconductors and metals: Electronic structure and process, *Ber. Bunsen-Gesellsch. Phys. Chem.* **1997**, 101(11), 1562–1572.

68 A. HENGLEIN, Mechanism of reactions on colloidal microelectrodes and size quantization effects, *Top. Curr. Chem.* **1988**, 143, 113–180.

69 A. HENGLEIN, Small particle research. Physicochemical properties of extremely small colloidal metal and semiconductor particles, *Chem. Rev.* **1989**, 89, 1861–1873.

70 Y. WANG, N. HERRON, Nanometer-sized semiconductor clusters. Materials synthesis, quantum size effects, and photophysical properties, *J. Phys. Chem.* **1991**, 95, 525–532.

71 B. LEVY, Photochemistry of nanostructured materials for energy applications. *J. Electroceram. I* **1997**, (3), 239–272.

72 H. WELLER, A. EYCHMULLER, Photo-chemistry and photo electrochemistry of quantized matter: Properties of semiconductor nanoparticles in solution and thin-film electrodes, in *Advances in Photochemistry*, ed. C. DOUGLAS, NECKERS, H. D. VOLMAN, G. VON BUNAU, John Wiley and Sons, New York, NY, USA **1995**, Vol. 20.

73 N. SERPONE, D. LAWLESS, E. PELIZZETTI, Subnanosecond characteristics and photophysics of nanosized TiO_2 particulates from Rp," = 10A to 34A: Meaning for heterogeneous photocatalysis, in *Fine Particles Science and Technology*, ed. E. PELIZZETTI, Kluwer Academic Publishers, Boston, MA, USA **1996**, pp. 657–673.

74 A. HAGFELDT, M. GRATZEL, Light-induced redox reactions in nanocrystalline systems, *Chem. Rev.* **1995**, 95, 49–68.

75 P. V. KAMAT, B. PATRICK, Photo-physics and photochemistry of quantized ZnO colloids, *J. Phys. Chem.* **1992**, 96, 6829–6834.

76 A. HENGLEIN, Q-particles: Size quantization effects in colloidal semiconductors, *Progr. Colloid Polym. Sci.* **1987**, 73, 1–4.

77 C. KORMANN, D. W. BAHNEMANN, M. HOFTMANN, Preparation and characterisation of quantum-size titanium dioxide, *J. Phys. Chem.* **1988**, 92, 5196–5201.

78 M. GRATZEL, *Heterogeneous Photochemical Electron Transfer*, CRC Press, Boca Raton, FL, **1989**.

79 M. ANPO, T. SHIMA, S. KODAMA, Y. KUBOKAWA, Photocatalytic hydrogenation of CHCCH with H2O on small particle TiO2. Size quantization effects and reaction intermediates, *J. Phys. Chem.* **1987**, 91, 4305–4310.

80 L. KAVAN, T. STOTO, M. GRATZEL, D. FITZMAURICE, V. SHKLOVER, Quantum size effects in nanocrystalline semiconducting TiO2 layers prepared by anodic oxidative hydrolysis of TiCl3, *J. Phys. Chem.* **1993**, 97, 9493–9498.

81 S. MARTIN, H. HERRMANN, W. CHOI, M. HOFFMANN, Photochemical destruction of chemical contaminants on quantum-sized semiconductor particles, *J. Sol. Energ.-T. ASME*, **1995**, 409–413.

82 A. J. HOFFMANN, G. MILLS, H. YEE, M. R. HOFFMANN, Q-sized CdS: Synthesis, characterisation, and efficiency of photoinitiation of polymerisation of several vinylic monomers, *J. Phys. Chem.* **1992**, 96, 5546–5552.

83 J. M. NEDELJKOVIC, M. T. NENODOVIC, O. I. MICIC, A. J. NOZIK, Enhanced photoredox chemistry in quantized semiconductor colloids, *J. Phys. Chem.* **1986**, 90, 12–13.

84 A. J. HOFFMANN, E. R. CARRAWAY, M. R. HOFFMANN, Photocatalytic production of H_2O_2 and organic peroxides on quantum-sized semiconductor colloids, *Environ. Sci. Technol.* **1994**, 28, 776–785.

85 Y. NOSAKA, N. OHTA, H. MIYAMA, Photochemical kinetics of ultrasmall semiconductor particles in solution: Effect of size on the quantum yield of electron transfer, *J. Phys. Chem.* **1990**, 94, 3752–3755.

86 G. P. LEPORE, C. H. LANGFORD, J. VICHOVA, A. VLCEK, Photochemistry and picosecond absorption spectra of aqueous suspensions of a polycrystalline titanium dioxide optically transparent in the visible spectrum, *J. Photochem. Photobiol. A, Chem.* **1993**, 75, 67–75.

87 A. J. HOFFMANN, H. YEE, G. MILLS, M. R. HOFFMANN, Photoinitiated polymerisation of methyl methacrylate using Q-sized ZnO colloids, *J. Phys. Chem.* **1992**, 96, 5540–5546.

88 C. WANG, C. Z. ZHANG, J. Y. YING, Photocatalytic decomposition of halogenated organics over nano-crystalline titania, *Nanostructured Mater.* **1997**, 9, 583–586.

89 Z. ZHANG, C.-C. WANG, R. ZAKARIA, J. Y. YING, Role of particle size in nanocrystalline TiO2-based photocatalysts, *J. Phys. Chem. B* **1998**, 102, 10871–10878.

90 S. E. PRATSINIS, *J. Aerosol Sci.* **1996**, 27, 153–154.

91 Z. X. DING, P. L. HU, G. Q. YUE, P. F. GREENFIELD, *Catal. Today* **2001**, 68, 173–182.

92 B. XIA, W. LI, B. ZHANG, X. YOUCHANG, *J. Mater. Sci.* **1999**, 34, 3505–3511.

93 R. R. BACSA, M. GRATZEL, *J. Am. Ceram. Soc.* **1996**, 79, 2185–2188.

94 C. ANDERSON, A. J. BARD, *J. Phys. Chem.* **1995**, 99, 9882–9885.

95 S. WATSON, D. BEYDOUN, J. SCOTT, R. AMAL, Preparation of nanosized crystalline TiO$_2$ particles at low temperature for photocatalysis, *J. Nanoparticle Res.* **2004**, 6, 193–207.

96 S. T. ARUNA, K. C. PATIL, Synthesis and properties of nanosized titania, *J. Mater. Synth. Process.* **1996**, 4, 175–179.

97 http://samms.pnl.gov/index.stm

98 F. SCHREIBER, Structure and growth of self-assembling monolayers, *Science* **2000**, 65, 151–156.

99 M. R. DE GUIRE, T. NIESEN, S. SUPOTHINA, J. WOLFF, J. BILL, C. N. SUKENIK, F. ALDINGER, A. H. HEUER, M. RUEHLE, Synthesis of Oxide and Non-Oxide Inorganic Materials at Organic Surfaces, *Z. Metallkunde*, **1998**, 89, 758–766.

100 J. BILL, R. C. HOFFMANN, T. M. FUCHS, F. ALDINGER, Deposition of Ceramic Materials from Aqueous Solution Induced by Organic Templates, *Z. Metallkunde*, **2002**, 93, 478–490.

101 P. MAJEWSKI, T. FUCHS, S. PRAKASH, Synthesis and properties of silane-based self-assembled-monolayers onto silica particles, *Mater. Forum* **2005**, 29, 489–493.

102 P. MAJEWSKI, T. FUCHS, S. PRAKASH, L. TRIBALAT, Synthesis of manganese oxide onto silica particles coated with self-assembled-monolayers, *J. Mater. Sci. Technol.* **2005**, 21, 29–32.

103 N. BALACHANDER, C. N. SUKENIK, Monolayer transformation by nucleophilic substitution: Applications to the creation of new monolayer assemblies, *Langmuir* **1990**, 6, 1622–1627.

104 R. J. COLLINS, C. N. SUKENIK, Sulfonate-functionalized, siloxane-anchored, self-assembled monolayers, *Langmuir* **1995**, 11, 2322–2324.

105 H. PIZEM, C. N. SUKENIK, U. SAMPATHKUMARAN, A. K. MCILWAIN, M. R. DE GUIRE, Effects of substrate surface functionality on solution-deposited titania films, *Chem. Mater.* **2002**, 13, 2476–2485.

106 G. E. FRYXELL, J. LIU, Designing surface chemistry in mesoporous silica, in *Adsorption at Silica Surfaces*, ed. E. PAPIRER, Marcel Dekker, New York, NY, USA, **2000**, pp. 665–688.

107 J. LIU, Y. SHIN, Z. NIE, J. H. CHANG, LI-Q. WANG, G. E. FRYXELL, W. D. SAMUELS, J. GREGORY, Exarhos, Molecular assembly in ordered mesoporosity: A new class of highly functional nanoscale materials, *J. Phys. Chem. A* **2000**, 104, 8328–8339.

108 J. LIU, G. E. FRYXELL, M. QIAN, LI-Q. WANG, Y. WANG, Interfacial chemistry in self-assembled nanoscale materials with structural ordering, *Pure Appl. Chem.*, **2000**, 72, 269–279.

109 (a) W. YANTASEE, Y. LIN, E. G. FRYXELL, B. BUSCHE, J. C. BIRNBAUM, Removal of heavy metals from aqueous solution using novel nanoengineered sorbents: Self-assembled carbamoylphosphonic acids on mesoporous silica, *Sep. Sci. & Technol.*, **2003**, 38, 3809–3825; (b) G. E. FRYXELL, Y. LIN, S. FISKUM, J. C. BIRNBAUM, H. WU, K. KEMNER, S. KELLEY, Actinide sequestration using self-assembled monolayers on mesoporous supports, *Environ. Sci. Technol.*, **2005**, 39, 1324–1331; (c) G. E. FRYXELL, H. WU, Y. LIN, W. J. SHAW, J. C. BIRNBAUM, J. C. LINEHAN, Z. NIE, K. KEMNER, S. KELLY, Lanthanide selective sorbents: Self-assembled monolayers on mesoporous supports (SAMMS), *J. Mater. Chem.*, **2004**, 14, 3356–3363.

110 S. V. MATTIGOD, G. E. FRYXELL, R. J. SERNE, K. E. PARKER, F. M. MANN, Evaluation of novel getters for adsorption of radioiodine from groundwater and waste glass leachates, *Radiochim. Acta*, **2003**, 91, 539–545.

111 W. YANTASEE, Y. LIN, G. E. FRYXELL, B. J. BUSCHE, J. C. BIRNBAUM, Removal of heavy metals from aqueous solution using novel nanoengineered sorbents: self-assembled carbamoylphosphonic acids on mesoporous silica. *Sep. Sci. Technol.*, **2003**, 38(15), 3809–3825.

112 G. E. FRYXELL, Y. LIN, H. WU, K. M. KEMNER, Environmental applications of self-assembled monolayers on mesoporous supports (SAMMS), in *Studies in Surface Science and Catalysis*, Vol. 141, ed. A. SAYARI, M. JARONIEC, Elsevier, Oxford, UK, **2003**, pp. 583–590.

113 Y. LIN, G. E. FRYXELL, H. WU, M. ENGLEHARD, Selective sorption of cesium using self-assembled monolayers on mesoporous supports (SAMMS) *Environ. Sci. Technol.*, **2001**, 35, 3962–3966.

114 G. E. FRYXELL, J. LIU, S. V. MATTIGOD, L. Q. WANG, M. GONG, T. A. HAUSER, Y. LIN, K. F. FERRIS, X. FENG, Environmental applications of interfacially modified mesoporous ceramics, in *Environmental Issues and Waste Management Technologies in the Ceramic and Nuclear Industries*, ed. G. T. CHANDLER, X. FENG, **2000**, Ceramics Transactions, American Ceramic Society, Westerville, OH, USA, Vol. 107, pp. 29–37.

115 G. E. FRYXELL, J. LIU, S. MATTIGOD, Environmental applications of self-assembled monolayers on mesoporous supports (SAMMS), *Mater. Technol.*, **1999**, 14, 188–191.

116 S. V. MATTIGOD, X. FENG, G. E. FRYXELL, J. LIU, M. GONG, Separation of complexed mercury from aqueous wastes using self-assembled mercaptan on mesoporous silica, *Sep. Sci. Technol.*, **1999**, 34, 2329–2345.

117 S. MATTIGOD, G. E. FRYXELL, X. FENG, J. LIU, Self-assembled monolayers on mesoporous supports for metal separation, in *Metal Separation Technologies Beyond 2000: Integrating Novel Chemistry with Processing*, ed. K. C. LIDDELL, D. J. CHAIKO, The Minerals, Metals and Materials Society, Warrendale, PA, USA, **1999**, pp. 71–79.

118 X. FENG, L. RAO, T. R. MOHS, J. XU, Y. XIA, G. E. FRYXELL, J. LIU, K. N. RAYMOND, Self-assembled monolayers on mesoporous silica, a super sponge for actinides, in *Environmental Issues and Waste Management Technologies IV*, ed. J. C. MARA, G. T. CHANDLER, **1999**, Ceramic Transactions, American Ceramic Society, Westerville, OH, USA, Vol. 93, pp. 35–42.

119 P. MAJEWSKI, Interaction of Functionalized Self-Assembled Monolayers with Pathogens in Water, The Nano and Bio-Newsletter, II, Ian Wark Research Inst., Adelaide, Australia, 2006.

120 P. MAJEWSKI, *Water Treatment based on Functionalised Organic Self-Assembled Monolayers*, Annual Report **2004**, Ian Wark Research Institute, University of South Australia, www.unisa.edu.au/iwri.

121 B. BRENNAN, Waterworks, *Sci. & Technol.*, **2001**, 79, 32–38.

122 S. O. OBARE, G. J. MEYER, Nanostructured materials for environmental remediation of organic contaminants in water, *J. Environ. Sci. Health, Part A: Toxic/Hazardous Substances Environ. Eng.*, **2004**, A39(10), 2549–2582.

123 C. H. Cooper, A. G. Cummings, M. Y. Starostin, C. P. Honsinger, Purification of fluids with nanomaterials containing defective carbon nanotubes, *PCT Int. Appl. WO 2004080578* (**2004**), 106 pp. CODEN: PIXXD2.

124 K. M. Paknikar, V. Nagpal, A. V. Pethkar, J. M. Rajwade, Degradation of lindane from aqueous solutions using iron sulfide nanoparticles stabilized by polymers, *Sci. Technol. Adv. Mater.*, **2005**, 6, 370–374.

125 X. Peng, Z. Luan, J. Ding, Z. Di, Y. Li, Binghui, Ceria nanoparticles supported on carbon nanotubes for the removal of arsenate from water, *Mater. Lett.*, 59(4), 399–403.

126 J. Y. Moran, D. B. Bursill, M. Drikas, H. Nguyen, A new technique for the removal of natural organic matter, Proceedings of the Australian Water & Wastewater Assoc., Water TECH Conference 1996, Australian Water & Wastewater Assoc., Sydney, Australia, pp. 428–432.

127 J. Y. Morran, M. Drikas, D. Cook, D. B. Bursill, Comparison of MIEX® treatment and coagulation on NOM character, *Water Sci. Technol.: Water Supply*, **2004**, 4(4), 129–137.

128 M. Drikas, J. Y. Morran, C. Pelekani, C. Hepplewhite, D. B. Bursill, Removal of natural organic matter – a fresh approach, *Water Sci. Technol.: Water Supply*, **2002**, 2(1), 71–79.

129 M. Drikas, J. Y. Morran, D. Cook, D. B. Bursill, Operating the MIEX® process with microfiltration or coagulation, Proceedings of the WQTC 2003, American Water Works Assoc., Denver, CO, USA, pp. 1–11.

130 H. Van Nguyen, D. Bruce Bursill, J. Young Morran, M. Drikas, V. L. Pearce, *US Patent No. 6669849*, **2003**.

10
Nanoparticles for the Photocatalytic Removal of Endocrine-disrupting Chemicals in Water

Heather M. Coleman

10.1
Introduction

There is currently much concern about the release into the aquatic environment of natural and synthetic oestrogens and compounds that can mimic oestrogens. Since the turn of the last century, new testing methods have allowed scientists to detect traces of at least 500 new synthetic chemicals in our bodies. Some of these chemicals are persistent; others that we are regularly exposed to are short-lived. The long-term effects of most of these chemicals are unknown but evidence is mounting that some of these substances, known as endocrine disrupters, could be wreaking havoc with human and animal hormones, reducing the chances of successful reproduction by lowering sperm counts and contributing to an increased incidence of several rare cancers and birth defects [1]. The range of substances reported to cause endocrine-disrupting effects is diverse, and continues to expand as the number of studies increases. Some are likely to be distributed widely in the environment, are long-lived, and can accumulate in the tissues of plants and animals. The term "Environmental oestrogens" includes all oestrogens that may affect the endocrine system (the communication system of glands, hormones and cellular receptors that control the body's internal functions) [2] and includes the natural oestrogens, synthetic oestrogens, xenoestrogens (oestrogen mimics) and phytoestrogens (plant oestrogens). However, the main compounds of concern are the natural and synthetic oestrogens that have been detected at significant biological levels in sewage effluent [3]. New methods for water treatment, as well as improvements to existing processes, are required because of more severe regulations resulting from increasing awareness of the urgent need to protect our environment. Concerning the new oxidation methods under development (usually termed "advanced oxidation technologies or processes", AOTs or AOPs), heterogeneous titanium dioxide photocatalysis appears very promising in terms of destroying organic micropollutants, leaving them in very low concentrations. The degradation of organic pollutants present in wastewaters using irradiated dispersions of TiO_2 nanoparticles is a fast growing field in basic and applied research.

This chapter describes the use of titanium dioxide photocatalysis for the degrada-

Nanotechnologies for the Life Sciences Vol. 5
Nanomaterials – Toxicity, Health and Environmental Issues. Edited by Challa S. S. R. Kumar
Copyright © 2006 WILEY-VCH Verlag GmbH & Co. KGaA, Weinheim
ISBN: 3-527-31385-0

tion of natural and synthetic oestrogens in water. The background to oestrogens in the environment is described in Section 10.2 and titanium dioxide (TiO_2) photocatalysis is discussed along with the main aim and objectives of the work in Section 10.3. Each objective is then addressed in the sections that follow (Sections 10.4–10.7) and conclusions are made and research needs identified (Section 10.8).

10.2
Background to Oestrogens in the Environment

The main substances of concern are the natural oestrogens 17β-oestradiol, oestrone and oestriol and the synthetic oestrogen 17α-ethynyloestradiol. This chapter involves the photocatalytic degradation of all of the oestrogens and particularly deals with 17β-oestradiol, the principal, natural and most potent oestrogen. Natural oestrogens are steroid hormones made primarily in the female ovaries and the male testes in humans and other animals. Known as the female hormones, oestrogens are found in greater amounts in females than males. These essential molecules influence growth, development and behavior (puberty), regulate reproductive cycles (menstruation, pregnancy) and affect many other body parts (bones, skin, arteries, the brain, etc.) [2]. Oestradiol is the principal, natural and most potent oestrogen, followed by oestrone and finally oestriol (Fig. 10.1).

17β-Oestradiol is a potent endogenous oestrogen responsible for the development of female secondary sex characteristics and reproduction [4]. As illustrated in Fig. 10.1, 17β-oestradiol and its derivatives oestrone and oestriol are 18-carbon steroids with a phenolic ring. The phenolic A ring is the structural component responsible for high affinity binding to the oestrogen receptor [4].

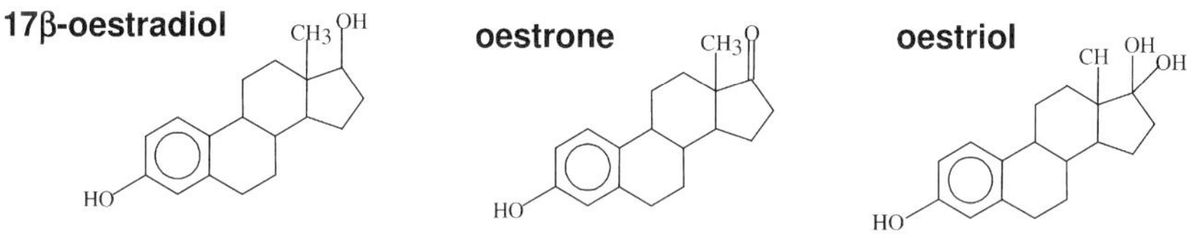

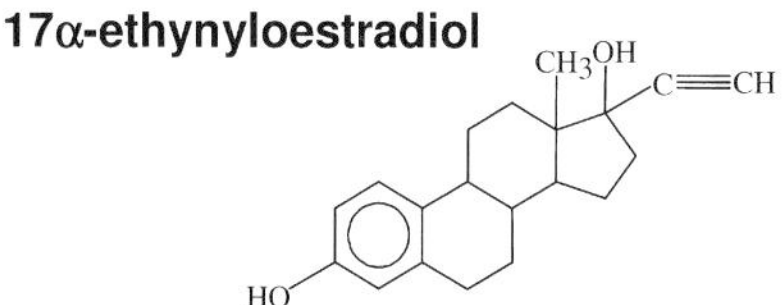

Fig. 10.1. Structures of the natural oestrogens (17β-oestradiol, oestrone and oestriol) and the synthetic oestrogen 17α-ethynyloestradiol.

17α-Ethynyloestradiol is a synthetic oestrogen used extensively in oral contraceptive formulations. It is structurally similar to 17β-oestradiol, with the exception of an ethynyl substitution at carbon-17 and has a higher affinity for the oestrogen receptor than 17β-oestradiol [4] (Fig. 10.1). Introducing an ethynyl group at the 17α-position of the oestradiol molecule produces a more stable compound. 17α-Ethynyloestradiol is the most frequently used oestrogen component in the contraceptive pill. The ethynyl group makes the D ring much more resistant to oxidation. Consequently, ethynyloestradiol is excreted up to 80% unchanged [5]. Oestrogen pharmaceutical products are used in both human and veterinary medicine [6], e.g., oral contraceptives, hormone replacement therapy and livestock yield improvement [4].

Exposure to endocrine-disrupting chemicals (EDCs) in the environment has been associated with abnormal thyroid function in birds and fish; decreased fertility in birds, fish, shellfish and mammals; decreased hatching success in fish, birds and turtles; demasculinization and feminization of male fish, birds and mammals; defeminization and masculinization of female fish, gastropods and birds; and alteration of immune function in birds and mammals [7]. Human illnesses linked to endocrine disruption include breast, prostate and testicular cancer, reproductive abnormalities such as declining sperm counts and malformed male genitals, learning and behavioral problems and immune system deficiencies.

The presence of low concentrations of natural and synthetic steroid oestrogens in the aquatic environment has been reported over the last 30 years [8–10], but only relatively recently was it realized that steroid oestrogens constitute the main oestrogenic component in domestic sewage treatment work (STW) effluents [9]. The major oestrogenic components of STW effluent include the natural oestrogens, 17β-oestradiol and oestrone, and the synthetic steroid oestrogen 17α-ethynyloestradiol [9]. These steroid oestrogens have been widely reported in sewage effluents [9, 11], rivers [12, 13] and spring water [14]. Steroid oestrogens are eliminated from the body mainly as biologically inactive forms, following their conjugation to water-soluble glucuronide and/or sulfate ester groups [15]. However, a large proportion of these steroid conjugates in wastewater are probably deconjugated before reaching the STW by faecal bacteria (such as *Escherichia coli*), which synthesize large quantities of the enzyme β-glucuronidase [16]. STWs typically remove in excess of 80% of the steroid content (or oestrogenic activity) present within the influent [17–20]. However, due to their extremely high biological potency, even trace amounts of steroid remaining in the effluent are capable of exerting biological effects on fish [21] and other aquatic organisms. The time needed for conventional biological methods to remove total organic carbon, combined with the relatively short hydraulic residency times of many STWs, means that many wastewater discharges still contain environmentally relevant levels of steroid oestrogens and their biotransformation products (Table 10.1). The need to reduce the output of steroid oestrogens in STW effluents is critical in countries with limited water resources, where effluent can become a major component of river flow in the summer months [22]. Therefore, there is a need to consider alternative strategies for the removal of trace amounts of steroid and steroid-like compounds pres-

Tab. 10.1. Oestrogen levels in STWs around the world [9, 23–25].

Location	Concentration (ng L^{-1})	
	17β-Oestradiol	**17α-Ethynyloestradiol**
United Kingdom (1998)	1–50	0–7
France, Paris (2003)	4.5–8.6	3.1–4.5
Netherlands (1999)	<0.6–12	<0.2–7.5
Germany (1999)	3	15
Canada (1999)	64	42
Sweden (1999)	1.1	4.5
Japan, Tokyo (2003)	15	9

ent in STW effluents before their discharge into rivers. There are also increasing regulatory pressures to reduce environmental levels of many EDCs.

10.2.1
Advanced Oxidation Techniques (AOTs)

In view of the growing medical use of synthetic steroids and the increasing use of birth-control pills, the synthetic ovulation-inhibiting hormones are expected to increase in concentration in wastewaters. While the concentrations in wastewater are bound to be extremely low at present, their high physiological activity at extremely low concentrations and their relatively greater stability in aqueous media than that of natural urinary hormones deserve consideration as possible contaminants of water to be processed for drinking. Consideration, therefore, needs to be given to the management strategies available in the light of such oestrogenic contamination, as the removal of these substances by current conventional water treatment methods is ineffective. A possible alternative treatment for the purification of water is titanium dioxide semiconductor photocatalysis, which is an advanced oxidation technology and a rapidly growing area of interest to both research workers and water purification companies [26]. Almost all these studies have been directed towards the oxidative degradation of organic pollutants in water [27]. While water purification techniques such as UV, ozonation or activated charcoal could significantly remove these microorganic contaminants, the high costs involved suggest that research into treatment optimization should receive more attention [3].

The common characteristic of all AOTs is the generation of very reactive free-radicals that oxidize pollutants. AOTs include thermal processes, H_2O_2 or ozone, and light-induced reactions such as homogeneous photolysis, UV/H_2O_2 photolysis, UV/ozone photolysis, heterogeneous photolysis and radiolysis [28]. Chemical oxidation technologies are useful in the oxidative degradation or transformation of a wide range of pollutants for the treatment of drinking water, groundwater, wastewater and contaminated soils. Chemical oxidation methods are especially applicable for: the treatment of hazardous organics present at low concentrations, such as in contaminated groundwaters; use as a pre-treatment step before biological treatment of low-volume, high strength wastewaters; treatment of wastewaters with constituents that are resistant to biodegradation methods or upset biological treatment reactors, such as cyanides and complex metals; and use as a post-treatment step following biological treatment to reduce aquatic toxicity. Chemical oxidation methods can also be combined with other treatment technologies to achieve optimum and cost-effective treatment technologies [29]. However, approaches for controlling the release of steroid oestrogens, or other compounds that are natural by-products of our very existence, is a unique challenge.

10.2.2
Ultraviolet Photolysis

Treatment by UV radiation alone was also investigated in this study as a comparison to photocatalysis. Direct ultraviolet photolysis was the first photochemical method used for pollutant degradation. Although several authors have proposed direct photooxidation with ultraviolet light for water treatment [30], there are several limitations to its general applicability. Direct photochemical degradation can be achieved only when the incident light [vacuum ultraviolet (VUV) light: $\lambda < 200$ nm or UV-light: $\lambda > 200$ nm] is absorbed by the pollutant. There are numerous reports of UV degradation of organics present in water, including fluorinated or chlorinated aliphatics, chlorinated hydrocarbons, trihalomethanes, dinitrotoluene, chlorophenols, pesticides, PCBs, chlorinated and nitrated aromatics, phenols, halogenated aliphatics, and other hazardous wastes [31]. Direct photolysis procedures are generally of low efficiency than procedures involving hydroxyl radical generation. However, photolysis of pollutants may be important in cases where hydroxyl radical reactions are very slow [30]. UVC disinfection is currently being used in some treatment facilities around the world (e.g., Orange County Water District, California, USA; Sydney Water, Australia; Essex and Suffolk Water, UK).

10.3
Nanoparticles for Water Treatment Applications

Chemists have known for decades that several semiconductor metal oxides are light-sensitive and can initiate redox reactions of adsorbates [32]. In 1972, Fujishima and Honda discovered the photocatalytic splitting of water on titanium dioxide electrodes [33]. This event marked the beginning of a new era in heterogeneous photocatalysis, stimulating a worldwide effort to characterize the physical features that control efficiency in these interfacial reactions [31]. Since then, research efforts in understanding the fundamental processes and in enhancing the photocatalytic efficiency of TiO_2 nanoparticles have come from extensive research performed by chemists, physicists and chemical engineers. In recent years, applications to environmental clean-up have been one of the most active areas in heterogeneous photocatalysis. This has been inspired by the potential application of TiO_2-based photocatalysts for the total destruction of organic compounds in polluted air and wastewaters [31]. The degradation of organic pollutants present in wastewaters using irradiated dispersions of TiO_2 is a fast growing field in basic and applied research. The development of this process to achieve complete mineralization of organic pollutants has been widely tested for a large variety of chemicals [31]. Carey et al. first reported, in 1976, the photocatalytic degradation of biphenyl and chlorophenyls in the presence of TiO_2 [34]. Since then, many applications using the TiO_2/UV process have been investigated. Other semiconductor dispersions have also been used for the photocatalytic degradation of pollutants [31]. There is an exhaustive amount of literature on titanium dioxide photocatalysis. Mills and Le Hunte published an excellent review in 1997 and estimated that over 2000 papers had been published since 1981 [35]. Over the years, reviews have been published [31, 36–39] and books written on the subject [40–43]. Commercial applications of titanium dioxide photocatalysis have been reviewed by Mills et al. [44].

10.3.1
Titanium Dioxide Photocatalysis

10.3.1.1 The Principle

Photocatalysis, as the name suggests, involves light and a catalyst to bring about a chemical reaction. Mills and Le Hunte define photocatalysis as the "acceleration of a photoreaction by the presence of a catalyst" [35]. In photocatalysis the catalyst is activated by the absorption of photons of light whose energy is used to overcome the activation energy. In this case the catalyst is a semiconductor and it provides a low energy activation pathway for the passage of electronic charge. The charge is effectively transferred from the reactant through the semiconductor (activated by a photon of light) to the product (a redox reaction) [41]. In TiO_2 photocatalysis for water purification the pollutants are usually organic compounds and, therefore, the overall process can be summarized by reaction Eq. (1) [27].

$$\text{Organic pollutant} + O_2 - TiO_2 \rightarrow CO_2 + H_2O \tag{1}$$

$$hv, \lambda \leq 400 \text{ nm}$$

The advantages of TiO_2 over other semiconductors, for photocatalytic treatment of water, are that it is an inexpensive readily available material, and it is non-toxic, insoluble and photostable. In addition, solar illumination is a possibility, organic compounds are completely oxidized, no expensive chemicals need be added and a wide spectrum of pollutants can be degraded [27]. The process can also be turned on and off at the flick of a switch. TiO_2 photocatalysis has potential application in the treatment of both waste and drinking water [45].

10.3.1.2 **Titanium Dioxide Nanoparticles as a Photocatalyst**

The initial process for heterogeneous photocatalysis of organic and inorganic compounds by semiconductors is the generation of electron–hole pairs in the semiconductor particles [31]. Several semiconductors have been evaluated as photocatalysts, including metal oxides (TiO_2, ZnO, SnO_2 and WO_3) and chalcogenides (ZnS, CdS). For oxidation reactions, the most useful and widely employed is titanium dioxide. TiO_2 is an intrinsic n-type semiconductor (due to an intrinsic oxygen deficiency, like ZnO). It is ionic and has a wider band gap (>3 eV) than non-ionic semiconductors. To promote an electron from the valence band to the conduction band, light of wavelength less than 400 nm is necessary to supply the required energy to cross the band gap [27]. TiO_2 nanoparticles have long been used in such applications as paint pigments and scratch-resistant optical coatings but can also display high activity for photocatalysis, which chemists and chemical engineers are beginning to exploit [46]. TiO_2 exhibits three different crystal forms: brookite, rutile and anatase. Rutile and anatase are the most stable and most common forms and their unit cells are shown in Fig. 10.2 [37]. They consist of chains of slightly distorted octahedrons of oxygen atoms around a titanium atom. Differences in the distortion of each octahedron by the assembly pattern of the octahedral chains cause differences in the mass densities and electronic band structure. For instance, the band gap for anatase is larger than that for rutile (3.23 and 3.02 eV respectively). The metastable anatase form transforms into the rutile form at high temperatures [37].

Many researchers have evaluated the photocatalytic activity of rutile and anatase

<table>
<tr><td>

<u>Rutile</u>

$d^{eg}_{Ti\text{-}O} = 1.949$ Å
$d^{AP}_{Ti\text{-}O} = 1.980$ Å

$a = 4.593$ Å
$c = 2.959$ Å

</td><td>

<u>Anatase</u>

$d^{eg}_{Ti\text{-}O} = 1.934$ Å
$d^{AP}_{Ti\text{-}O} = 1.980$ Å

$a = 3.784$ Å
$c = 9.515$ Å

</td></tr>
</table>

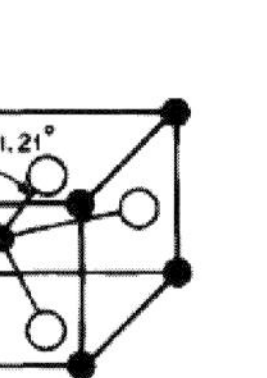

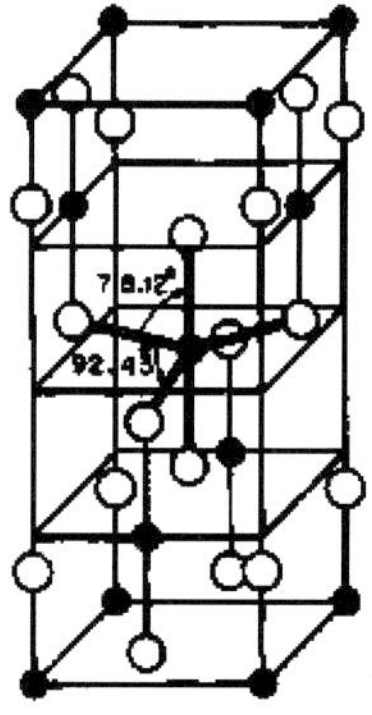

$E_g = 3.1$ eV
$\rho = 4.250$ g/cm^3
$\Delta G^o_f = -212.6$ kcal/mole

$E_g = 3.3$ eV
$\rho = 3.894$ g/cm^3
$\Delta G^o_f = -211.4$ kcal/mole

Fig. 10.2. Structure of rutile and anatase [37].

and, in general, rutile is claimed to be less active than anatase or even inactive [27, 47]. This is because anatase has a much larger surface area than rutile. Anatase requires photons having energies greater than 3.2 eV ($\lambda \sim 380$ nm) to excite an electron from the valence band to the conduction band. The separated electron–hole pairs can then be made available for oxidation–reduction reactions. The reduction potential for $^{\cdot}OH$ has been assigned a value of $+2.85$ V. It is, therefore, thermodynamically favorable for the hole site formed in the valence band of TiO_2 to oxidize water to $^{\cdot}OH$ and for the separated electron promoted to the conduction band to reduce oxygen at -0.13 V. The production of an OH radical from water, and the reduction of oxygen, requires a semiconductor with a band gap above 3 eV. Hence, anatase is an ideal photocatalyst for use in aqueous solutions [27, 35]. The most widely used form of TiO_2, which was used in this research, with high activity, is actually a mixture of the anatase and rutile forms (ca. 70–80% anatase). It is produced in particulate form (30 nm diameter crystalline size, aggregated together into 0.1 μM macroscopic particle sizes) by the Degussa Corporation and is called Degussa P25. Its high activity is suggested to be because the conduction band of anatase is more positive than that of rutile, and the light-promoted electrons may pass from rutile to anatase, enhancing the separation of holes and electrons [48, 49].

10.3.1.3 Mechanism of TiO$_2$ Photocatalysis

Figure 10.3 shows a schematic of a TiO_2 particle, illustrating the mechanism of TiO_2 photocatalysis. When a photon of light of sufficient energy ($E \geq E_{bg}$) strikes a TiO_2 particle the energy of the photon is absorbed and used to promote an electron (e^-) from the valence band to the conduction band. This movement of an electron leaves a hole (h^+) in the valence band. These species (h^+ and e^-) produced by the absorption of light can either recombine or migrate to the surface of the TiO_2

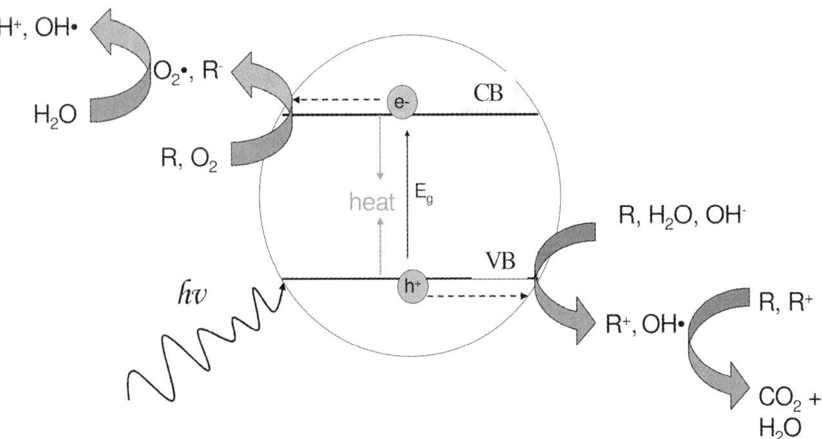

Fig. 10.3. Schematic of a titanium dioxide particle, illustrating the mechanism of titanium dioxide photocatalysis (VB = valence band, CB = conduction band, E_g = band gap energy, e^- = electron, h^+ = hole, R = organic).

particle where they can react with other species at the interface. The holes can directly oxidize organic species adsorbed onto the TiO_2 particle or can give rise to hydroxyl radicals ($^{\bullet}OH$) by reacting with water or OH^-. These highly reactive hydroxyl radicals then attack organic compounds present at or near the surface [27]. Electrons promoted to the conduction band must be removed rapidly from the TiO_2 to prevent recombination with the holes and allow the mechanism to continue. Usually the electrons are passed on to molecular oxygen at the interface [31].

The source of OH^- may be related to the nature of the surface-bound water associated with anatase. The hydroxyl radical can oxidize organic contaminants. This process finally affords carbon dioxide and water if the reaction proceeds to completion [45]. The strong oxidizing power of the photogenerated holes, together with the chemical inertness and non-toxicity of TiO_2, has made it an attractive photocatalyst [46]. Photocatalysis has seen explosive growth, particularly during the past ten years. The general mechanism of photocatalysis on the TiO_2 surface involves the oxidation of surface hydroxyl groups, which participate in the photocatalytic oxidation process. Although direct oxidation of substrates by photo-generated holes is possible, the involvement of $^{\bullet}OH$ in the oxidation process has gained much experimental support. To drive the photocatalytic reaction and maintain charge neutrality, oxygen undergoes reduction in aerated aqueous media, yielding O_2^- and H_2O_2, which in turn participate in further oxidation processes [46]. The band-gap model has proven very useful in explaining the mechanism of semiconductor-catalyzed oxidative degradation of organic material in aqueous systems [36, 37, 50]. The first event in the photocatalytic process is absorption of a photon of ultraband gap energy to produce the electron–hole pair on a particle [Eq. (2)].

$$TiO_2 + h\nu \rightarrow e^- + h^+ \tag{2}$$

The electron in the conduction band can be transferred to adsorbed H^+, O_2 or a chlorinated pollutant (RX) [Eqs. (3)–(6)], initiating various reactions [30].

$$e^- + O_2 \rightarrow O_2^{-\bullet} \tag{3}$$

$$O_2^- + H^+ \rightleftharpoons HOO \xrightarrow{HOO} H_2O_2 + O_2 \tag{4}$$

$$e^- + H_2O_2 \rightarrow H_2O_2^{\bullet-} \rightarrow HO^- + HO^{\bullet} \tag{5}$$

$$e^- + RX \rightarrow RX^{\bullet-} \rightarrow R^{\bullet} + X^- \tag{6}$$

The hole (h^+) in the valence band can react with surface-bound water, hydroxide groups, and anions (A^-) to give the $HO^{\bullet}$ (or $A^{\bullet}$) radical, and with organic substrate (RH) to give radical cations ($RH^{\bullet+}$) [Eqs. (7)–(10)] [30, 31].

$$h^+ + HOH \rightarrow HOH^{\bullet+} \rightarrow HO^{\bullet} + H^+ \tag{7}$$

$$h^+ + HO^- \rightarrow HO^{\bullet} \tag{8}$$

$$h^+ + A^- \rightarrow A^{\bullet} \tag{9}$$

$$h^+ + RH \rightarrow RH^{\bullet+} \rightarrow R^{\bullet} + H^+ \tag{10}$$

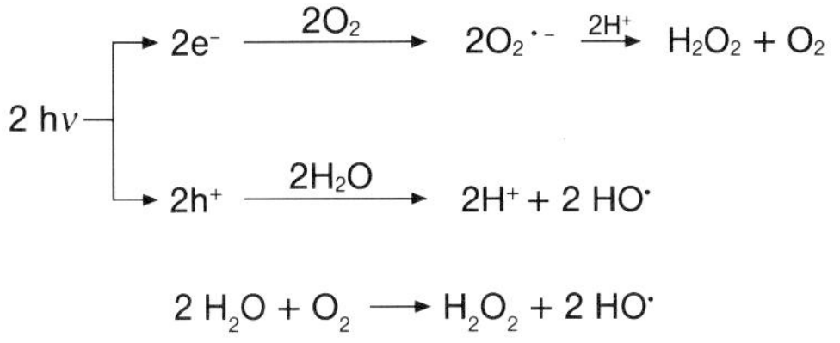

$$2\,h\nu \begin{cases} 2e^- \xrightarrow{\ 2O_2\ } 2O_2{}^{\cdot-} \xrightarrow{\ 2H^+\ } H_2O_2 + O_2 \\[2ex] 2h^+ \xrightarrow{\ 2H_2O\ } 2H^+ + 2\,HO^{\cdot} \end{cases}$$

$$2\,H_2O + O_2 \longrightarrow H_2O_2 + 2\,HO^{\cdot}$$

Scheme 10.1.

Once formed, oxidative intermediates, mainly hydroxyl radicals, can react with the organic contaminant to initiate a sequence of reactions that lead to complete oxidative mineralization. Scheme 10.1 summarizes the material balance of H_2O_2 and $HO^{\cdot}$ formation [31].

The main aim of this work was to investigate the degradation of the natural oestrogens 17β-oestradiol, oestrone and oestriol and the synthetic oestrogen 17α-ethynyloestradiol in water, using titanium dioxide photocatalysis.

The main objectives were:

- To determine if oestrogens in water can be degraded by titanium dioxide photocatalysis and UV radiation.
- To determine if all oestrogenic activity is removed after treatment with photocatalysis and UV radiation.
- To investigate the effect of varying reaction conditions on the photocatalytic reaction, i.e., initial concentration of pollutant and light intensity.
- To compare photocatalysis with UVA and UVC radiation.

These objectives were investigated and are discussed in Sections 10.4–10.7.

10.4
Photocatalytic Degradation of 17β-Oestradiol in Water over an Immobilized TiO$_2$ Catalyst

Initial work involved investigating the degradation of the principal most potent natural oestrogen 17β-oestradiol in water by photocatalysis monitored using high-performance liquid chromatography (HPLC) with fluorescence detection in a quartz water-jacketed reactor [51].

Given its low solubility in water, 17β-oestradiol (Sigma > 98%) was initially dissolved in acetonitrile (Labscan) and then diluted with water to the desired concentration. TiO$_2$ (Degussa P25) was immobilized on Ti-6Al-4V alloy by an electrophoretic method described previously [52]. A 1 cm^2 area of support was coated with a catalyst loading of 1.5 mg cm^{-2}. The photocatalytic reactor was a water-jacketed quartz cell that held 8 mL of solution [52]. The supported TiO$_2$ film was submerged in the reactant solution and irradiated through the wall of the quartz

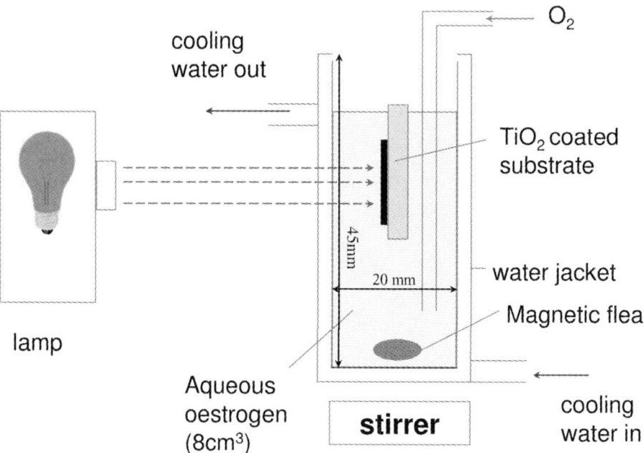

Fig. 10.4. Quartz water-jacketed reactor [52].

reactor using a 150-W xenon lamp (Sage Analytical) with stabilized power supply (Applied Photophysics) (Fig. 10.4). The incident light was passed through a borosilicate glass filter prior to the reactor to remove $\lambda < 300$ nm. The solution was purged with oxygen (BOC) before and during illumination and the reactor was thermostatted at 20 ± 2 °C. The photonic flux entering the reactor ($\lambda = 300$–400 nm) was determined to be 8×10^{-8} Einsteins s^{-1} cm^{-2} by ferrioxalate actinometry [53] with a 300–400 nm band pass filter (Speirs Robertson). 17β-Oestradiol degradation was monitored using HPLC with fluorescence detection. The HPLC conditions were: 25 cm ODS column (Hypersil), P2000 solvent delivery pump (Thermo-separation Products), mobile phase 40:60% acetonitrile:water, flow rate of 1 mL min^{-1}. The fluorescence detector (Perkin Elmer LS30) was linked to an $x - y$ plotter. Excitation and emission wavelengths used were $\lambda_{ex} = 280$ nm and $\lambda_{em} = 315$ nm.

In the presence of the illuminated TiO$_2$ film, 3 μmol dm^{-3} (0.8 mg L^{-1}) 17β-oestradiol was 50% degraded in 40 min and 98% degraded in 3.5 hours. When the samples were illuminated in the absence of titanium dioxide, there was about 11% degradation in 40 min, and 44% in 3.5 h, showing that some direct photolysis takes place. A semi-log plot of initial concentration versus irradiation time was linear, indicating overall pseudo-first order kinetics. The experiment was repeated with a range of initial concentrations from 0.05 to 3 μmol dm^{-3}. The initial rates for each concentration were determined from the pseudo-first-order rate constants and initial concentrations. The data were then fitted to the Langmuir–Hinshelwood kinetic rate model, which has been applied to the initial rates of photocatalytic destruction of many organic compounds [36]. The rate law is shown in Eq. (11), where R_i is the initial rate of disappearance of substrate and C_i is the initial concentration, k is the reaction rate constant and K is the Langmuir adsorption constant.

$$R_i = -dC_i/dt = kKC_i/(1 + KC_i) \tag{11}$$

From a plot of $1/R_i$ versus $1/C_i$, with the slope equal to $1/kK$ and intercept equal to $1/k$, k and K were determined as 4.4×10^{-2} µmol dm^{-3} min^{-1} and 3.47×10^{-1} dm^3 µmol^{-1}, respectively.

This initial study showed that the principal natural oestrogen 17β-oestradiol is readily degraded by semiconductor photocatalysis under oxygen on immobilized TiO$_2$ powder and that photocatalysis was much more effective than UV light alone. The initial rate kinetics fit the Langmuir–Hinshelwood model. Micromolar concentrations of aqueous 17β-oestradiol were 98% degraded in 3.5 h by photocatalysis over the titanium dioxide powder immobilized on Ti-6Al-4V alloy. The degradation kinetics were fitted to a Langmuir–Hinshelwood model with $k = 0.044$ µmol dm^{-3} min^{-1} and $K = 0.347$ dm^3 µmol^{-1}. The pseudo-first order rate constant was 0.016 min^{-1}. It is very important to determine if all oestrogenic activity is removed from water samples after treatment with photocatalysis since this is the main concern. This was, therefore, investigated in the same quartz water-jacketed reactor and monitored using a yeast screen bioassay as a test for oestrogenicity (Section 10.5).

10.5
Rapid Loss of Oestrogenicity of Natural and Synthetic Oestrogens in Water by Photocatalysis and UVA Photolysis Monitored using a Yeast Screen Bioassay

The presence of low levels of natural and synthetic steroid oestrogens in the aquatic environment, and their biological and oestrogenic effects on aquatic organisms, are presently issues of concern. In this study, we investigated the temporal removal of oestrogenic activity of the potent and environmentally relevant steroid oestrogens by photocatalysis over an immobilized titanium dioxide catalyst. We used a recombinant yeast assay to measure oestrogenic activity. Application of photocatalysis to remove steroid compounds and their oestrogenic activity within STW effluent released into the aquatic environment is discussed.

Stock solutions of 17β-oestradiol, oestriol, oestrone and 17α-ethynylestradiol (100 mg L^{-1}) (>98%; Sigma-Aldrich) were prepared in ethanol and then diluted 10 000-fold to a working stock concentration of 10 µg L^{-1} using sterile double-distilled water (0.01% ethanol final concentration). These working stocks were prepared shortly before the experiments, and were added directly to the reactor. The working stock concentration was chosen on the basis that a 10-µL aliquot would produce a maximal response in the yeast screen (concentration of 500 ng-steroid L^{-1} in the assay) without the need for further sample manipulation. Moreover, any removal of steroid during the reactions would result in a readily detectable loss in the assay response, as the steroid concentrations would fall within the linear part of the dose–response curve of the assay.

The quartz water-jacketed reactor with titanium dioxide immobilized on titanium alloy (Fig. 10.4) was used with a 125-W Philips high-pressure mercury lamp placed 3 cm from the reactor. Some 8 mL of freshly prepared steroid (10 µg L^{-1}) in sterile water was transferred to the reactor and allowed to equilibrate for 1 h before irradiation with or without the TiO_2 catalyst (photocatalysis and photolysis, respectively). Duplicate samples (10 µL) were removed at intervals starting at time 0 (before UV light exposure), and then regularly throughout the experiments. The 10 µL aliquots were assayed directly in the recombinant yeast oestrogen assay. Details of the oestrogen-inducible expression system in yeast (validation) and preparation of the medium components have been described previously [54]. In brief, this yeast expresses the human oestrogen receptor (hERα) and contains expression plasmids containing oestrogen-responsive sequences that control expression of the reporter gene Lac-Z (coding for the enzyme β-galactosidase). In the presence of oestrogens, β-galactosidase is produced and is secreted into the medium, where it breaks down the yellow chromogenic substrate chlorophenol red-β-D-galactopyranoside (CPRG) into a red product that can be measured by absorbance at 540 nm. Duplicate water samples (10 µL) from each run were transferred to a 96-well optically flat microtiter plate (Linbro/Titertek, ICN FLOW, Bucks UK) at frequent time intervals throughout the experiments. Seeded yeast medium (190 µL) containing CPRG was then added to the microtiter plate and the plates were then sealed with autoclave tape, shaken vigorously for 2 min on a titer plate shaker, and incubated for 72 h at 32 °C. Each plate contained duplicate rows of samples at each time point, a row of blanks (200 µL assay medium only), a row containing sterile water (solvent control) and a row containing a serial dilution of the appropriate steroid substrate (positive control). After incubation, the plates were shaken and allowed to settle for 1 h, after which the absorbance was read at 540 nm (for color) and at 620 nm (for turbidity) using a Titertek Multiscan MCC/340 plate reader. To correct the oestrogenic response of each test chemical for turbidity (A_{corr}), the correction shown in Eq. (12) was applied to the data in each well.

$$A_{corr} \text{ Chem} = A_{540} \text{ Chem} - (A_{620} \text{ Chem} - A_{620} \text{ Blank}) \tag{12}$$

Nominal steroid concentrations in the reactor were derived from the appropriate steroid standard curve, using the mean corrected absorbance values from each duplicate sample.

17β-Oestradiol, oestrone and 17α-ethynyloestradiol were found to be oestrogenic *in vitro*, and their relative potencies agreed with previous findings [54]. Limits of detection for the measurement of oestrogenic activity within the reactor were 53 ng L^{-1} for 17β-oestradiol and 17α-ethynyloestradiol, and 100 ng L^{-1} for oestrone. Oestriol did not produce an oestrogenic response in the yeast screen bioassay, indicating that it has insufficient oestrogenic activity to be detected at this level.

Results from the water control experiments were negative, indicating that both the reactor and the sterile water were free of oestrogenic contamination before commencing the experiments (data not shown). Similarly, in control experiments (8.5 h duration) with steroids in the absence of UV light the level of oestrogenic activity was unchanged over that time (data not shown). Therefore, no discernable adsorption of oestrogenic substrate on to the surface of the reactor, immobilized titanium dioxide or magnetic flea had occurred. There was also no difference in activity between standards prepared in ethanol and those prepared in sterile water.

Tab. 10.2. Time (minutes) taken for 50%, 90% and 100% removal of oestrogenic activity by photocatalysis and UVA photolysis.

Steroid oestrogen	Photocatalysis			UV radiation		
	50%	90%	100%	50%	90%	100%
17β-oestradiol	10	24	55	195	248	485
Oestrone	7	18	60	68	195	360
17α-Ethynyloestradiol	8	27.5	50	23	72	120

Photocatalysis was the most effective method of inactivating all three steroid oestrogens, with virtually all oestrogenic activity being removed within 55 min. In contrast, UVA photolysis took 9× longer for oestradiol, 6× longer for oestrone and 2.4× longer for ethynyloestradiol. For all three oestrogens, 50% of their oestrogenic activity was removed by photocatalysis within 10 min, and 100% within 1 h (Table 10.2 and Figs. 10.5–10.7). The decay rates (measured as loss of oestrogenic activity)

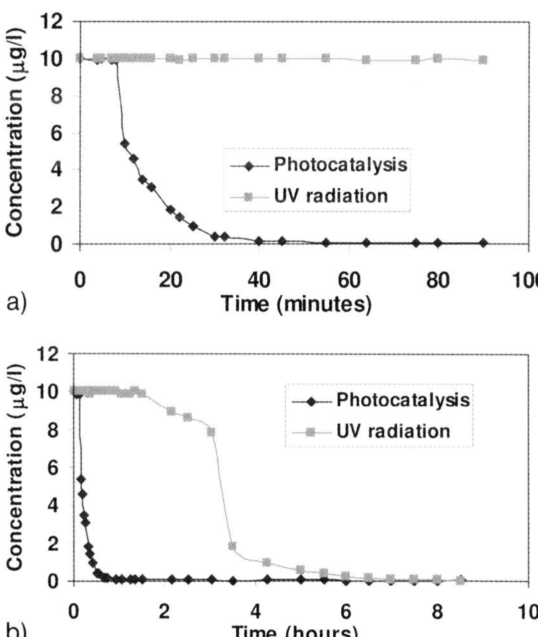

Fig. 10.5. Photocatalysis and UVA photolysis of 17β-oestradiol monitored using the yeast screen bioassay. Temporal removal of oestrogenic activity is shown over 90 min (a) and 8.5 h (b) reaction time.

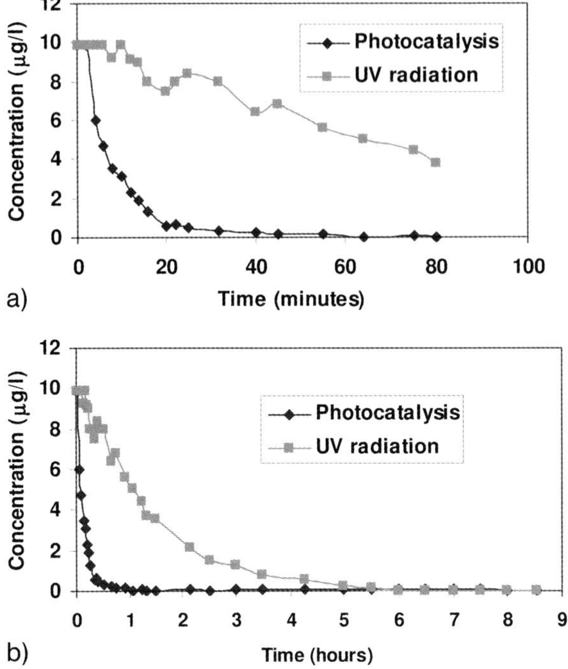

Fig. 10.6. Photocatalysis and UVA photolysis of oestrone monitored using the yeast screen bioassay. Temporal removal of oestrogenic activity is shown over 80 min (a) and 8.5 h (b) reaction time.

were similar for all three oestrogens (Table 10.3). In contrast, there were substantial differences in decay rates of the three steroids by UVA photolysis (no TiO_2 catalyst), with 17α-ethynylestradiol decaying considerably faster than either of the two other steroids (Table 10.3).

We have directly compared temporal changes in the oestrogenic activity of aqueous solutions (initially 10 µg L^{-1}) of 17β-oestradiol, oestrone and 17α-ethynyloestradiol following both UVA photolysis and TiO_2 photocatalytic degradation. A recombinant oestrogen assay based on yeast was used to measure the biological activity of the reaction mixtures. This yeast screen assay was previously shown to be highly specific for a range of steroid oestrogens and their metabolites, and the relative potencies of the steroids used in this study were found to be consistent with previous findings [54]. The yeast oestrogen screen detects both the parent compounds and any intermediate degradation products that bind to, and activate, the oestrogen receptor, regardless of their identity. Thus, this approach provides a real indication of the ability of the various treatments to affect the biological activity of the reaction mixture over time; information that cannot be obtained by analytical chemistry alone. This is particularly important given the rela-

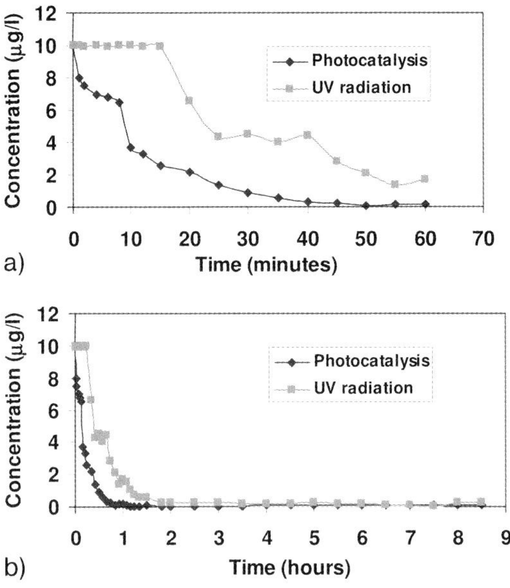

Fig. 10.7. Photocatalysis and UVA photolysis of 17α-ethynyloestradiol monitored using the yeast screen bioassay. Temporal removal of oestrogenic activity is shown over 60 min (a) and 8.5 h (b) reaction time.

tively persistent and poorly characterized nature of the intermediate products of endocrine disrupters generated during photocatalytic degradation [55, 56].

We have demonstrated that the oestrogenic activity of 17β-oestradiol, oestrone and 17α-ethynyloestradiol was eliminated at the same rate during photocatalysis, whereas variable (and much slower) removal rates of oestrogenic activity occurred by UVA photolysis, with the order 17α-ethynyloestradiol > oestrone > 17β-oestradiol. This compares favorably with our work in a quartz coil reactor monitored using fluorescence detection (Section 10.6) [57]. The ability of UVA to

Tab. 10.3. Kinetic data for photocatalysis and UVA photolysis of oestrogens.

Steroid	Photocatalysis pseudo-first order rate constant (min^{-1})	UVA photolysis pseudo-first order rate constant (min^{-1})
17β-Oestradiol	0.106	0.011
Oestrone	0.086	0.012
17α-Ethynyloestradiol	0.086	0.035

remove all three steroids is important given that surface water systems are exposed to natural sunlight, and this may provide a mechanism for the removal of oestrogenic effects. Also, some water treatment plants employ UVC radiation to disinfect potable water, which may help eliminate oestrogenic compounds. The effect of UVC disinfection on degrading oestrogens in water is discussed further in Section 10.7. Although 17α-ethynyloestradiol appears to be more susceptible to UVA degradation than the natural oestrogens 17β-oestradiol and oestrone, the ethynyl group of 17α-ethynyloestradiol also increases its resistance to bacterial oxidation, and therefore its persistence during STW aerobic digestion [5]. The similarity in removal rates of oestrogenic activity during TiO$_2$ photocatalysis supports a similar mechanism of degradation for the steroid compounds. This may occur via extraction of the benzylic hydrogen to form the ·CH radical, which combines with oxygen, or via attack of the hydroxyl group to form the quinone. Indeed, Ohko and colleagues previously reported that photocatalysis of 17β-oestradiol starts via oxidation of the phenol moiety [58], which is critical for receptor binding and for conferral of oestrogenicity to all steroid oestrogens [59].

Given that most known xenoestrogens are phenolic chemicals, and benzene rings are rapidly photodegraded by preferential hydroxyl radical attack [55, 58], it is likely that photocatalysis may quickly remove the capacity of these chemicals to bind to, and activate, the oestrogen receptor. The measurement of total organic carbon (TOC) may therefore underestimate the true capacity of photocatalysis to inactivate endocrine-disrupting chemicals (EDCs), where early reaction stages are key. However, the production of intermediate products with different types of biological activity must be considered. For example, testosterone-like species were identified by GC/MS during the photocatalysis of 17β-oestradiol [58], although any putative androgenic activity of these degradation products was not investigated.

Oestrone is a major and relatively persistent biotransformation product of oestradiol during aerobic digestion [11]. The fact that photocatalysis was able to eliminate oestrone as quickly as 17β-oestradiol is therefore encouraging, given that oestrone is only around 2-fold less oestrogenic than 17β-oestradiol. However, 17α-ethynyloestradiol is probably the most important steroid pollutant, given that it is biologically active in fish at concentrations as low as 0.1 ng L^{-1} [60, 61]. The greater potency of 17α-ethynyloestradiol *in vivo* depends on the 17-ethynyl group, which increases its longevity in the body by reducing the rate of metabolism at C-16 and C-17 compared with endogenous steroids [62]. As photocatalysis was able to eliminate all three steroid oestrogens at a similar rate, it may be a valuable process for reducing the impact of this persistent and highly active group of compounds in the aquatic environment. The initial concentrations of steroids used in our reactor were three to four orders of magnitude greater than reported environmental levels, while the limits of detection for the steroids within the reactor approached levels reported for oestradiol and oestrone measured in influents to STWs [3].

In summary, we have demonstrated rapid temporal removal of the biological activity in aqueous solutions of oestrone, 17β-oestradiol and 17α-ethynyloestradiol by photocatalysis over an immobilized TiO$_2$ catalyst. Photocatalysis was much more efficient than UV light alone. Similar rates of removal for all these oestrogens support an identical mechanism of action that is likely to involve oxidation of the phenol moiety of the steroids. Photocatalytic treatment of wastewaters could therefore serve to further decrease the oestrogenic steroid load entering the aquatic environment. In a more detailed study on photocatalysis of the natural and synthetic oestrogens in water we employed fluorescence detection, and also investigated the effects of initial pollutant concentration and light intensity (Section 10.6).

10.6
Photocatalytic Degradation of 17β-Oestradiol, Oestriol and 17α-Ethynyloestradiol in a Quartz Coil Reactor Monitored using Fluorescence Spectroscopy

Photocatalytic degradation of the natural oestrogens 17β-oestradiol and oestriol and the synthetic oestrogen 17α-ethynyloestradiol in water were investigated. Reactions were carried out in a quartz coil reactor coated internally with titanium dioxide (Degussa P25). Degradation by UV light alone was also investigated. Fluorescence spectroscopy was used to monitor the reactions. The effect of both initial concentration and light intensity on photocatalysis and photolysis of 17β-oestradiol in water were also investigated.

A quartz coil reactor and a Hanovia 125-W medium pressure mercury lamp blanketed in nitrogen were used [57]. Coils were prepared from a 1 m length of 2 mm ID quartz tubing with TiO_2 immobilized onto the quartz coil as follows. The coil was first cleaned with chromic acid, rinsed thoroughly and dried. It was then filled with a 20% solution of hydrofluoric acid and allowed to stand for about 30 min, after which it was washed thoroughly with distilled water. The coil was filled with TiO_2 suspension (1% suspension of Degussa P25 TiO_2 powder, sonicated for 20 min) and allowed to stand for a few minutes, drained and then dried by passing warm air through it. A thin white film of TiO_2 was formed on the inside surface of the coil. This procedure was repeated 2–3 times to ensure that the inside surface of the coil is completely covered. After the final drying, distilled water was continuously pumped through the coil to wash away any loose powder [63]. The wavelength of light emitted from the Hanovia lamp also includes the UVB and UVC regions of the spectrum. Experiments with TiO_2 are therefore photolysis as well as photocatalysis. A solution (3 μmol dm^{-3}; ∼0.8 mg L^{-1}) of the oestrogen under study was made up in MilliQ water from a stock solution in acetonitrile. The oestrogen solution was passed through the quartz coil reactor at different flow rates, giving different retention times, and a sample was collected and analyzed by fluorescence detection, λ_{ex} 230 nm, λ_{em} 310 nm. This experiment was repeated using an identical quartz coil without TiO_2, i.e., photolysis of the oestrogen solution. This procedure was carried out at least twice for each oestrogen. The effect of initial concentration on the reaction with 17β-oestradiol was investigated by repeating the procedure at varying concentrations (3, 2, 1, 0.5, 0.25, and 0.1 μmol dm^{-3}) (0.8–0.03 mg L^{-1}). The effect of light intensity on photocatalysis and photolysis of 17β-oestradiol was investigated by placing the reactor at different distances from the

Tab. 10.4. Average values for rate constant, k, initial rate and half-life for photocatalysis and photolysis of 17β-oestradiol, oestriol and 17α-ethynyloestradiol.

	k (min^{-1})	Initial rate (μmol dm^{-3} min^{-1})	Half-life (min)	R^2
17β-Oestradiol				
Photocatalysis	0.174	0.522	2.095	0.934
Standard error	0.016		0.282	0.005
Photolysis	0.134	0.402	3.45	0.920
Standard error	0.004		0.259	0.018
Oestriol				
Photocatalysis	0.156	0.468	3.775	0.972
Standard error	0.001		0.018	0.003
Photolysis	0.093	0.280	6.65	0.966
Standard error	0.006		0.318	0.010
17α-Ethynyloestradiol				
Photocatalysis	0.231	0.694	1.55	0.907
Standard error	0.006		0.106	0.001
Photolysis	0.195	0.585	1.775	0.926
Standard error	0.011		0.018	0.007

lamp (3.25, 5, 7.5 and 10 cm, corresponding to light intensities of 2509, 1060, 470 and 265 mW, respectively).

Table 10.4 gives average values obtained for the rate constant, k, the initial rate and the half-life for photocatalysis and photolysis of 17β-oestradiol, oestriol and 17α-ethynyloestradiol.

The fact that the lamp used in this work (125-W Hanovia medium-pressure mercury) emits radiation in the UVA, UVB and UVC region of the spectrum some photolysis as well as photocatalysis takes place. The results show that both photocatalysis and photolysis are effective in degrading all three oestrogens in water. Plots of time against $\ln(C/C_0)$ for the photocatalysis and photolysis of 17β-oestradiol confirm our earlier work [51, 64] (Sections 10.4 and 10.5) that photocatalytic and photolysis of 17β-oestradiol and 17α-ethynyloestradiol follow pseudo-first order reactions. More recently, other workers [58, 65–67] have reported similar findings. This work shows that oestriol follows similar behavior to the other natural oestrogens and the synthetic oestrogen. The sets of results are consistent with each other, with a good correlation in each case. This is confirmed by the low standard errors of the average values for the rate constant, k, initial rate and correlation coefficient (R^2) (Table 10.4). The average rate constants for photocatalysis and photolysis show that the photocatalytic degradation of 17β-oestradiol and oestriol is almost

1.5× faster than degradation by UV light alone. However, photocatalysis and photolysis of 17α-ethynyloestradiol occur at almost the same rate (Table 10.4).

Table 10.4 also shows that 17α-ethynyloestradiol has the fastest rate for photocatalysis, followed by 17β-oestradiol and finally oestriol (1.5× slower than 17α-ethynyloestradiol). 17α-Ethynyloestradiol also degrades the fastest by photolysis, at almost the same rate as photocatalysis, degrading 1.5× faster than 17β-oestradiol and over twice as fast as oestriol. '17α-ethynyloestradiol seems to be a less stable molecule than 17β-oestradiol and oestriol under photocatalytic and photolytic conditions. The addition of the ethynyl group possibly causes the molecule to be less stable and degrade more rapidly under these conditions due to the triple bond of the ethynyl group which would absorb UV light more easily. Oestriol seems to be the most stable molecule of the three oestrogens, degrading at the slowest rate. The addition of the OH group may stabilise the phenolic ring and help resist breakdown by photocatalysis and UV light. The mechanism of degradation of the oestrogens may occur via extraction of the benzylic hydrogen to form the CH˙ radical which combines with oxygen, or via attack of the hydroxyl group to form the quinine (as outlined in section 5). Ohko *et al.* previously reported that the photocatalysis reaction with 17β-oestradiol starts via the phenol moiety and also confirmed that 17β-oestradiol in water is completely mineralised as a result of the photocatalytic reactions and suggested a mechanism for the reaction, identifying 10ε-17β-dihydroxy-1,4-estradien-3-one and testosterone-like species as intermediate products [58]'.

Initial work on photocatalysis of 17β-oestradiol [51] in a batch reactor with TiO_2 immobilized on Ti alloy gave a first-order rate constant of 0.016 min^{-1} (Section 10.4). This is much lower than the rate constant obtained in this work (0.174 min^{-1}), indicating that the quartz coil reactor is a much more efficient system, owing to the increased surface area. The trend obtained here for photocatalysis compares favorably with work [64] detailed in Section 10.5 where photocatalytic degradation of 17β-oestradiol, 17α-ethynyloestradiol and oestrone were monitored using a yeast screen bioassay as a test for oestrogenicity. The oestrogenic activity of 17β-oestradiol, oestrone and 17α-ethynyloestradiol was eliminated at the same rate during photocatalysis whereas variable (and much slower) removal rates occurred by UVA photolysis, with the order 17α-ethynyloestradiol > oestrone > 17β-oestradiol (0.086–0.106 min^{-1}). The rates of photolysis in this work (0.134 min^{-1} for 17β-oestradiol and 0.195 min^{-1} for 17α-ethynyloestradiol) are much higher than for this previous work (0.011 and 0.035 min^{-1}, respectively) due to the different lamps used. The lamp used in this work (125-W Hanovia medium-pressure mercury) emits radiation in UVA, UVB and UVC region of the spectrum, indicating that some photolysis is taking place as well as photocatalysis. The increase may also be due to the increased area for reaction in the coil reactor.

Table 10.5 gives the average values for the rate constant, *k*, initial rate and half-life calculated for each concentration.

Initial concentration against rate was plotted for the results obtained above for both photocatalysis and photolysis of 17β-oestradiol (Figs. 10.8 and 10.9, respectively).

Tab. 10.5. Average values of rate constant, k, initial rate and half-life for photocatalysis and photolysis of 17β-oestradiol at different initial concentrations.

Concentration (μmol dm^{-3})	k (min^{-1})	Initial rate (μmol dm^{-3} min^{-1})	Half-life (min)	R^2
Photocatalysis				
0.1	0.086	0.009	7.4	0.925
0.25	0.141	0.036	4.062	0.944
0.5	0.164	0.082	3.188	0.953
1	0.224	0.224	2.062	0.942
2	0.188	0.376	2.125	0.898
3	0.174	0.522	2.095	0.934
Photolysis				
0.1	0.125	0.012	2.7	0.785
0.25	0.137	0.034	2.55	0.844
0.5	0.187	0.094	1.44	0.862
1	0.200	0.200	2.0	0.944
2	0.138	0.276	2.812	0.898
3	0.134	0.402	3.45	0.920

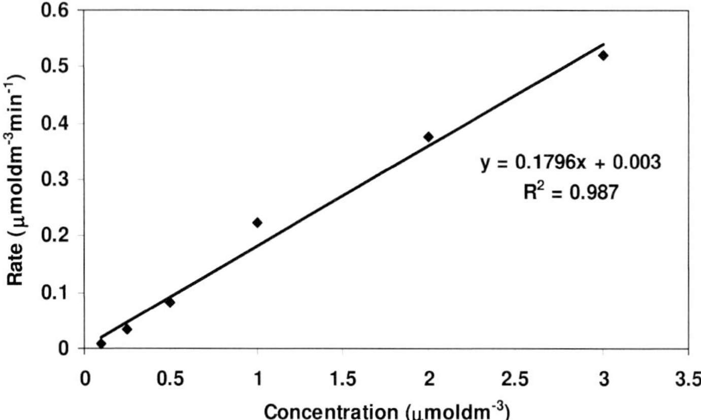

Fig. 10.8. Initial concentration against rate for photocatalysis of 17β-oestradiol monitored using fluorescence.

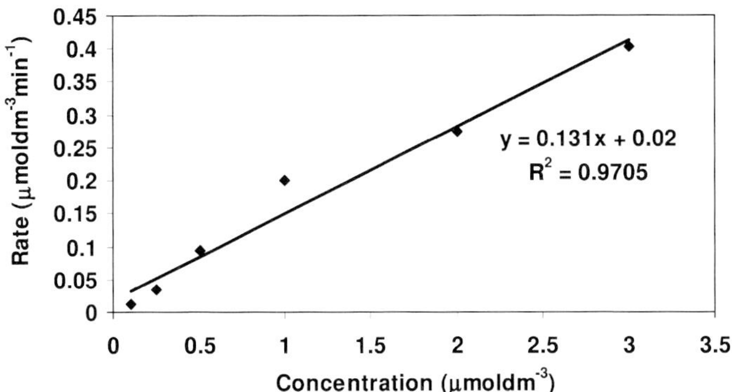

Fig. 10.9. Initial concentration against rate for photolysis of
17β-oestradiol monitored using fluorescence.

The results are consistent with each concentration with good correlation (Table
10.5). The slope of the best-fit line is 0.180 for photocatalysis and 0.131 for photolysis (Figs. 10.8 and 10.9, respectively). The results demonstrate that, as the initial
concentration of 17β-oestradiol increases, the reaction rate increases proportionally
for both photocatalysis and photolysis, confirming that the reactions are first order
[51, 57]. The slope of the graph for photocatalysis is 1.5× that of photolysis, indicating that photocatalysis coupled with photolysis occurs at 1.5× the rate of photolysis alone.

Interestingly, the effect of initial concentration on photocatalysis follows a different pattern than shown previously [51] where the Langmuir–Hinshelwood model
applied for the same concentration range. In this model, adsorption of the reactant
on the surface of the catalyst is considered. The rate increases with initial concentration of 17β-oestradiol and then levels off at higher concentrations. This is because, at higher concentrations, the surface of the TiO_2 becomes saturated with
oestradiol molecules and the reaction rate reaches a maximum. In this work, photolysis occurs at a rate comparable to that of photocatalysis. This competing reaction also consumes 17β-oestradiol, reducing the number of molecules available
for saturation of the titanium dioxide surface. This observation may be important
in industrial applications of the photocatalytic breakdown of organic pollutants in
water. Photocatalysis and photolysis could be applied together to increase rates of
reaction at high concentrations of pollutants.

Table 10.6 summarizes results obtained for the effect of light intensity on both
photocatalysis and photolysis of 17β-oestradiol. Good correlation was obtained for
all the results.

Light intensity (I) is an important parameter to consider in photocatalysis
($R \propto I$), especially for industrial applications in terms of cost. Increasing the light
intensity affects the rate of the reaction by increasing the number of charge carriers generated in the semiconductor. Most researchers have found different effects
at different levels of light intensity [68].

Tab. 10.6. Average values for k, initial rate, half-life and R^2 for effect of light intensity on photocatalysis and photolysis of 17β-oestradiol.

Distance (cm)	Light intensity (mW)	k (min^{-1})	Initial rate (μmol dm^{-3} min^{-1})	Half-life (min)	R^2
Photocatalysis					
3.25	2509	0.500	1.501	0.362	0.926
5	1060	0.324	0.972	1.25	0.939
7.5	470	0.236	0.708	1.625	0.882
10	265	0.174	0.522	2.095	0.934
Photolysis					
3.25	2509	0.400	1.200	0.1	0.876
5	1060	0.364	1.092	0.875	0.973
7.5	470	0.229	0.687	1.7	0.918
10	265	0.134	0.402	3.45	0.920

- At low light intensities, the rate increases in proportion to the light intensity, i.e., $R \propto I$, $n = 1$.
- At intermediate light intensities the rate only varies with the square root of intensity [69, 70], i.e., $R \propto I^{1/2}$, $n = 1/2$.
- At high light intensities the rate of photodegradation is independent of light intensity, i.e., $R \propto I^0$, $n = 0$.

For photocatalysis, as light intensity increases the rate increases due to the increased number of oxidizing species produced. The rate increases with light intensity to a power n (the gradient of a linear log–log plot, [Eq. (14)]). At low light intensities the rate of initial degradation increases directly in proportion to light intensity (gradient $\sim$ 1), suggesting that few oxidizing species are lost through recombination processes. At high light intensities the rate of initial degradation increases in proportion to I to the power of 0, i.e., the rate becomes independent of light intensity and the expected rate-limiting factor is mass transfer. At intermediate light intensities the rate only varies with the square root of intensity [69, 70] and hence efficiency suffers. This was attributed by Egerton and King [71] to energy wasting recombination reactions between electrons and holes and by Kormann [72] to bimolecular combination of hydroxyl radicals.

Increased intensity always results in an increase in the volumetric reaction rate until the mass transfer limit is encountered. However, once intermediate light intensities are reached any increase in I will not lead to a proportional increase in rate and industrially may not be worth the extra cost. The I^1 to $I^{0.5}$ rate transition is

Tab. 10.7. Dependency of rate of photo-catalysis on light intensity for various organic substrates.

Substrate	n	Ref.
Phenol	0.6	74
Phenol	0.5	69
4-Chlorophenol	0.7	47
Formaldehyde	0.4	75
Methyl orange	0.2	76
Dichlorophenoxyacetic acid	0.5	77
Phenoxyacetic acid	0.5	78
2,4-Dichlorophenol	0.5	79
Cyanide	0.5	80
Salicyclic acid	0.6	81

said to depend on the catalyst material [73]. The relationship between rate and light intensity can therefore be represented by Eq. (13).

$$\text{Rate} = k_1 I^n \tag{13}$$

The rate constant k will include terms for the extent of recombination, substrate concentration, etc. This constant will be different for different transition regions. The linear form of this equation is:

$$\ln(\text{rate}) = \ln k + n \ln(I) \tag{14}$$

A plot of ln(rate) against ln(light intensity) will yield a straight line for each transition region, with a slope equal to n. For the photocatalysis of 17β-oestradiol, such a plot gave a best fit line with a slope, n, of 0.461, which is approximately 0.5, indicating intermediate light intensities. Other workers found similar results when investigating light intensities for other substrates. Table 10.7 shows the literature values of n for different organic compounds.

Dionysiou et al. have reported the photocatalytic degradation of 4-chlorobenzioc acid as a function of light intensity, using a rotating disk photocatalytic reactor [82]. They found that the rate of degradation followed a linear dependency with incident light intensity and attributed it to the existence of low local values of incident light intensity on the illuminated disk. Ohko et al. have investigated the effect of light intensity on the degradation of propan-2-ol and the effect of concentration [83]. They suggested that for small concentrations of contaminant it is more beneficial to operate the process at low light intensities and at the range where the rates are not mass-transport controlled. At high contaminant concentration, the reaction

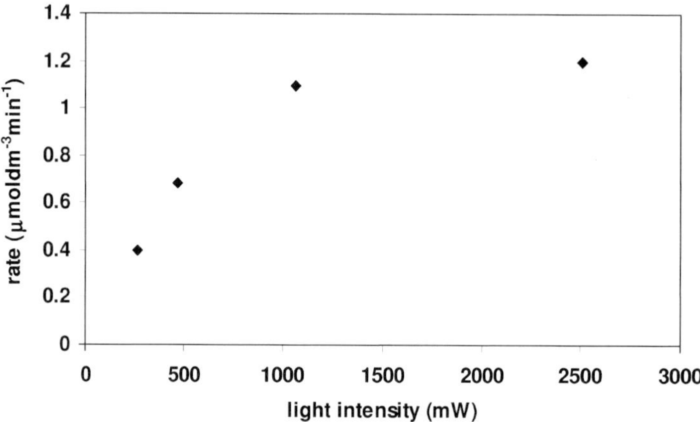

Fig. 10.10. Rate against light intensity for photolysis of 17β-oestradiol monitored using fluorescence.

may be light-limited but light utilization efficiency will be higher. This was also reported by Minero [84], who recommended that, in solar photocatalytic applications, concentration of the solar light is unnecessary. Industrially, these are important points to consider if photocatalysis were applied for the treatment of real environmental water samples where oestrogens and other EDCs are present in very low concentrations and also in situations where solar illumination is used as the UV source. Figure 10.10 shows the plot of rate against light intensity for the photolysis of 17β-oestradiol.

The relationship is linear at low light intensities up to a certain point (intermediate light intensity) where the rate starts to level off (high light intensities). This suggests that the rate is proportional to light intensity up to a certain point, where it then becomes independent of light intensity. This could be interesting from an industrial viewpoint, where a combination of photocatalysis and UV radiation could be used to increase the rate of reaction instead of increasing the UV light alone, where the rate levels off at high light intensities.

Photocatalysis and photolysis are effective for the degradation of the three oestrogens 17β-oestradiol, oestriol and 17α-ethynyloestradiol in water. Photocatalysis coupled with photolysis is much more effective in degrading the oestrogens than photolysis alone. The reactions follow pseudo-first order kinetics. 17α-Ethynyloestradiol degrades the fastest for both photocatalysis and photolysis followed by 17β-oestradiol and then oestriol. This was attributed to the triple bond of the ethynyl group, which absorbs UV light more easily. The rate varies linearly with initial concentration both for photocatalysis combined with photolysis and for photolysis of 17β-oestradiol in water. Photocatalysis degrades 17β-oestradiol at twice the rate of photolysis. The rate was proportional to the square root of light intensity for photocatalysis of 17β-oestradiol. The relationship between light intensity and degradation rate by photolysis is linear up to a point, when it then starts to level off. The ability of UV light to remove all steroids is important given that surface water systems are exposed to natural sunlight, and this may provide a mechanism for the removal of oestrogenic effects. Also, some water treatment plants employ UVC radiation to disinfect potable water, which would help to eliminate oestrogenic compounds in the water. An investigation into the efficiency of UVC disinfection for the removal of the natural and synthetic oestrogen in water was made and compared with photocatalysis and UVA light alone. This work is outlined in section 7 below.

10.7
Comparison of Photocatalysis with UVA and UVC Radiation for the Degradation of Natural and Synthetic Oestrogens in Water

Our previous studies have shown that TiO_2 photocatalysis is effective for the degradation of natural and synthetic oestrogens in water and that it is more effective than UV light alone. Some water treatment companies currently use UVC disinfection to remove trace organics as a final stage in the water treatment process. UVC radiation uses high intensity light at 253 nm, which is energy intensive and therefore very expensive. In this work we study the effect of UVC disinfection for the removal of the natural and synthetic oestrogens in water and compare it with photocatalysis and UVA light alone.

Photocatalysis experiments were carried out in a spiral Pyrex reactor (85 mL volume), with TiO_2 immobilized onto the inside wall, with a black light blue fluorescent lamp (NEC, 15 W, maximum emission at ~350 nm, emission range 300–400 nm) fitted through the centre of the coil. TiO_2 was immobilized onto the reactor walls as described previously [57] (Section 10.6). Experiments with UVA light alone were carried out in the same reactor free from TiO_2. A similar set-up was used for the UVC radiation experiments except the spiral reactor (85 mL volume) was made from quartz and free from TiO_2 and was used with a slim line germicidal lamp (UV Air Pty Ltd, emission $\lambda = 253$ nm). The reactors were connected to a peristaltic pump (Masterflex ® Quick-Load, Cole-Palmer Instrument Co.) by Masterflex flexible tubing to enable solution circulation through the reactor at 150 mL min^{-1} and to an on-line fluorescence spectrometer (Perkin Elmer LS-45 with FLWinLab software), which monitored degradation of the oestrogens at $\lambda_{ex} = 230$ nm and $\lambda_{em} = 310$ nm.

A standard solution of 3 µmol dm^{-3} of oestrogen (i.e., 0.82 mg L^{-1} 17β-oestradiol, 0.86 mg L^{-1} oestriol and 0.89 mg L^{-1} 17α-ethynyloestradiol) was made up in MilliQ water from stock solutions in acetonitrile (final pH 5.5). Oestrogen (120 mL) was then pumped through the reactor set-up at 150 mL min^{-1} for approximately 5 min to allow the solution to equilibrate before the experiment was started. The UV lamp was then turned on and the "timedrive" method (FLWinLab software), which measured fluorescence intensity over time, was started where measurements were taken every second. This procedure was carried out for each oestrogen in duplicate for TiO_2 photocatalysis, UVA radiation and UVC radiation.

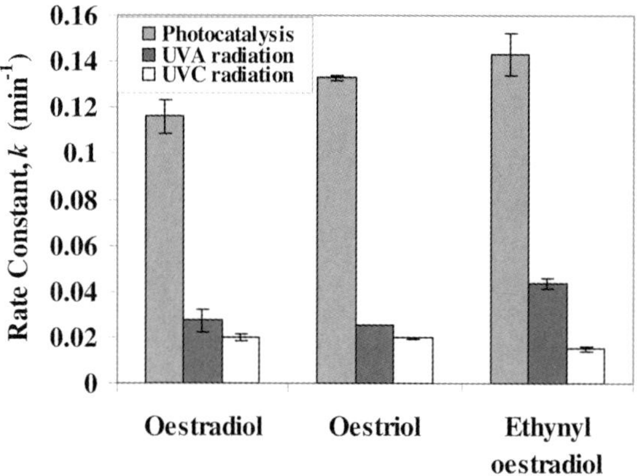

Fig. 10.11. Degradation by photocatalysis, UVA and UVC radiation of natural and synthetic oestrogens in water in a spiral reactor monitored using fluorescence spectroscopy.

Figure 10.11 shows a histogram of the rate constants for photocatalysis of the natural and synthetic oestrogens over TiO_2/UVA compared to UVA and UVC radiation alone. Table 10.8 presents the first-order rate constants for each reaction with standard errors.

Clearly, from Fig. 10.11 and Table 10.8, photocatalysis is much more effective than UVA or UVC radiation for all of the oestrogens. The rates are up to 5× greater for photocatalysis than for UVA light alone and up to 9× faster (17α-ethynyloestradiol) than with UVC disinfection. Photocatalytic rates are quite comparable for all three oestrogens when standard errors are taken into account, although rates for 17α-ethynyloestradiol and oestriol are slightly higher than for 17β-oestradiol. UVA radiation rates for the natural oestrogens 17β-oestradiol and oestriol are comparable. However, UVA radiation for the synthetic oestrogen 17α-ethynyloestradiol is significantly higher. This agrees with previous work in a

Tab. 10.8. Kinetic data for photocatalysis, UVA photolysis and UVC photolysis of oestrogens in water.

Steroid	First order rate constants, k (min^{-1})		
	Photocatalysis	UVA photolysis	UVC photolysis
17β-Oestradiol	0.116 ± 0.007	0.028 ± 0.005	0.0201 ± 0.002
Oestriol	0.133 ± 0.001	0.025	0.196
17α-Ethynyloestradiol	0.143 ± 0.009	0.044 ± 0.002	0.0151 ± 0.001

stirred-tank batch reactor with a 125-W high-pressure mercury lamp (Section 10.5) [64] and a photocatalytic quartz coil reactor and 125-W medium-pressure mercury lamp (Section 10.6) [57] and, as mentioned earlier, may be because the triple bond of the ethynyl group present would absorb UV light more easily. Rates for UVA and UVC radiation are comparable for 17β-oestradiol and oestriol. However, UVA and UVC radiation rates for 17α-ethynyloestradiol differ significantly, with UVA radiation being almost 3× faster than UVC radiation. 17α-Ethynyloestradiol may absorb better in the UVA region of the spectrum than in the UVC region. The blue-black lamp (UVA radiation) covers a broader spectrum ($\lambda = 300$–400 nm with maximum emission at $\lambda = 350$ nm), and so also covers part of the visible and the UVB spectrum, whereas the UVC lamp emits radiation only at $\lambda = 253$ nm. The UV spectra of 17β-oestradiol, oestriol and 17α-ethynyloestradiol show that they absorb at $\lambda = 230$ nm and $\lambda = 280$ nm and that absorption is much more intense at $\lambda = 230$ nm. Both peaks are broad and there may be some absorption at $\lambda = 300$ nm for the $\lambda = 280$ nm peak. The rate of photons absorbed per reactor volume for each lamp was calculated using ferrioxalate actinometry [85]. Table 10.9 shows the results.

Table 10.9 shows that the UVC lamp is much more intense than the UVA lamp. However, UVA is sufficient to breakdown the oestrogens in water. This is encouraging from an industrial viewpoint since UVC disinfection requires much more energy and is therefore more costly. Photocatalysis, however, is still much more efficient than UV light alone, with rates up to 9× faster.

Our initial work on the photocatalysis of 17β-oestradiol [51] (Section 10.4) in a batch reactor with TiO_2 immobilized on Ti alloy gave a first-order rate constant of 0.016 min^{-1}. This is much lower than the rate constant obtained in this work (0.116 min^{-1}), which is comparable with more recent work in a quartz coil reactor (0.174 min^{-1}) (Section 10.6) [57], indicating that a coil reactor is a much more efficient system due to the increased surface area. The trend obtained here for photocatalysis compares favorably with previous work (Sections 10.5 and 10.6) [57, 64] where 17β-oestradiol and 17α-ethynyloestradiol were degraded at the same rate during photocatalysis whereas variable and slower degradation rates occurred by UV radiation, with the order 17α-ethynyloestradiol > 17β-oestradiol (0.086–0.106 min^{-1}). Nakashima et al. [86] reported rate constants of 0.033 and 0.050 min^{-1}

Tab. 10.9. Photons absorbed per reactor volume for each of the lamps measured using ferrioxalate actinometry (μEinsteins s^{-1} L^{-1} ± standard error).

Lamp	Photons absorbed (μEinsteins s^{-1} L^{-1})
UVC lamp ($\lambda = 253$ nm)	74 ± 3
UVA lamp ($\lambda = 350$ nm)	57 ± 1

for 17β-oestradiol in two different types of reactors with TiO$_2$ immobilized on PTFE mesh sheets, and, in more recent work [87], 0.15 and 0.12 min^{-1} for 17β-oestradiol and oestrone, respectively.

Titanium dioxide photocatalysis is an effective method for the degradation of the natural oestrogens 17β-oestradiol and oestriol and the synthetic oestrogen 17α-ethynyloestradiol in water in an immobilized TiO$_2$ spiral coil reactor. It is much more efficient than UVA or UVC radiation alone (up to 9× faster). Industrially, this is very important as costs and energy requirements can be drastically reduced. Particle-mediated photocatalytic techniques can improve the performance of water treatment systems employing UV by reducing the energy requirements for the UV stage. The successful and efficient degradation of these compounds will potentially provide cheaper and cleaner means of removing them from groundwater, wastewaters and drinking water. Because the only energy source is near-UV light, the technology should, ultimately, be able to function on sunlight alone.

10.8
Overall Conclusions and Identification of Research Needs

The main aim of this work was to investigate the degradation of natural oestrogens (17β-oestradiol, oestrone and oestriol) and a synthetic oestrogen (17α-ethynyloestradiol) in water using titanium dioxide photocatalysis. This was carried out using different chemical and biological analytical techniques.

The first objective was to determine if oestrogens could be degraded by photocatalysis and UV radiation. Initial studies in a batch reactor monitored using HPLC and fluorescence detection showed that 17β-oestradiol in water could be degraded by photocatalysis and was much more effective than UV light alone. The reaction followed first-order reaction kinetics and Langmuir–Hinshelwood behavior was observed.

The second objective was to determine if oestrogenic activity is removed from the oestrogenic water samples after treatment with photocatalysis and UV radiation since the oestrogenic activity of these substances is the main concern. This was carried out using a yeast screen bioassay. It was found that photocatalysis and UV radiation can remove all oestrogenic activity from water samples containing the natural oestrogens 17β-oestradiol and oestrone and the synthetic oestrogen 17α-ethynyloestradiol. Again, photocatalysis is much more efficient than UV light alone. The rates of reaction for the three oestrogens studied were very similar for photocatalysis. For UV radiation, 17α-ethynyloestradiol degraded the fastest followed by oestrone and 17β-oestradiol.

The work with fluorescence spectroscopy also fulfilled the first objective, showing that all the natural and synthetic oestrogens in water could be degraded by photocatalysis. The results demonstrated that 17α-ethynyloestradiol degraded the fastest followed by 17β-oestradiol and oestriol for both photocatalysis and UV radiation combined and for UV light alone. Direct photolysis reactions are generally of low efficiency compared with procedures involving hydroxyl radical generation.

However, here there was no major difference in the rates for photocatalysis and UV light alone. This was due to the lamp used, which emitted in the UVA, UVB and UVC range of the spectrum. The reactions also follow pseudo-first order kinetics. 17α-Ethynyloestradiol degrades the fastest for both photocatalysis and photolysis followed by 17β-oestradiol and oestriol. This was attributed to the triple bond of the ethynyl group in 17α-Ethynyloestradiol, which absorbs UV light more easily.

The third objective was to investigate the effect of varying reaction conditions on the photocatalytic reaction, i.e., initial concentration and light intensity. The relationship between initial concentration and rate is linear for both photocatalysis combined with photolysis and for photolysis of 17β-oestradiol in water. Photocatalysis degrades 17β-oestradiol at twice the rate of photolysis. The rate was proportional to the square root of light intensity for photocatalysis of 17β-oestradiol. The relationship between light intensity and the rate of degradation by photolysis is linear up to a point; it then starts to level off.

The final objective was to compare photocatalysis with UVA and UVC disinfection. Here, photocatalysis was found to be much more efficient than UV light alone. Titanium dioxide photocatalysis is an effective method for the degradation of the natural oestrogens 17β-oestradiol and oestriol and the synthetic oestrogen 17α-ethynyloestradiol in water in an immobilized TiO_2 spiral coil reactor. The spiral reactor was much more efficient than a batch reactor with TiO_2 immobilized on Ti alloy due to the increased surface area of the former. Photocatalysis was much more efficient than UVA or UVC radiation alone (up to 9$\times$ faster). This is very important industrially, where costs and energy requirements can be drastically reduced. Particle-mediated photocatalytic techniques can improve the performance of water treatment systems employing UV by reducing the energy requirements for the UV stage. The fact that solar illumination can be used in these systems is an added advantage for applications in countries with a hot and sunny climate. As the importance of water reuse and water recycling increases, an effective technique is needed to remove trace organics in our water supply. TiO_2 photocatalysis can be used to develop a safe, cost-efficient water treatment process.

Based on the work presented in this chapter, several research areas and needs can be proposed:

1. Improvements in chemical and biological analysis for the detection of steroid oestrogens at ng L^{-1} levels and for real water sample analysis.
2. Investigations into the application of photocatalysis for the removal of steroid oestrogens in real water samples, e.g., treated sewage effluent samples where other substances are present.
3. Improvement in the engineering design of photocatalytic reactor systems with a view to incorporating the photocatalytic treatment stage into existing water treatment systems.

In summary, the challenge is for water treatment scientists to design and apply titanium dioxide nanoparticles for the photocatalytic breakdown of micropollutants such as steroid oestrogens in real water samples.

The principles of organic photoelectrochemistry elaborated over the past decade clearly show that interfacial electron transfer to surface adsorbates can result in oxidative degradation, often leading to complete mineralization of organic pollutants. The wide band gap and high chemical stability of TiO_2 nanoparticles gives them an extremely broad reactivity range. Irradiated TiO_2 nanoparticles efficiently degrade nearly every significant functional group, including the most environmentally hazardous and persistent substances. Since our first report of the use of photocatalysis for the degradation of oestrogens in water [51], its potential use as a method for reducing discharges of EDCs into the aquatic environment has been of increasing interest [57, 64–66, 86–92]. Indeed, the degradation of nonylphenol polyethoxylate surfactants [90] and their biotransformation products [65, 66, 88], Bisphenol-A [83, 86, 89, 92–94], phthalates [93], atrazine [95], resorcinol [96] and amitrole [97] using TiO_2 photocatalysts have been reported. However, the significant proportion of the overall oestrogenic activity of many effluent discharges [9] is due to the steroid oestrogens. Photocatalytic degradation of 17β-oestradiol [51, 57, 64, 83, 86–88, 91, 98, 99], oestrone [51, 57, 64, 98, 99] and 17α-ethynylestradiol [51, 57, 64, 87, 88, 91, 98, 99] has been shown. Solar photocatalytic degradation has also been investigated for the degradation of bisphenol A [100]. TiO_2 nanoparticles are the most widely used for the removal of EDCs in water. However, there is one report of the use of a visible-light-driven $BiVO_4$ photocatalyst for the degradation of nonylphenols in water by solar radiation [101]. However, titanium dioxide nanoparticles have been shown to be the most effective photocatalyst for water treatment applications. TiO_2 as a suspension, coating, or immobilized catalyst exhibits promising potential for environmental amelioration, particularly as continuing research into the engineering optimization of photocatalyst dispersal builds on the framework of earlier investigations [32]. Few studies have evaluated the efficacy of TiO_2-assisted photodegradation in the treatment of mixtures of contaminants and actual wastewaters. The process is unlikely to be used in the treatment of high-strength industrial wastewaters or for the large-scale direct clean-up of contaminated soils. Loss of efficiency due to competing substrates in the case of mixtures, interference by dissolved anions and cations, which cause significant reductions in rates of photodegradation, and light interference by high concentrations of soils are likely problems in industrial wastewaters and soils [29]. The literature indicates that oestrogens play an important role in the development of reproductive abnormalities and other health problems. Hopefully, growing awareness that reproductive function may be at risk will stimulate both the basic and clinical research within this field that have received relatively little attention [102]. Biologists, chemists and engineers should continue to work in this area to fully understand the environmental implications of these compounds and in the disposal, monitoring and removal of them. There is a need for the development of an effective technique for the removal of these compounds and other trace organic contaminants in our water supply. Titanium dioxide nanoparticles and UV light to produce the photocatalytic process could be a solution to the problem of EDCs and other micropollutants in the aquatic environment.

References

1 CARMICHAEL, H., Sex offenders, *Chem. Br.* **1998**, 34(10), 25–28.

2 *Environmental Estrogens and other Hormones*, Center for Bioenvironmental Research of Tulane and Xavier Universities, New Orleans, L.A., website; http://www.tmc.tulane.edu/ecme/eehome/.

3 JOHNSON, A.C., SUMPTER, J.P., Removal of endocrine-disrupting chemicals in activated sludge treatment works, *Environ. Sci. Technol.* **2001**, 35(24), 4697–4703.

4 ARCAND-HOY, L.D., NIMROD, A.C., BENSON, W.H., Endocrine-modulating substances in the environment: estrogenic effects of pharmaceutical products, *Int. J. Toxicol.* **1998**, 17, 139–158.

5 TURAN, A., Excretion of natural and synthetic estrogens and their metabolites: Occurrence and behaviour in water, in *Expert Round Endocrinitically Active Chemicals in the Environment*, ed. A. GILES Umweltbundesamt, Berlin, **1996**, pp. 15–20.

6 BITMAN, J., CECIL, H.C., Estrogenic activity of DDT analogs and polychlorinated biphenyls. *J. Agric. Food Chem.* **1970**, 18(6), 1108–1112.

7 COLBURN, T., VOM SAAL, F.S., SOTO, A.M., Developmental effects of endocrine-disrupting chemicals in wildlife and humans, *Environ. Health Perspect.* **1993**, 101(5), 378–384.

8 BARONTI, C., CURINI, R., D'ASCENZO, G., DI CORCIA, A., GENTILI, A., SAMPERI, R., Monitoring natural and synthetic estrogens at activated sludge sewage treatment plants and in a receiving river water, *Environ. Sci. Technol.* **2000**, 34, 5059–5066.

9 DESBROW, C., ROUTLEDGE, E.J., BRIGHTY, G.C., SUMPTER, J.P., WALDOCK, M., Identification of estrogenic chemicals in STW effluent. I. Chemical fractionation and in vitro biological screening, *Environ. Sci. Technol.* **1998**, 32, 1549–1558.

10 TABAK, H.H., BLOOMHUFF, R.N., BUNCH, R.L., Steroid hormones as water pollutants. II. Studies on the persistence and stability of natural and synthetic ovulation-inhibiting hormones in untreated and treated wastewaters, *Develop. Ind. Microbiol.* **1970**, 11, 497–519.

11 TERNES, T.A., STUMPF, M., MUELLER, J., HABERER, K., WILKEN, R.-D., SERVOS, M., Behaviour and occurrence of estrogens in municipal sewage treatment plants – I. Investigations in Germany, Canada and Brazil, *Sci. Total Environ.* **1999**, 225, 81–90.

12 STUMPF, M., TERNES, T.A., HABERER, K., BAUMANN, W., Determination of natural and synthetic estrogens in sewage plants and river water, *VomWasser* **1996**, 87, 251–261.

13 ISOBE, T., SHIRAISHI, H., YASUDA, M., SHINODA, A., SUZUKI, H., MORITA, M., Determination of estrogens and their conjugates in water using solid-phase extraction followed by liquid chromatography-tandem mass spectrometry, *J. Chromatogr. A* **2003**, 984, 195–202.

14 RURAINSKI, R.D., THEISS, H.J., ZIMMERMANN, W., The occurrence of natural and synthetic oestrogens in drinking water, *Das Gas-und Wasserfach wasser/abwasser* **1977**, 118, 288–291.

15 ADLERCREUTZ, H., FOSTIS, T., BANNWART, C., HÄMÄLÄINEN, E., BLOIGU, S., OLLUS, A., Urinary estrogen profile determination in young Finnish vegetarian and omnivorous women, *J. Steroid Biochem.* **1986**, 24, 289–296.

16 DRAY, J., TILLIER, F., DRAY, F., ULLMANN, A., Etude de l'hydrolyse des metabolites urinaires de différentes hormones steroids par la β-glucuronidase de *Eschericia coli*, *Ann. l'Institut Pasteur* **1972**, 123, 853–857.

17 KIRK, L.A., TYLER, C.R., LYE, C.M., SUMPTER, J.P., Changes in estrogenic and androgenic activities at different stages of treatment in wastewater

treatment works, *Environ. Toxicol. Chem.* **2002**, 21(9), 72–79.

18 JOHNSON, A.C., BELFROID, A., DI CORCIA, A., Estimating steroid oestrogen inputs into activated sludge treatment works and observations on their removal from the effluent, *Sci. Total Environ.* **2000**, 256, 163–173.

19 KORNER, W., BOLZ, U., SUSSMUTH, W., HILLER, G., SCHULLER, W., HANF, V., HAGENMAIER, H., Input/output balance of estrogenic active compounds in a major municipal sewage plant in Germany, *Chemosphere* **2000**, 40, 1131–1142.

20 SHORE, L.S., GUREVITZ, M., SHEMESH, M., Estrogen as an environmental pollutant, *Bull. Environ. Contam. Toxicol.*, **1993**, 51, 361–366.

21 ROUTLEDGE, E.J., SHEAHAN, D., DESBROW, C., BRIGHTY, G.C., WALDOCK, M., SUMPTER, J.P., Identification of estrogenic chemicals in STW effluent. 2. *In vivo* responses in trout and roach, *Environ. Sci. Technol.* **1998**, 32, 1559–1565.

22 WILLIAMS, R.J., JURGENS, M.D., JOHNSON, A.C., Initial predictions of the concentrations and distribution of 17(R)-oestradiol, oestrone and ethinyloestradiol in 3 English rivers, *Water Res.* **1999**, 33, 1663–1671.

23 LARSSON, D.G.J., ADOLFSSON-ERICI, M., PARKKONEN, J., PETTERSON, M., BERG, A.H., OLSSON, P.-E., FORLIN, L., Ethinyloestradiol – an undesired fish contraceptive?, *Aquatic Toxicol.* **1999**, 42, 91–97.

24 EC, *European Workshop on the Impact of Endocrine Disrupters on Human Health and Wildlife: Report of the Proceedings*, Weybridge, UK, European Commission, **1997**.

25 SVENSON, A., ALLARD, A.S., EK, M., Removal of estrogenicity in Swedish municipal sewage treatment plants, *Water Res.* **2003**, 37, 4433–4443.

26 LITCHEN, N.N., DONG, J., VIJAYAKUMAR, K.M., Photopromoted TiO$_2$-catalyzed oxidative decomposition of organic pollutants in water and in the vapor phase, *Water Pollut. Res. J., Can.* **1992**, 27, 203–210.

27 MILLS, A., DAVIES, R.H., WORSLEY, D., Purification of water by semiconductor photocatalysis, *Chem. Soc. Rev.* **1993**, 22(6), 417–425.

28 LITTER, M.I., Heterogeneous photocatalysis. Transition metal ions in photocatalytic systems, *Appl. Catal. B: Environ.* **1999**, 23, 89–114.

29 VENKATADRI, R., PETERS, R.W., Chemical oxidation technologies: Ultraviolet light/hydrogen peroxide, Fenton's reagent, and titanium dioxide-assisted photocatalysis, *Hazardous Waste Hazardous Mater.* **1993**, 10(2), 107–149.

30 PROUSEK, J., Advanced oxidation processes for water treatment, *Chem. Listy* **1996**, 90, 307–315.

31 LEGRINI, O., OLIVEROS, E., BRAUN, A.M., Photochemical processes for water treatment, *Chem. Rev.* **1993**, 93, 671–698.

32 FOX, M.A., Photocatalysis: Decontamination with sunlight, *Chemtech* **1992**, 11, 680–685.

33 FUJISHIMA, A., HONDA, K., Electrochemical photolysis of water at a semiconductor electrode, *Nature*, **1972**, 238, 37–38.

34 CAREY, J.H., LAWRENCE, J., TOSINE, H.M., Photodechlorination of PCB's in the presence of titanium dioxide in aqueous suspensions, *Bull. Environ. Contam. Toxicol.* **1976**, 16(6), 697–701.

35 MILLS, A., LE HUNTE, S., An overview of semiconductor photocatalysis, *J. Photochem. Photobiol. A, Chem.* **1997**, 108, 1–35.

36 HOFFMANN, M.R., MARTIN, S.T., CHOI, W., BAHNEMANN, D.W., Environmental applications of semiconductor photocatalysis, *Chem. Rev.* **1995**, 95, 69–96.

37 LINSEBIGLER, A.L., LU, G., YATES, J.T., JR., Photocatalysis on TiO$_2$ surfaces: Principles, mechanisms, and selected results, *Chem. Rev.* **1995**, 95, 735–758.

38 FUJISHIMA, A., RAO, T.N., TRYK, D.A., Titanium dioxide photocatalysis, *J. Photochem. Photobiol. C: Photochem. Rev.* **2000**, 1, 1–21.

39 CARP, O., HUISMAN, C.L., RELLER, A., Photoinduced reactivity of titanium dioxide, *Progr. Solid State Chem.* **2004**, 32, 33–177.

40 Schiavello, M. (ed.), *Heterogeneous Photocatalysis*, John Wiley and Sons Chichester, UK, **1997**.

41 Ollis, D.F., Al-Ekabi, H., *Photocatalytic Purification and Treatment of Water and Air*, Elsevier Amsterdam, The Netherlands, **1993**.

42 Serpone, N., Pelizetti, E., *Photocatalysis: Fundamentals and Applications*, John Wiley and Sons, New York, **1989**.

43 Pelizzetti, E., Schiavello, M., *Photochemical Conversion and Storage of Energy, Proceedings of the Eighth International Conference on Photochemical Conversion and Storage of Solar Energy*, IPS-8, held July 15–20, 1990, Palermo, Italy, Kluwer Academic Publishers, Dordrecht, **1991**.

44 Mills, A., Lee, S.K., A web-based overview of semiconductor photochemistry-based current commercial applications, *J. Photochem. Photobiol. A: Chem.* **2002**, 152, 233–247.

45 Wold, A., Photocatalytic properties of TiO$_2$, *Chem. Mater.* **1993**, 5, 280–283.

46 Fujishima, A., Rao, T.N., Recent advances in heterogeneous TiO$_2$ photocatalysis, *Proc. Indian Acad. Sci. (Chem. Sci.)* **1997**, 109(6), 471–486.

47 Mills, A., Morris, S., Davies, R.J., Photomineralisation of 4-chlorophenol sensitised by titanium dioxide: A study of the intermediates, *J. Photochem. Photobiol. A: Chem.* **1993**, 70(2), 183–191.

48 Schindler, K.-M., Kunst, M., Charge-carrier dynamics in titania powders, *J. Phys. Chem.* **1990**, 94, 8222–8226.

49 Bickley, R.I., Gonzalez-Carreno, T., Lees, J.S., Palmisano, L., Tilley, R.D., A structural investigation of titanium dioxide photocatalysts, *J. Solid State Chem.* **1991**, 92, 178.

50 Kamat, P.V., Photochemistry on non-reactive and reactive (semiconductor) surfaces, *Chem. Rev.* **1993**, 93, 267.

51 Coleman, H.M., Eggins, B.R., Byrne, J.A., Palmer, F.L., King, E., Photocatalytic degradation of 17β-oestradiol on immobilised TiO$_2$, *Appl. Catal. B: Environ.* **2000**, 24(1), 1–5.

52 Byrne, J.A., Eggins, B.R., Brown, N.M.D., McKinney, B., Rouse, M., Immobilisation of TiO$_2$ powder for the treatment of polluted water, *Appl. Catal. B: Environ.* **1998**, 17, 25–36.

53 Calvert, J.G., Pitts, J.N., *Photochemistry*, Wiley, New York, **1973**.

54 Routledge, E.J., Sumpter, J.P., Estrogenic activity of surfactants and some of their degradation products assessed using a recombinants yeast screen, *Environ. Toxicol. Chem.* **1996**, 15(3), 242–248.

55 Ohko, Y., Ando, I., Niwa, C., Tatsuma, T., Yamahura, T., Nakashima, T., Kubota, Y., Fujishima, A., Degradation of Bisphenol A in water by TiO$_2$ photocatalyst, *Environ. Sci. Technol.* **2001**, 35, 2365–2368.

56 Horikoshi, S., Hidaka, H., Non-degradable triazine substrates of atrazine and cyanuric acid hydrothermally and in supercritical water under the UV-illuminated photocatalytic cooperation, *Chemosphere* **2003**, 51(2), 139–142.

57 Coleman, H.M., Abdullah, M., Eggins, B.R., Palmer, F.L., Photocatalytic degradation of 17®-oestradiol, oestriol and 17α-ethynyloestradiol in water monitored using fluorescence spectroscopy, *Appl. Catal. B: Environ.* **2005**, 55(1), 23–30.

58 Ohko, Y., Iuchi, K.-I., Niwa, C., Tatsuma, T., Nakashima, T., Iguchi, T., Kubota, Y., Fujishima, A., 17®-Estradiol degradation by TiO$_2$ photocatalysis as a means of reducing estrogenic activity, *Environ. Sci. Technol.* **2002**, 36, 4175–4181.

59 Brzozowski, A., Pike, A.C.W., Dauter, Z., Hubbard, R.E., Bonn, T., Engstrom, O., Ohman, L., Greene, G.L., Gustafsson, J.-A., Carlquist, M., Molecular basis of agonism and antagonism in the oestrogen receptor, *Nature* **1997**, 389(16), 753–758.

60 Purdom, C.E., Hardiman, P.A., Bye, V.J., Eno, N.C., Tyler, C.R., Sumpter, J.P., Estrogenic effects of effluent from sewage treatment works, *Chem. Ecol.* **1994**, 8, 275–285.

61 Thorpe, K.L., Cummings, R.I.,

HUTCHINSON, T.H., SCHOLZE, M., BRIGHTY, G., SUMPTER, J.P., TYLER, C.R., Relative potencies and combination effects of steroidal estrogens in fish, *Environ. Sci. Technol.* **2003**, 37, 1142–1149.

62 GUENGERICH, F.P., Minireview: Metabolism of 17α-ethynylestradiol in humans, *Life Sci.* **1990**, 47, 1981–1988.

63 ABDULLAH, M.I., EEK, E., Automatic photocatalytic method for the determination of dissolved organic carbon (DOC) in natural waters, *Water Res.* **1996**, 30(8), 1813–1822.

64 COLEMAN, H.M., ROUTLEDGE, E.J., SUMPTER, J.P., EGGINS, B.R., BYRNE, J.A., Rapid loss of estrogenicity of steroid estrogens by UVA photolysis and photocatalysis over an immobilised titanium dioxide catalyst, *Water Res.* **2004**, 38(14–15), 3233–3240.

65 IKE, M., ASANO, M., BELKADA, F.D., TSUNOI, S., TANAKA, M., FUJITA, M., Degradation of biotransformation products of nonylphenol ethoxylates by ozonation and UV/TiO2 treatment, *Water Sci. Technol.* **2002**, 46, 127–132.

66 KOHTANI, S., MAKINO, S., KUDO, A., TOKUMURA, K., ISHIGAKI, Y., MATSUNGA, T., NIKAIDO, O., HAYAKAWA, K., NAKAGAKI, R., Photocatalytic degradation of 4-n-nonylphenol under irradiation from solar simulator: Comparison between BiVO$_4$ and TiO$_2$ photocatalysts, *Chem. Lett.* **2002**, 7, 660–661.

67 LIU, B., LIU, X., Direct photolysis of estrogens in aqueous solutions, *Sci. Total Environ.* **2004**, 320(2–3), 269–274.

68 OLLIS, D.F., *Photochemical Conversion and Storage of Solar Energy*, ed. PELIZETTI, E., SCHIAVELLO, E., Kluwer Academic Publishers, Dordrecht, **1991**, p. 593.

69 OKAMOTO, K., YAMAMOTO, Y., TANAKA, H., TANAKA, M., ITAYA, A., Heterogeneous photocatalytic decomposition of phenol over anatase powder, *Bull. Chem. Soc. Jpn.* **1985**, 58, 2015.

70 KORMANN, C., BAHNEMANN, D.W., HOFFMANN, M.R., Photolysis of chloroform and other organic molecules in aqueous titanium dioxide suspensions, *Environ. Sci. Technol.* **1991**, 25, 494–500.

71 EGERTON, T.A., KING, C.J., The influence of light intensity on photoactivity in titanium dioxide pigmented systems, *J. Oil Colour Chem. Assoc.* **1979**, 62(10), 386–391.

72 KORMANN, C., Ph.D. thesis, California Institute of Technology, Pasadena CA, **1989**.

73 MATTHEWS, R.W., ABDULLAH, M., LOW, G.K.-C., Photocatalytic oxidation for total organic carbon analysis, *Anal. Chim. Acta* **1990**, 233(2), 171–179.

74 WEI, T.-Y., WAN, C.-C., Kinetics of photocatalytic oxidation of phenol on titanium oxide (TiO$_2$) surface, *J. Photochem. Photobiol. A: Chem.* **1992**, 69(2), 241–249.

75 SHIN, E.-M., SENTHURCHELVAN, R., MUNOZ, J., BASAK, S., RAJESHWAR, K., Photolytic and photocatalytic destruction of formaldehyde in aqueous media, *J. Electrochem. Soc.* **1996**, 143(5), 1562–1570.

76 CHEN, L.C., CHOU, T.C., Kinetics of photodecolorization of methyl orange using titanium dioxide as catalyst, *Ind. Eng. Chem. Res.* **1993**, 32, 1520–1527.

77 TRILLAS, M., PERAL, J., DOMENECH, X., Redox photodegradation of 2,4-dichlorophenoxyacetic acid over TiO$_2$, *Appl. Catal. B: Environ.* **1995**, 5, 377–387.

78 TRILLAS, M., PERAL, J., DOMENECH, X., Photo-oxidation of phenoxyacetic acid by TiO$_2$-illuminated catalyst, *Appl. Catal. B: Environ.* **1993**, 3(1), 45–53.

79 SERRA, F., TRILLAS, M., GARCIA, J., DOMENECH, X., Titanium dioxide-photocatalyzed oxidation of 2,4-dichlorophenol, *J. Environ. Sci. Health, Part A: Environ. Sci. Eng.* **1994**, 29(7), 1409–1421.

80 AUGUGLIARO, V., LODDO, V., MARCI, G., PALMISANO, L., LOPEZ-MUNOZ, M.J., Photocatalytic oxidation of cyanides in aqueous titanium dioxide suspensions, *J. Catal.* **1997**, 166(2), 272–283.

81 MILLS, A., HOLLAND, C.E., DAVIES, R.H., WORSLEY, D., Photomineralization of salicylic acid: A kinetic study,

J. Photochem. Photobiol. A: Chem. **1994**, 83(3), 257–263.

82 DIONYSIOU, D.D., SUIDAN, M.T., BAUDIN, I., LAINE, J.-M., Oxidation of organic contaminants in a rotating disk photocatalytic reactor: Reaction kinetics in the liquid phase and the role of mass transfer based on the dimensionless Damköhler number, *Appl. Catal. B: Environ.* **2002**, 38(1), 1–16.

83 OHKO, Y., IKEDA, K., RAO, T.N., HASIMOTO, K., FUJISHIMA, A., Photocatalytic reaction kinetics on TiO2 thin films under light-limited and mass transport-limited conditions, *Z. Physik. Chem. (Muenchen)* **1999**, 213, 33–42.

84 MINERO, C., Kinetic analysis of photoinduced reactions at the water semiconductor interface, *Catal. Today* **1999**, 54(2–3), 205–216.

85 HATCHARD, C.G., PARKER, C.A., A new sensitive chemical actinometer. II. Potassium ferrioxalate as a standard chemical actinometer, *Proc. Royal Soc. (London)* **1956**, A235, 518–536.

86 NAKASHIMA, T., OHKO, Y., TRYK, D.A., FUJISHIMA, A., Decomposition of endocrine-disrupting chemicals in water by use of TiO$_2$ photocatalysts immobilized on polytetrafluoro-ethylene mesh sheets, *J. Photochem. Photobiol. A: Chem.* **2002**, 151, 207–212.

87 NAKASHIMA, T., OHKO, Y., KUBOTA, Y., FUJISHIMA, A., Photocatalytic decomposition of estrogens in aquatic environment by reciprocating immersion of TiO$_2$-modified polytetrafluoroethylene mesh, *J. Photochem. Photobiol. A: Chem.* **2003**, 160, 115–120.

88 TANIZAKI, T., KADOKAMI, K., SHINOHARA, R., Catalytic photodegradation of endocrine disrupting chemicals using titanium dioxide photosemiconductor thin films, *Bull. Environ. Contam. Toxicol.*, **2002**, 68, 5, 732–739.

89 FUKAHORI, S., ICHIURA, H., KITAOKA, T., TANAKA, H., Photocatalytic decomposition of bisphenol A in water using composite TiO$_2$-zeolite sheets prepared by a papermaking technique, *Environ. Sci. Technol.*, **2003**, 37(5), 1048–1051.

90 HORIKOSHI, S., WATANABE, N., ONISHI, H., HIDAKA, H., SERPONE, N., Photodecomposition of a nonylphenol polyethoxylate surfactant in a cylindrical photoreactor with TiO$_2$ immobilized fiberglass cloth, *Appl. Catalal. B: Environ.*, **2002**, 37(2), 117–129.

91 ONARI, Y., YOSHIOKA, O., IWASAKI, S., TAKAHASHI, M., Decompositions of endocrine disrupting chemicals by advanced oxidations such as oxidation with ozone, photo-catalysis and ultraviolet light irradiation, *Mie-ken Kagaku Gijutsu Shinko Senta Hoken Kankyo Kenkyubu Nenpo*, **2002**, 4, 98–101 (in Japanese).

92 WATANABE, N., HORIKOSHI, S., KAWABE, H., SUGIE, Y., ZHAO, J., HIDAKA, H., Photodegradation mechanism for bisphenol A at the TiO$_2$/H$_2$O interfaces, *Chemosphere* **2003**, 52, 851–859.

93 OOKA, C., YOSHIDA, H., HORIO, M., SUZUKI, K., HATTORI, T., Adsorptive and photocatalytic performance of TiO$_2$ pillared montmorillonite in degradation of endocrine disruptors having different hydrophobicity, *Appl. Catal. B: Environ.* **2003**, 41, 313–321.

94 LEE, J.-M., KIM, M.-S., KIM, B.-W., Photodegradation of bisphenol-A with TiO$_2$ immobilized on the glass tubes including the UV light lamps, *Water Res.* **2004**, 38, 3605–3613.

95 HORIKOSHI, S., HIDAKA, H., Non-degradable triazine substrates of atrazine and cyanuric acid hydro-thermally and in supercritical water under the UV-illuminated photo-catalytic cooperation, *Chemosphere* **2003**, 51, 139–142.

96 LAM, S.W., CHIANG, K., LIM, T.M., AMAL, R., LOW, G.K.-C., Effect of charge trapping species of cupric ions on the photocatalytic oxidation of resorcinol, *Appl. Catal. B: Environ.* **2005**, 55, 123–132.

97 WATANABE, N., HORIKOSHI, S., KAWASAKI, A., HIDAKA, H., Formation of refractory ring-expanded triazine intermediates during the photocatalyzed mineralization of the endocrine disruptor mintrole and related trazole derivatives at UV-irradiated TiO_2/H_2O interfaces, *Environ. Sci. Technol.* **2005**, 39, 2320–2326.

98 MITAMURA, K., NARUKAWA, H., MIZUGUCHI, T., SHIMADA, K., Degradation of estrogen conjugates using titanium dioxide as a photocatalyst, *Anal. Sci.* **2004**, 20, 3–4.

99 COLEMAN, H.M., HON, H., AMAL, R., LESLIE, G., WEHNER, M., FITZSIMMONS, S., Removal of oestrogenic and carcinogenic substances from water using alternative water treatment methods, Proceedings of the *Australian Water Association Speciality Conference on Contaminants of Concern*, Rydges Lakeside Hotel, Canberra, Australia, 22–23 June **2005**.

100 KANECO, S., RAHMAN, M.A., SUZUKI, T., KATSUMATA, H., OHTA, K., Optimization of solar photocatalytic degradation conditions of bisphenol A in water using titanium dioxide, *J. Photochem. Photobiol. A: Chem.* **2004**, 163, 419–424.

101 KOHTANI, S., KOSHIKO, M., KUDO, A., TOKUMURA, K., ISHIGAKI, Y., TORIBA, A., HAYAKAWA, K., NAKAGAKI, R., Photodegradation of 4-alkylphenols using $BiVO_4$ photocatalyst under radiation with visible light from a solar simulator, *Appl. Catal. B: Environ.* **2003**, 46, 573–586.

102 CARLSEN, E., GIWERCMAN, A., KEIDING, N., SKAKKEBAEK, N.E., Declining semen quality and increasing incidence of testicular cancer: Is there a common cause?, *Environ. Health Perspect.* **1995**, 103(7), 137–139.

11
Nanosensors for Environmental Applications

Wan Y. Shih and Wei-Heng Shih

11.1
Introduction

11.1.1
Overview

Current water-born pathogen sensor development relies on colony growth or fluorescent-based techniques, which are not *in situ*, rapid, or sensitive. In our laboratory, we have developed piezoelectric microcantilever sensors that can perform rapid, *in situ*, in-water pathogen detection with sensitivities well above that of current techniques. We have shown that using PZT/glass cantilevers of sub-millimeter length with a 2 mm glass tip, which exhibited 5×10^{-11} g Hz^{-1} mass detection sensitivity, *in situ* quantification of *Salmonella typhimurium* was achieved with a concentration limit of 10^3 cells mL^{-1}; this is lower than the infectious dosage, 10^5 cells mL^{-1}, which is also the concentration limit of commercial ELISA, QCM, and nanowire-based sensors. Furthermore, we have developed two types of miniaturized piezoelectric cantilevers for even better sensitivity. With 500 μm long PMN-PT/Cu microcantilevers, fabricated from freestanding PMN-PT films, that exhibited 3×10^{-13} g Hz^{-1} detection sensitivity we achieved a better than 50 spores mL^{-1} detection limit in 1 mL of *Bacillus anthracis* suspension. With PZT/SiO$_2$ microcantilevers less than 50 μm long, fabricated by silicon-based microfabrication techniques, the detection sensitivity is expected to reach better than 10^{-16} g Hz^{-1} and further lower the concentration limit. In addition to ultrasensitive, rapid, *in situ* detection, piezoelectric microcantilevers use simple electrical measurements, which is ideal for portable simultaneous array sensing in environmental applications.

11.1.2
Sensor

Current sensing technologies rely on fluorescence [1, 2], laser [3] or fiber-optics-based methods [4], quartz crystal microbalance [5], electrochemical enzyme immu-

Nanotechnologies for the Life Sciences Vol. 5
Nanomaterials – Toxicity, Health and Environmental Issues. Edited by Challa S. S. R. Kumar
Copyright © 2006 WILEY-VCH Verlag GmbH & Co. KGaA, Weinheim
ISBN: 3-527-31385-0

noassays [6], amplification schemes such as polymerase chain reaction (PCR) [7–9], and binding to nanometer-size metal particles [10]. Most of the techniques are neither direct nor quantitative and are slow. They do not lend themselves to multiplexing and high throughput. Development of direct sensing technologies relies heavily on silicon-based microcantilevers [11–15] due to their availability and ease of integration with existing silicon-based methodologies. Binding of target antigens to the antibody on the cantilever surface is directly detected by monitoring the cantilever's resonance frequency shift due to the mass of the adsorbed target antigens. Silicon-based microcantilevers offer high detection sensitivity, $\Delta m/\Delta f \sim 10^{-12}$ g Hz^{-1} [16, 17], where Δm and Δf denote, respectively, the mass change and corresponding resonance frequency change due to the binding of target molecules. However, all silicon-based microcantilevers rely on complex external optical components for deflection detection and an external driver for actuation generation. Moreover, immersing silicon-based microcantilevers in water reduces the Q factor (ratio of the resonance peak frequency relative to the resonance peak width at half peak height) to about one, prohibiting silicon-based microcantilevers from in-water detection [18]. Silicon-based microcantilevers cannot have high resonance peaks in water because they are not piezoelectric. They rely on a vibration driver located at the cantilever base to generate deflections at the cantilever tip, which is ineffective. In comparison, piezoelectric sensors use electrical means for detection and do not have the bulkiness and complexity of silicon-based sensors. However, current piezoelectric biosensors are based on quartz crystal microbalances (QCM) [19], which are disk devices with a mass detection sensitivity of 10^{-8} g Hz^{-1}, about $10\,000\times$ less sensitive than the silicon-based microcantilevers. A QCM is about 1–3 cm in size, and silicon microcantilevers require laser alignment. Both are unfit for high-throughput array environmental applications.

In this chapter, we describe a new type of biosensor: array piezoelectric microcantilever sensors (PEMS) that can simultaneously detect multiple antigens *in situ* both in water and in-air with high sensitivity for water-borne pathogen applications. Section 11.1.3 provides an introduction to piezoelectric cantilever sensors (PECS). Section 11.2 gives a brief theoretical description of PECS without (Section 11.2.1) and with (Section 11.2.2) a nonpiezoelectric tip. Section 11.3 gives various examples of *in situ*, in-water biodetection as well as in-air nerve-gas simulant detection. Section 11.4 describes the miniaturization approaches of piezoelectric cantilever and the detection sensitivity enhancement with a reducing sensor size.

Current commercial enzyme-linked immunosorbent assays (ELISA), which use optical means for detection, offer a concentration limit of 10^5 cells mL^{-1} for *Salmonella typhimurium*, about the same as the infection dosage. Other nanobiosensors such as nanowire-based sensors that also use electrical measurements for detection offer a concentration sensitivity of about 10^5 cells mL^{-1} in *Escherichia coli* detection [20]. In comparison, current 400 μm long piezoelectric lead magnesium niobate–lead titanate/copper (PMN-PT/Cu) microcantilevers have already achieved 50 spores mL^{-1} concentration sensitivity in a 1 mL *Bacillus anthraces* (which is about the same size as *E. coli* and *S. typhimurium*) suspension, far exceeding the concentration limit of ELISA and that of nanowire sensors. Although silicon-based nanocantilevers offer similar detection sensitivity as piezoelectric microcantilevers they require optical measurements in air [17] or in vacuum [21], which prohibits their use for portable, *in situ*, multiplexed detection. As we have demonstrated, in addition to their high sensitivities, piezoelectric microcantilevers offer the advantages of *in situ*, real-time, multiplexed detection, and are thus more suitable for environmental applications.

11.1.3
Piezoelectric Cantilever Sensors (PECS)

Piezoelectric cantilever sensors (PECS) are a new type of mass sensors we have developed that uses electrical means for detection and can be miniaturized for better mass detection sensitivity [22]. By monitoring the resonance frequency shifts we have demonstrated PECS for rapid, label-free, *in situ* quantitative detection of pathogens with simple all-electrical measurements. We have shown that millimeter size PECS could detect *S. typhimurium* at concentrations below 5000 cells mL^{-1} without flow or concentration, which is lower than the infection dosage level (10^5 cells mL^{-1}) and more sensitive than enzyme-linked immunosorbent assays (ELISA) and array biosensors [23]. Piezoelectric sensors have the advantage of both driving and sensing the mechanical resonance electrically. Receptors are coated on the piezoelectric device surface to bind the molecules of interest [24, 25]. The change in mass due to the binding of the target molecules shifts the mechanical resonance frequency of the device.

11.2
Theory of PECS

A piezoelectric cantilever is a flexural transducer that consists of a piezoelectric layer, e.g., lead zirconate titanate (PZT) bonded to a nonpiezoelectric layer, e.g., stainless steel. Figure 11.1(a) and (b) show, respectively, a schematic of a piezoelectric cantilever of a uniform thickness (unimorph) and a piezoelectric cantilever with a nonpiezoelectric extension. The nonpiezoelectric extension may also be narrower than the piezoelectric section [26]. Bending vibrations can be generated by applying a small alternating-current (ac) voltage (<1 V) across the thickness of the piezoelectric layer. The stress generated by the bending vibration in the piezoelectric layer in turn induces a measurable piezoelectric voltage in phase with the applied voltage. A piezoelectric cantilever's mechanical resonance frequency and resonance strength can be measured by monitoring the maximum of the real part or the phase angle of the complex electrical impedance. Monitoring a cantilever's resonance frequency has many useful applications. In mass detection, a piezoelectric cantilever's resonance frequency shift is measured to quantify the small mass attached to the cantilever surface [27]. In liquid property characterization, both the resonance peak frequency and the resonance peak width of PECS inserted in the liquid are measured to simultaneously determine the liquid's viscosity and density

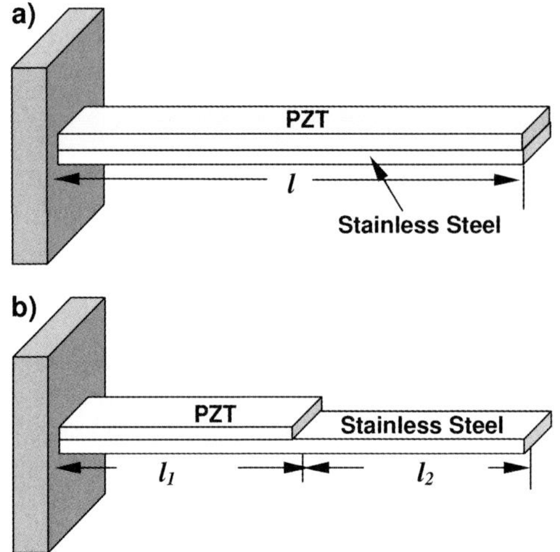

Fig. 11.1. Schematic of a piezoelectric cantilever of (a) uniform thickness (unimorph) and (b) with a nonpiezoelectric extension.

[28]. In liquid–solid transition detection, an abrupt resonance frequency shift with respect to temperature is measured to identify the transition [29]. A piezoelectric cantilever has the advantage of having a built-in piezoelectric layer to generate strong vibrations that can better withstand damping in water. The ability to withstand damping in water makes piezoelectric cantilevers excellent biosensors for direct and label-free detection. To this end, recent experiments have demonstrated real-time, in-water, direct detection of cells where the mass of the attached cells and the cell concentration in the solution were deduced from the cantilever resonance frequency shift and the resonance frequency shift with time, respectively.

11.2.1
Unimorph

When an ac voltage is applied to a unimorph cantilever, vibration occurs. The *n*th-mode flexural resonance frequency is related to the bending modulus per unit width, the length, and the mass per unit area, *m*, of the cantilever as

$$f'_n = \frac{v_n^2}{2\pi} \sqrt{\frac{K}{M_e + \Delta m}} \tag{1}$$

where v_n^2 is the dimensionless *n*th-mode eigen value, K and M_e the effective spring constant and the effective mass of the cantilever, respectively. For an added mass $\Delta m \ll M_e$, the resonance frequency shift, Δf_n, due to the added mass, Δm, is given by Eq. (2).

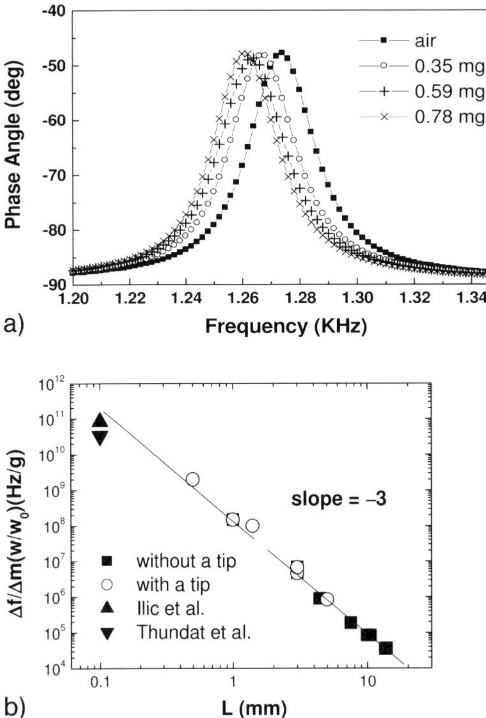

b)

Fig. 11.2. (a) Phase-angle versus frequency spectrum of a 12 mm long piezoelectric cantilever with a 8 mm long stainless steel extension. As the mass of aluminum foil on the cantilever increases, the resonance frequency shifts to lower values. (b) Resonance frequency shift per unit mass, $\Delta f / \Delta m$, with a normalized width versus cantilever length. Open circles and filled squares denote cantilevers with and without a stainless steel tip, respectively.

$$\Delta f_n = f_n' - f_n \cong -\frac{1}{2} f_n \frac{\Delta m}{M_e} \tag{2}$$

The resonance frequency shift per unit loaded mass is therefore

$$\frac{\Delta f_n}{\Delta m} \cong -\frac{1}{2} \frac{f_n}{M_e} = -\frac{1}{2} \left(\frac{v_n^2}{2\pi\sqrt{12}} \frac{h}{L^2} \sqrt{\frac{\tilde{E}}{\tilde{\rho}}} \right) \frac{1}{0.236 Lwh\tilde{\rho}} = -\frac{v_n^2}{4\pi} \frac{1}{L^3 w} \left(\frac{1}{0.236\sqrt{12}\tilde{\rho}} \sqrt{\frac{\tilde{E}}{\tilde{\rho}}} \right) \tag{3}$$

where h, L, w, $\tilde{E}$, $\tilde{\rho}$, are the total thickness, length, width, effective Young's modulus, and effective density of the cantilever, respectively. To verify Eq. (3), measurements were carried out with cantilevers of 0.4 and 0.2 cm in width with a PZT layer 0.25 mm thick bonded to a stainless steel 0.1 mm thick. The length of the cantilevers was varied from 1.37 cm to 0.44 cm by changing the clamp position. These resonant frequencies were identified and monitored by measuring the electrical impedance spectrum with an impedance analyzer (Agilent 4294A, Agilent, Palo Alto, CA). Figure 11.2 shows the experimental results. Note that off resonance the cantilever was a capacitor, exhibiting a phase angle around $-90°$. At resonance, the flexural motion gave rise to a peak in the real part of the impedance, and hence a peak in the phase angle due to the direct piezoelectric effect. As an example, Fig. 11.2(a) shows the first-mode resonance frequency spectrum, i.e., phase angle versus frequency, of a cantilever of $L = 1.37$ cm and $w = 0.4$ cm without loading ($\blacksquare$), loaded with an aluminum foil of 3.5×10^{-4} g ($\circ$), 5.9×10^{-4} g ($+$), and 7.8×10^{-4} g ($\times$) at the cantilever tip. Clearly, the resonance peak shifted to a lower frequency as the mass of the aluminum foil increased. Presently, a 1 cm long cantilever has a $\Delta f / \Delta m = 3.2 \times 10^5$ Hz g^{-1}. To compare with Eq. (3), the normalized $\Delta f / \Delta m (w/w_0)$ versus cantilever length, L, is plotted in Fig. 11.2(b), where w_0 was 2 mm. Clearly, the log–log plot of the normalized $\Delta f / \Delta m (w/w_0)$ yielded a slope of -3, validating Eq. (3). Also plotted in Fig. 11.2(a) are data points obtained with PZT cantilevers with a nonpiezoelectric tip described below. Extrapolating from Fig. 11.2(b), for a cantilever 10 μm long, its $\Delta f / \Delta m$ will approach 10^{17} Hz g^{-1}. A more detailed description of the application of a piezoelectric cantilever as a mass sensor and how the sensitivity changes with the length, width, and height of the cantilever can be found in Yi et al. [27].

11.2.2
PECS with a Nonpiezoelectric Extension

A piezoelectric cantilever of uniform thickness (Fig. 11.1a) exhibits the strongest peak intensity in the first mode and the intensity of higher-order resonance peaks decreases with an increasing order [30, 31]. The ratio of the nth resonance frequency to the first resonance frequency is also predicted by the solution of the vibration wave equation of a uniform beam. Surprisingly, a piezoelectric cantilever with a nonpiezoelectric extension (Fig. 11.1b) could exhibit a higher second-mode peak intensity than the first-mode one and the ratio of the nth-mode resonance frequency to the first-mode one could be quite different from that of a uniform beam.

Theoretically, we considered a piezoelectric cantilever with a nonpiezoelectric extension as two distinctive sections, (1) the piezoelectric section and (2) the nonpiezoelectric extension section, with each section possessing a different thickness, elastic modulus and mass density distribution in the thickness direction. We solved the cantilever's flexural vibration equation analytically to derive a transcendental equation that could be solved numerically to obtain the cantilever's flexural vibration wave forms and its resonance spectrum.

Examples of spectra in region I where the first peak was always higher than the second peak are shown in Fig. 11.3(a) with cantilevers of a 8 mm and 9.2 mm PZT section and a 4 mm nonpiezoelectric extension length ($l_1/l_2 = 2.0$ and 2.3, respectively). Figure 11.3(b) shows examples of spectra in region II with cantilevers of a 2, 4, 6, 8 and 10 mm PZT section and an 8 mm nonpiezoelectric extension length

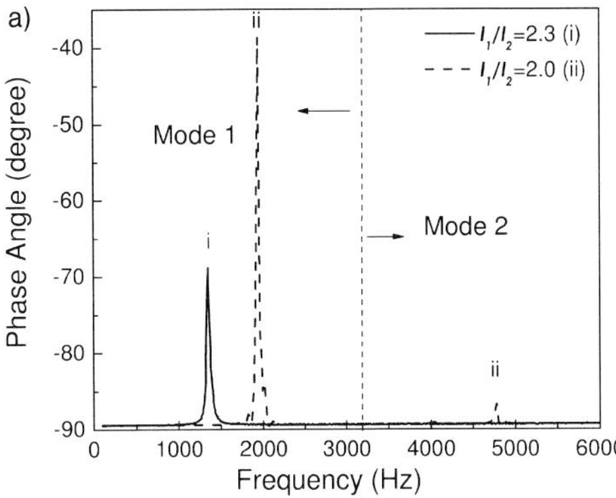

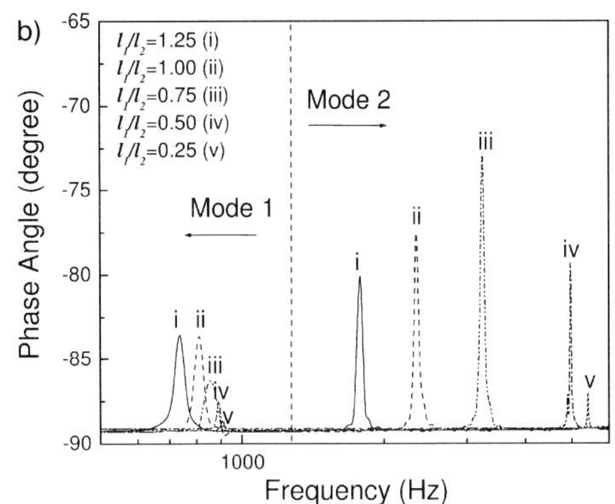

Fig. 11.3. Phase-angle versus frequency of cantilevers with (a) $l_1/l_2 \geq 1.5$ [$l_2 = 4$ mm and $l_1 = 8$ (dashed line) and 9.2 mm (solid line)] and (b) $l_1/l_2 \leq 1.5$ [$l_2 = 8$ mm and $l_1 = 10$ (i), 8 (ii), 6 (iii), 4 (iv), and 2 mm (v)].

(l_1/l_2 ranging from 0.25 to 1.25). Clearly, the ratio l_1/l_2 played an important role in determining the resonance peak intensity ratio. This came about because the resonance electrical impedance signals were solely determined by the vibration amplitude within the piezoelectric layer. How the peak height of a higher-mode resonance compared to the peak height of the first-mode depended mainly on how the higher-mode vibration amplitude within piezoelectric layer compared to that of the first mode, which could be strongly affected by the ratio of the length of the piezoelectric section to that of the nonpiezoelectric extension, l_1/l_2. To illustrate this point, calculated $h_1(x)$ and $h_2(x)$ versus x are plotted in Fig. 11.4(a) for the cantilever with $l_1 = 9.2$ mm and $l_2 = 4$ mm ($l_1/l_2 = 2.3$) whose spectrum

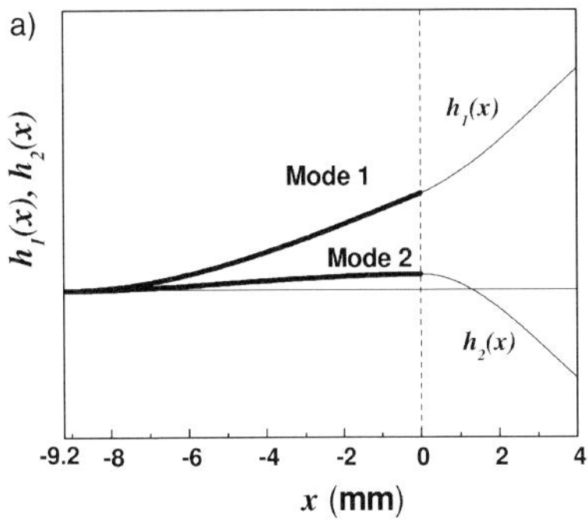

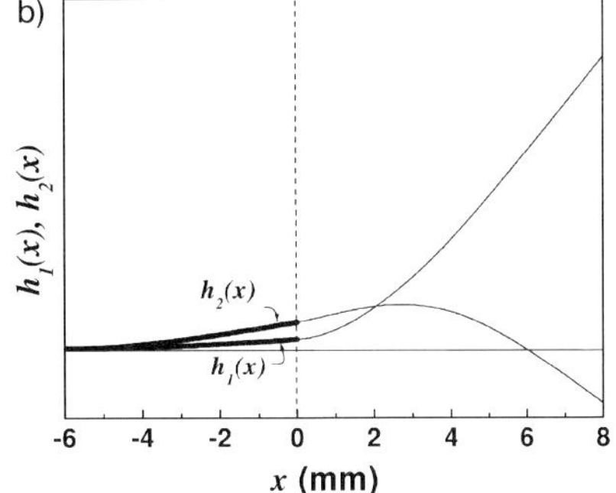

Fig. 11.4. Calculated amplitudes, $h_1(x)$ and $h_2(x)$, versus x for a cantilever with (a) $l_1 = 9.2$ mm and $l_2 = 4$ mm ($l_1/l_2 = 2.3$) and (b) $l_1 = 6$ mm and $l_2 = 8$ mm ($l_1/l_2 = 0.75$).

was shown as the solid line in Fig. 11.3(a), and in Fig. 11.4(b) for the cantilever with $l_1 = 6$ mm and $l_2 = 8$ mm ($l_1/l_2 = 0.75$) whose spectrum is labeled (iii) in Fig. 11.3(b). Figure 11.3(a) shows that the second-mode vibration amplitude, $h_2(x)$, exhibited a nodal point near $x = l_1$. Note that the second peak of this cantilever was absent in Fig. 11.3(a), as is consistent with the presence of the nodal point at $x = l_1$ shown in Fig. 11.4(a), which indicated that there was little bending stress and therefore little induced piezoelectric voltage near $x = l_1$. In contrast, for $l_1/l_2 = 0.75$, the second-mode vibration amplitude within the piezoelectric section ($x \leq l_1$) was higher than that of the first mode, indicating a higher stress and hence a higher piezoelectric response, although the first-mode vibration amplitude was higher than the second mode towards the free end ($x = l_1 + l_2$) of the nonpiezo-electric extension section.

11.3
Examples of Detections

11.3.1
Immobilization and In-solution Quantification of Yeast Cells

A PZT/stainless steel cantilever with a 3 mm long, 4 mm wide, and 0.3 mm thick PZT layer bonded to a 0.1 mm thick stainless steel with a 4 mm long stainless steel tip thinly coated with a poly-L-lysine (Sigma P8920, 0.1% w/v) layer was used for yeast detection. Poly-L-lysine is positively charged, to which negatively charged yeast is attracted and immobilized. The stainless steel tip was immersed in yeast suspensions of 1 and 2 g L^{-1} with a 3 mm dipping depth. The in-water resonance peak intensity of this cantilever is similar to its resonance peak intensity in air (Fig. 11.5a) with a Q factor about 30, making the in-water measurements as accurate and reliable as the in-air ones. Figure 11.6(a) shows the optical micrographs of yeast cells on the stainless steel tip after 60 min of immersion in the yeast suspension. A scanning electron microscopic (SEM) micrograph provided a close-up of the immobilized yeast cells (Fig. 11.6b). Adsorption of the yeast cells indeed caused a shift in the cantilever resonance frequency. Figure 11.5(b) shows a plot of the resonance frequency shift versus time. Note that the resonance frequency changed more slowly in 1 than in 2 g L^{-1}, eventually reaching the same saturated Δf in both concentrations. The resonance frequency change with time was used to characterize the cell concentration. The adsorption of cells was limited by the diffusion of cells to the cantilever surface. At short times, e.g., $t < 10$ min for 2 g L^{-1} and $t < 40$ min for 1 g L^{-1} (Fig. 11.5b), the adsorbed amount and hence the resonance frequency shift should be proportional to $t^{1/2}$ and the slope of Δf versus $t^{1/2}$ should be proportional to concentration [3]. For 1 g L^{-1} (2 g L^{-1}), at $t < 40$ min ($t < 10$ min), Δf is linear with $t^{1/2}$ and the slope for 2 g L^{-1} is indeed twice that for 1 g L^{-1}, indicating that one can use Δf versus t to quantify the concentration. More detailed results on yeast detection are presented in a recent paper by Yi et al. [32].

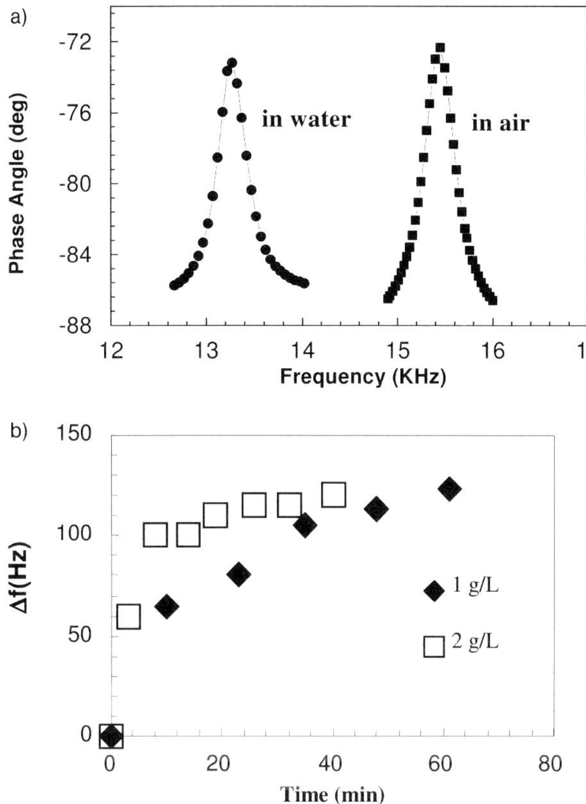

Fig. 11.5. (a) First-mode resonance frequency spectra of the yeast-detecting cantilever; (b) resonance frequency shift versus time after the cantilever was immersed in the 1 g L^{-1} (filled diamonds) and 2 g L^{-1} (open squares) suspension.

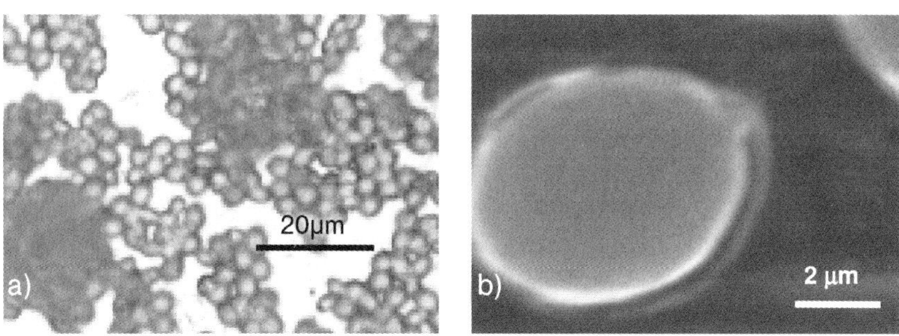

Fig. 11.6. (a) Optical micrograph of yeast cells immobilized on the poly-L-lysine layer coated on the cantilever stainless steel tip surface after 60 min and (b) SEM micrograph of immobilized yeast cells.

11.3.2
Detection of Binding of Biotinylated Polystyrene Spheres to Immobilized Avidin

We have demonstrated that a piezoelectric cantilever could detect the binding of a target protein (biotin) to an immobilized protein (avidin) at the cantilever tip. We used a PZT/stainless steel cantilever, 1.3 mm long, 2 mm wide and with a 0.127 mm thick PZT layer bonded to a 0.05 mm thick stainless steel foil, and a 2.87 mm long stainless steel tip coated with a gold layer by sputtering. The cantilever was rigidly set in an epoxy-filled glass tube that served as the clamp [insert (a) Fig. 11.7]. Avidin was immobilized on the gold surface via 3-mercaptopropionic acid (MPA). The carboxyl groups of MPA bind to avidin covalently while the sulfur binds to the gold surface chemically [33–35]. The avidin-coated tip was then vertically immersed in a 0.125 wt% suspension of biotin-coated polystyrene microspheres of 2 µm diameter (Polysciences, Warrington, PA) with a 2.2 mm dipping depth. The resonance frequency of the cantilever decreased with time and the resonance frequency shift with time is shown in Fig. 11.7. The resonance frequency shift increased with time and reached about 240 Hz shift at $t = 40$ min. An optical micrograph of the immobilized biotinylated microspheres on the avidin-coated stainless steel cantilever tip taken at 20 min is given in insert (b) of Fig. 11.7, which shows about 10% surface coverage and corresponds to about 1×10^{-6} g microspheres adsorbed on the cantilever tip, indicating that the cantilever has a mass detection sensitivity, $\Delta m / \Delta f$, of about 4×10^{-9} g Hz^{-1}.

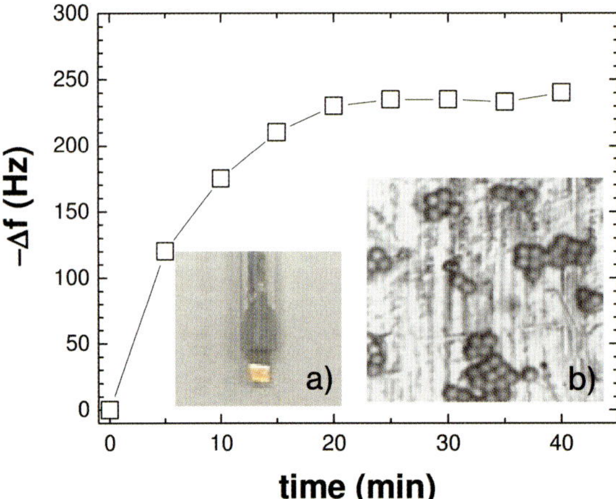

Fig. 11.7. Resonance frequency shift with time of a PZT/stainless steel cantilever in a biotinylated polystyrene suspension. Inserts: (a) photograph of the cantilever. The PZT layer was 1.3 mm long, 3.35 mm wide. The 2.87 mm long gold-coated stainless steel tip was coated with avidin. (b) Optical micrograph of immobilized biotinylated spheres after 20 min.

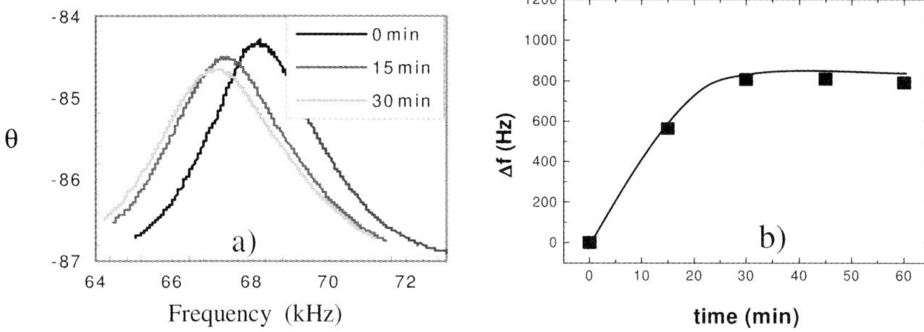

Fig. 11.8. (a) Resonance spectra at different times due to the immobilization of avidin on the stainless steel tip and (b) the resultant resonance frequency shift versus time.

11.3.3
Detection of Avidin Immobilization at the Cantilever Tip

To detect the immobilization of avidin we used an even *shorter* cantilever that had a PZT layer *0.7 mm long*, 2.27 mm wide, and 0.127 mm thick bonded to a 0.05 mm thick stainless steel foil. The cantilever has a 2 mm long gold-coated stainless steel tip. The freshly sputtered gold surface was first coated with MPA as described above and then immersed in a 0.1 mg mL^{-1} avidin solution. The dipping depth was 1.3 mm. Immobilization of avidin on the cantilever tip was detected by monitoring the resonance frequency shift of the cantilever at around 68 kHz. Figure 11.8(a) shows the resonance spectrum at different times and Fig. 11.8(b) shows the frequency shift with time after the cantilever tip was immersed in the avidin solution. The saturated adsorption amount of avidin was known to be about 3 ng mm^{-2} [36, 37]. This indicated that the mass detection sensitivity of this 0.7 mm long cantilever was $\Delta m/\Delta f \sim 1 \times 10^{-10}$ g Hz^{-1}. The saturated adsorption amount of avidin yields a resonance frequency change of 40 Hz with a conventional 5 MHz QCM. The present piezoelectric cantilever's 800 Hz resonance frequency change is already 20× better than that of a 5 MHz QCM.

11.3.4
Salmonella typhimurium Detection

Salmonella is a water-born/food-born pathogen that causes more than 500 deaths each year [38]. It is also a category B bioterrorism agent that can seriously contaminate water resources. There is no device at present that can actually monitor a space and perform an analysis of the air or water within that space in real-time. However, such a device would dramatically decrease the "lag time between release of an agent and its detection," and thus enable more prompt treatment of those individuals possibly exposed to a particular bioterrorism agent in addition to reducing the total number of individuals affected [39].

A piezoelectric microcantilever was fabricated by bonding a layer of lead zirconate titanate (PZT) 0.127 mm thick, 2 mm long, and 1.5 mm wide to a titanium foil 0.127 mm thick with a titanium tip 3 mm long (Fig. 11.1b). The PZT layer had a piezoelectric coefficient d_{31} of -320 pm V^{-1} and a Young's modulus Y_{11} of 62.5 GPa. The titanium foil has a Young's modulus of 103 GPa. The salmonella strain was *Salmonella typhimurium* and the antibody used was an antibody to Salmonella Common Structural Antigens (CSA-1) (both from Krikegaard & Perry Laboratories, Inc., Gaithersburg, Maryland). This antibody is specifically designed to recognize several species of bacteria in the *Salmonella* genera, including our *typhimurium* species. The CSA-1 antibody was immobilized on the tip of a PZT-Ti microcantilever by means of a linking molecule. Titanium was chosen as the non-piezoelectric layer as a result of its good mechanical properties as well as its outstanding biocompatibility [40]. This biocompatibility is demonstrated by its widespread use as a material for implants in the human body.

Though common methods of immobilization on surfaces rely on MPA or 3-aminopropyltriethoxysilane (APS) followed by activation with 1-ethyl-3-(3-dimethylaminopropyl)carbodiimide (EDC) and *N*-hydroxysulfosuccinimide (NHS) [41], we used a much simpler method adapted from silica-silane chemistry. Since a passivated titanium surface exhibits similar surface chemistry to that of silica [42], a bifunctional linker, glycidoxypropyltrimethoxysilane (GOPTS), was used as the linking molecule [43, 44]. GOPTS presents, on one end, a trialkyloxysilane group, like APS, which allows its incorporation on activated silica or silica-like surfaces. Once GOPTS is deposited on a surface, it exposes epoxy moieties that are unstable and therefore very reactive toward nucleophilic groups such as amines, thiols, and alcohols. These groups will promptly react, causing opening of the bond angle strained epoxide ring. The main advantage of GOPTS over other bifunctional linkers is that it does not require an activation step, whereas activation with EDC/NHS is required for standard carboxylic functionalities. In our work, the trimethoxysilane end of GOPTS was bound to the titanium surface of the microcantilever and CSA-1 antibody was subsequently immobilized on the sensor under basic conditions, through addition of its amine groups to epoxide rings exposed on the surface.

Following functionalization of the microcantilever with antibody, several detections of the *S. typhimurium* cells were performed by lowering the cantilever tip into a vial containing the cell suspension by means of a micropositioner, such that approximately two-thirds of the tip was submerged. This allowed the resonance frequency shift to be monitored with time as described above.

In the present detection set-up, the titanium tip was only partially immersed in the solution to avoid wetting of the PZT layer by the solution. As a result, there was an upward background resonance shift due to a receding water level along with the resonance frequency shift due to cell binding. Consequently, a typical cell-detection resonance frequency shift versus time plot exhibited an initial decrease, reaching a minimum that was followed by an almost linear rise (solid curve

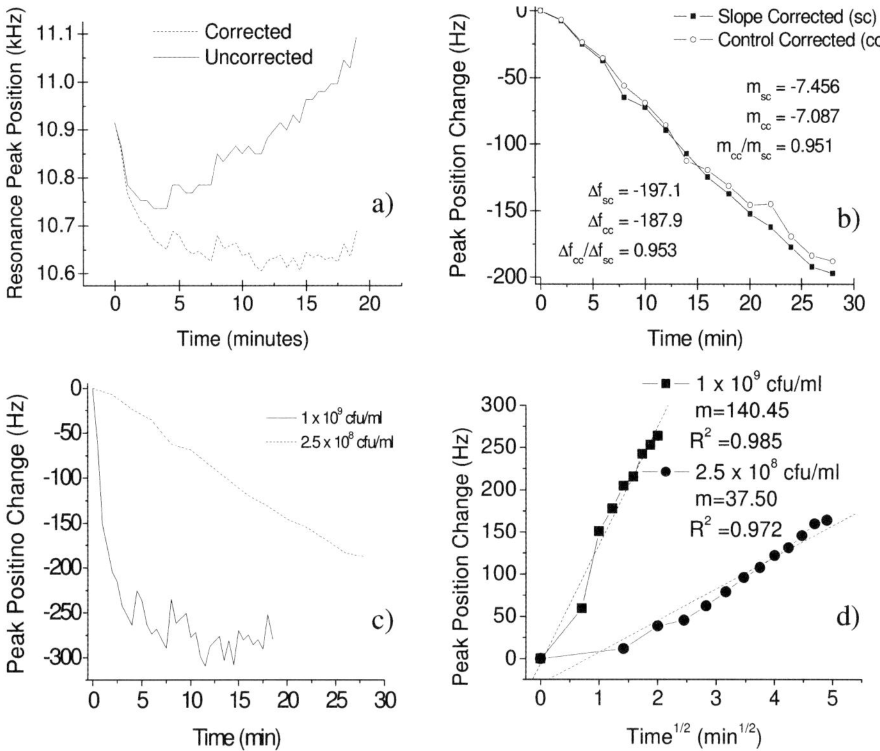

Fig. 11.9. (a) Uncorrected salmonella detection data with the corresponding corrected data. (b) Comparison of slope corrected vs. control corrected detection. (c) Detection of 1×10^9 and 2.5×10^8 cells mL^{-1} *Salmonella typhimurium* suspensions. (d) Kinetic analysis of *S. typhimurium* binding.

in Fig. 11.9a). To address the issue of background resonance frequency shift, we carried out three different checks. The first was to obtain the rate of upward background shift in water prior to the detection experiment. The second was to obtain the rate of upward background shift using the later stage of the resonance frequency shift that showed an almost linear increase. The third check was to use a two-cantilever array to perform the detection. One cantilever was treated with the GOPTS as detailed above, while the other cantilever was left passivated but untreated with GOPTS. Then, both cantilevers were simultaneously subjected to the same series of steps associated with a detection process: (a) exposure to the antibody, (b) rinse in phosphate-buffered saline (PBS) and (c) dipping in the cell suspension. We found that the first two corrections were practically the same, indicating that the upward shift in the later stages of detection was indeed a manifest of receding water level. In Fig. 11.9(a), the corrected resonance frequency shift due to cell binding is shown as the dashed line after subtracting the background upshift from the initial data (solid line). Figure 11.9(b) shows two curves from the same detection of a 2.5×10^8 cells mL^{-1} *S. typhimurium* suspension. The *control corrected* data represent subtraction of the upward drift based on the control cantilever that was not coated with the GOPTS (and thus was not functionalized with the CSA-1 antibody). The *slope corrected* data represent the removal of the linear drift simply by taking a linear regression of the detection data after cell adhesion on the sensor tip had ceased. As can be seen in the figure, the ratio of the two slopes is nearly unity, as is the ratio of the total resonant frequency shifts. Furthermore, as mentioned above, the cantilever used in Fig. 11.9(a) was also monitored in water alone and the same linear up-drift was obtained, thus providing further substantiation of the post-cell adhesion up-drift subtraction method. Based on these findings, all subsequent cell detection experiments used only the post-cell adhesion linear drift data to correct for the upward drift. This greatly simplified the experiments and thus allowed for more efficient data collection.

Meanwhile, PZT/gold-coated glass cantilever sensors with a 0.127 mm thick, 0.5 mm long and 2 mm wide PZT layer bonded to a gold coated glass layer of 0.150 mm in thickness that had a 2 mm long gold-coated glass tip were also tested in *S. typhimurium* detection. The gold-coated glass surface was coated with MPA, which was treated in a solution of 5 mg mL^{-1} EDC and 5 mg mL^{-1} NHS in distilled water for 30 min followed by dipping in 1 μL of a 0.1 mg mL^{-1} solution of CSA-1 in a phosphate-buffered saline (PBS) solution at pH 7.4 for 30 min. The antibody-coated cantilevers were then used to detect *S. typhimurium* at various concentrations for up to 30 min. Figure 11.10 gives the resultant Δf versus time plots from the PZT/gold-coated cantilevers at various *S. typhimurium* concentrations. As can be seen, Δf decreases sharply initially and reached a plateau after 10 min. The detections were quite repeatable as each curve was the average of 3–4 independent measurements. The standard deviations of Δf were about 10–20%. Table 11.1 lists

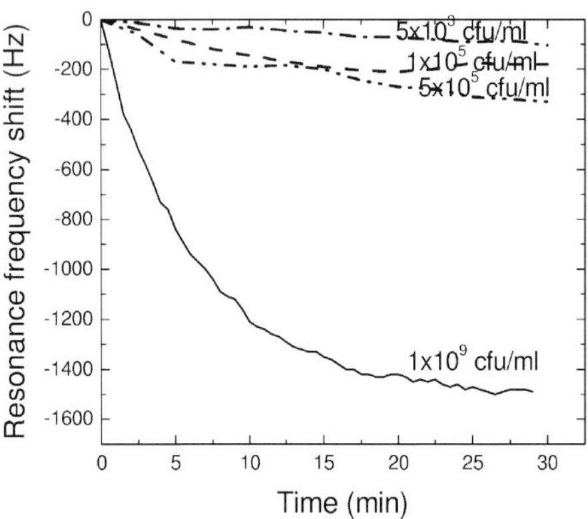

Fig. 11.10. Resonance frequency shift versus time of the PZT/gold-coated glass cantilever in various *Salmonella typhimurium* concentrations.

Table 11.1. Resonance frequency shift of a PZT/Au-coated glass cantilever with 5×10^{-11} g/Hz sensitivity in various concentrations of *Salmonella t.*

Salmonella concentration (cells/ml)	Δf (PZT/Au-coated glass)
10^9	1500
5×10^5	330
1×10^5	180
5×10^3	90

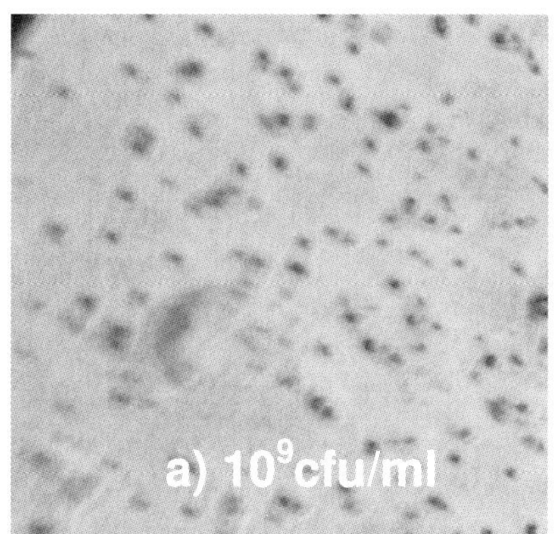

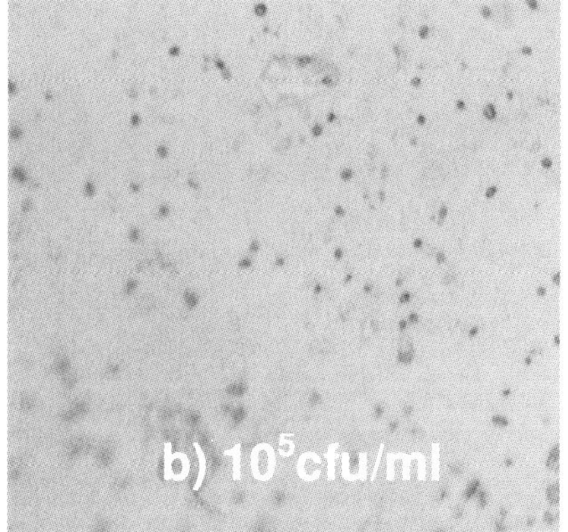

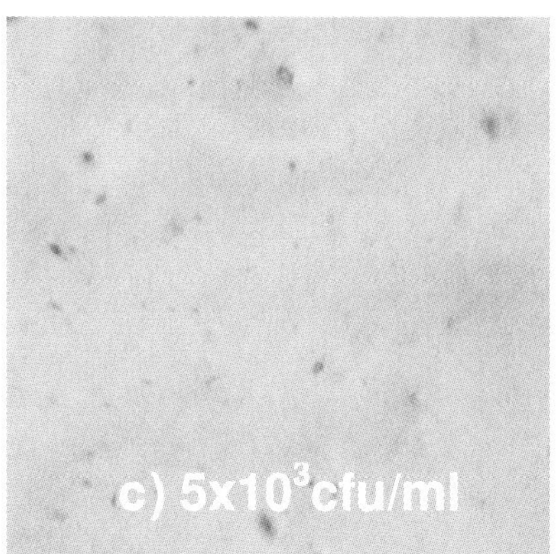

Fig. 11.11. Optical micrographs of a PZT/gold-coated cantilever surface after 30 min detection in a (a) 1×10^9 (b) 1×10^5 and (c) 5×10^3 cells mL^{-1} *Salmonella typhimurium* suspensions.

the average Δf at $t = 30$ min for various *Salmonella* concentrations, as obtained from the PZT/gold coated glass cantilever. As can be seen, at 5×10^3 cells mL^{-1}, the Δf obtained by the PZT/gold coated glass cantilevers at $t = 30$ min was 90 Hz, well above the standard deviation of 20 Hz, and at 1×10^3 cells mL^{-1} the obtained Δf was no longer meaningful compared to the standard deviation. This puts the detection limit of the current PZT/gold-coated glass cantilever at about the 1×10^3 cells mL^{-1} range.

To validate the results shown in Fig. 11.10, the PZT/gold-coated glass cantilever surface was examined by optical microscopy after 30 min of detection. Figure 11(a)–(c) shows SEM micrographs of captured *Salmonella* cells on the PZT/gold-coated glass cantilever surface at $t = 30$ min from 1×10^9, 1×10^5, and 5×10^3 cells mL^{-1} suspensions, respectively. With the mass of each cell being 2×10^{-12} g and by counting the number of cells in the micrographs, the mass of the captured cells on the cantilever surface was estimated to be 5×10^{-8}, 3×10^{-9}, and 1×10^{-9} g for 1×10^9, 1×10^5, and 5×10^3 cells mL^{-1} suspensions, respectively. With Δf at 30 min being 970, 300 and 100 Hz for the 1×10^9, 1×10^5, and 5×10^3 cells mL^{-1} suspensions, respectively, $\Delta m/\Delta f$ was 5×10^{-11} g Hz^{-1}, which is consistent with the 5×10^{-11} g Hz^{-1} as calibrated with a quartz crystal microbalance (QCM) in the antibody immobilization.

11.3.5
Nerve Gas Simulant Detection

The piezoelectric cantilever sensors we have developed are quantitative and suitable for use in air. As an example of detecting chemical agents, we studied the de-

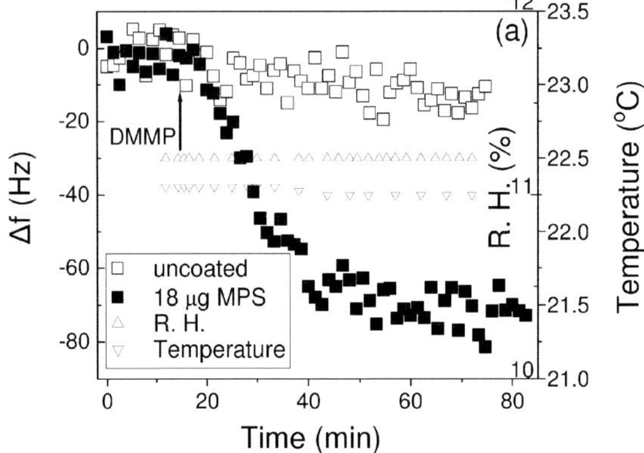

Fig. 11.12. Resonance frequency shift versus time of the uncoated (open squares) and 18 μg (filled squares) microporous silica (MPS)-coated cantilever after spraying 10 mL liquid DMMP in a 66 L chamber. Also shown are the relative humidity (open triangles) and temperature (inverted open triangles) in the chamber during the test. The arrow indicates the injection of DMMP.

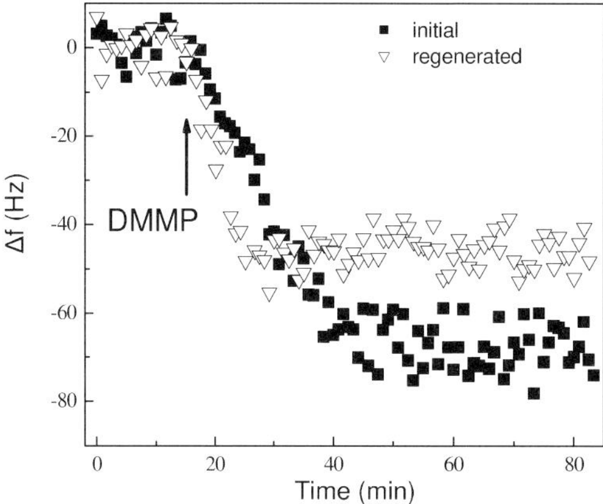

Fig. 11.13. Resonance frequency shift versus time during the initial DMMP test (filled squares) and after regeneration (inverted open triangles).

tection of the nerve gas simulant dimethyl dimethyl phosphonate (DMMP) using a PZT cantilever that had a PZT layer 1 mm long, 2.1 mm wide, and 0.127 mm thick bonded to a 0.05 mm thick stainless steel foil. The cantilever had a stainless steel tip 2.4 mm long and 0.05 mm thick coated with ultrahigh surface area (800 m^2 g^{-1} and pore size 10 Å) microporous SiO_2. The coated cantilever had a lower resonance frequency ($\sim$80 Hz shift) due to the mass of the microporous SiO_2 coating. The SiO_2-coated cantilever was placed in a closed chamber with a constant humidity of 11.2% and a constant temperature of 22.2 °C, as shown in Fig. 11.12 for DMMP tests. When 5 mL of DMMP was injected and sprayed into the chamber, the cantilever resonance frequency decreased with time and saturated at about 70 Hz change (■ in Fig. 11.12). Note that without the silica coating, the resonance frequency was stable (□), indicating that silica-coated cantilevers can indeed be sensitive DMMP sensors.

After the initial test, the cantilever was regenerated in air for 5 days. In the second DMMP test, the resonance frequency of the regenerated cantilever decreased at a similar rate as in the initial test, except that the resonance frequency shift saturated at 50 Hz instead of 70 Hz (Fig. 11.13). The smaller saturated resonance frequency shift for the regenerated cantilever was probably due to incomplete removal of DMMP. A clear advantage of oxide adsorbent is that it is readily reusable due to the physical nature of the adsorption.

11.4
Piezoelectric Cantilever Miniaturization

As shown in Fig. 11.2(b), miniaturized cantilevers can achieve higher mass detection sensitivity. We have developed two types of piezoelectric microcantilevers: one is fabricated from freestanding lead magnesium niobate-lead titanate (PMN-PT) films developed in our laboratory, exhibiting up to 2×10^{-14} g Hz^{-1} sensitivity; the other is fabricated using silicon microfabrication techniques with lead zirconate titanate thin films on silicon wafers that can be as small as 50 μm long, offering better than 10^{-16} g Hz^{-1} sensitivity.

11.4.1
PMN-PT/Cu Microcantilevers

The PMN-PT/Cu microcantilevers consisted of a highly piezoelectric layer fabricated from freestanding lead magnesium niobate-lead titanate (PMN-PT) films bonded to a copper layer by electroplating. The microcantilever shape was achieved by wire-saw cutting. As an example, the top-view optical micrograph, cross-section scanning electron microscopy (SEM) micrograph, and a typical resonance spectrum of 500 μm long, 700 μm wide PMN-PT/Cu microcantilevers made of 22 μm thick PMN-PT and 5 μm thick copper are shown in Fig. 11.14(a)–(c) respectively. The spectrum shown in Fig. 11.14(c) exhibits multiple resonance peaks with

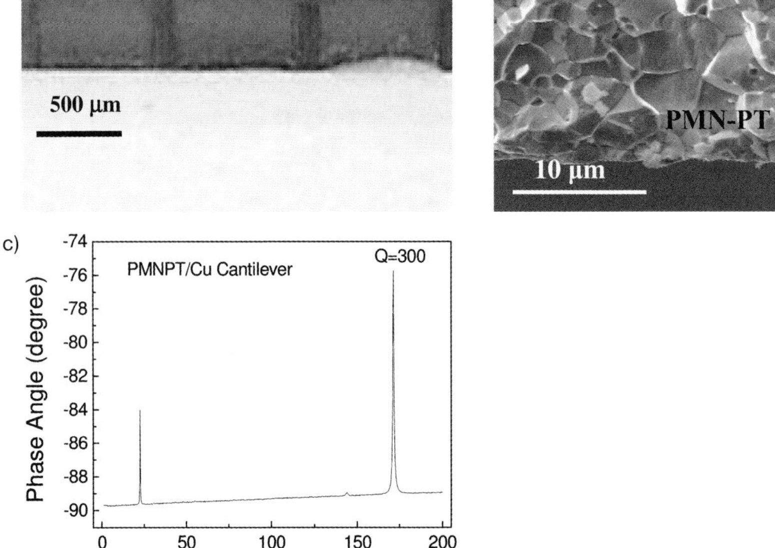

Fig. 11.14. (a) Optical micrograph and (b) SEM cross-section micrograph of 500 μm long, 700 μm wide PMN-PT/Cu microcantilevers with 22 μm thick PMN-PT and 5 μm thick Cu on a glass substrate, and (c) a typical resonance spectrum. Note that because of the highly piezoelectric PMN-PT layer we were able to retain the higher-mode resonance peaks with $Q > 300$ for better detection sensitivities. This PMN-PT PEMS exhibited femtogram mass detection sensitivity, $\Delta m / \Delta f = 3 \times 10^{-13}$ g Hz^{-1}.

Tab. 11.2. Resonance frequency shift of a PMN-PT/Cu microcantilever with 3×10^{-13} g Hz^{-1} sensitivity compared with a PZT/Au-coated glass cantilever with 5×10^{-11} g Hz^{-1} sensitivity. Note: the PMN-PT/Cu microcantilever lowered the BA concentration limit by two orders of magnitude.

BA concentration (cells mL^{-1})	Δf (PMN-PT/Au-Cu)[a]	Δf (PZT/Au-coated glass)
10^5		1200
2×10^4		600
5×10^3	6800	110
2×10^3		50
500	5700	
50	100	

[a] PMN-PT/Cu microcantilever had copper on one side and was cold coated on the other side for antibody immobilization.

$Q > 300$ (Q is the ratio of the resonance frequency to the resonance peak width at half the peak height). The mass detection sensitivity of this PMN-PT/Cu microcantilever was determined to be 3×10^{-13} g Hz^{-1}, with the second resonance peak at 170 kHz, as calibrated by QCM. The PMN-PT/Cu microcantilever was gold coated on both sides for antibody immobilization using the MPA self-assembly monolayer approach (Section 11.3.4). Because the PMN-PT/Cu microcantilevers were about two orders of magnitude more sensitive than the PZT PECS described in Sections 11.3.3–11.3.5, they realized an even lower concentration limit. For example, Table 11.2 lists the resonance frequency shifts of PMN-PT microcantilevers at various *Bacillus anthracis* (BA) concentrations in BA spores detection. Also listed are the resonance frequency shifts of a PZT/Au-coated glass cantilever with a sensitivity of 5×10^{-11} g Hz^{-1} (similar to that used in *S. typhimurium* detection in Section 11.3.4). Clearly, the more sensitive PMN-PT/Cu microcantilevers lowered the detection concentration limit to 50 cells mL^{-1} – two orders of magnitude lower than that of the PZT/Au-coated glass cantilevers.

11.4.2
PZT/SiO$_2$ Microcantilevers and PZT/SiO$_2$-Si$_3$N$_4$ Nanocantilevers

Alternatively, we have successfully fabricated PZT/SiO$_2$ microcantilevers by depositing 1.6 mm thick PZT .lms on a Pt/TiO$_2$/SiO$_2$/Si substrate with a novel sol–gel process and repeated spin coating. The Pt/TiO$_2$/SiO$_2$ substrate was necessary to prevent unwanted interfacial reactions and diffusions. The precursor solution contained 50% excess lead to compensate for lead loss during repeated heat treatment. Figure 11.15(a) shows a SEM micrograph of a 1.6 μm thick PZT film. The film was further made into piezoelectric microcantilevers by integrating the piezoelectric films with the microfabrication process. Examples of a 50 μm long PZT/SiO$_2$ mi-

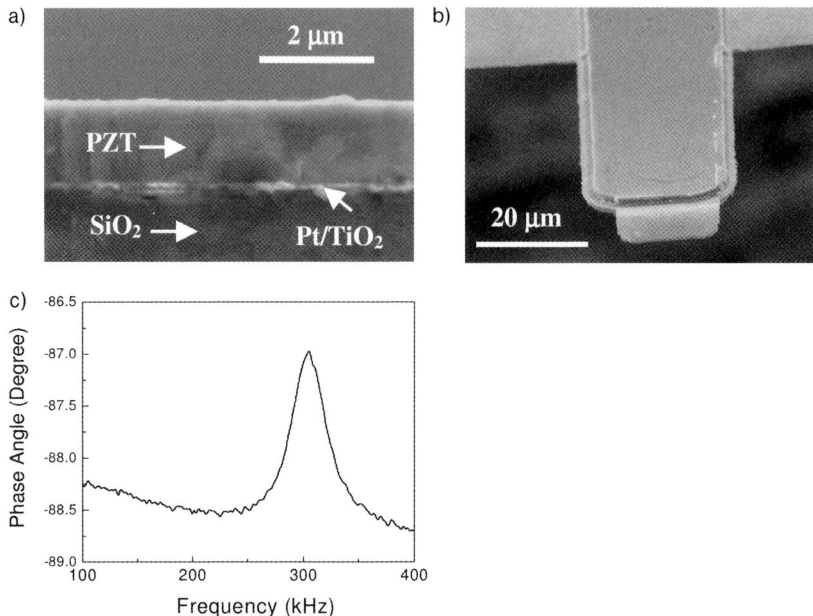

Fig. 11.15. SEM cross-section of (a) a 1.6 µm thick PZT film and (b) a 50 µm long, 20 µm wide PZT/SiO$_2$ microcantilever with a 10 µm long SiO$_2$ tip. (c) Resonance frequency spectrum of the PZT/SiO$_2$ microcantilever in (b).

crocantilever and its resonance frequency spectrum are shown in Fig. 11.15(b) and 11.15(c), respectively. From our earlier experimental and theoretical work, we expect such piezoelectric microcantilevers to have better than 10^{-16} g Hz^{-1} sensitivity. Furthermore, in the making are PZT/SiO$_2$-Si$_3$N$_4$ piezoelectric nanocantilever sensors (PENS) with a 20 µm long PZT PZT/SiO$_2$-Si$_3$N$_4$ section and a nano-size SiO$_2$-Si$_3$N$_4$ tip that is less than 300 nm in width and less than 1 µm in length. Theoretical calculation indicates that such PZT PZT/SiO$_2$-Si$_3$N$_4$ PENS will exhibit better than 10^{-18} g Hz^{-1} sensitivity (the mass of a single protein or DNA).

11.5
Conclusions

We have developed piezoelectric microcantilever sensors of different sizes and types that can perform rapid, *in situ*, in-water pathogen detection with sensitivities well above that of current techniques. Both theoretical and experimental studies were carried out to characterize the sensors. We showed that using PZT/glass cantilevers of sub-millimeter length with a 2 mm glass tip that exhibited 5×10^{-11} g Hz^{-1} mass detection sensitivity, *in situ* quantification of S. typhimurium was achieved with a concentration limit of 10^3 cells mL^{-1}, lower than the infectious dosage, 10^5 cells mL^{-1}, which is also the concentration limit of ELISA and QCM. Furthermore, using 500 μm long miniaturized PMN-PT/Cu microcantilevers that exhibited 3×10^{-13} g Hz^{-1} detection sensitivity, we showed that in a liquid volume of less than 1 mL the detection concentration limit was lowered to below 50 cells mL^{-1}. With PZT/SiO_2 microcantilevers less than 50 μm long and PZT/SiO_2-Si_3N_4 PENS less than 20 μm long with a SiO_2-Si_3N_4 less than 300 nm wide and less than 1 μm long, we expect the detection sensitivity to reach better than 10^{-16} g Hz^{-1} and 10^{-18} g Hz^{-1}, respectively and the detection concentration to be further lowered.

Acknowledgment

This work is supported in part by the National Aeronautics and Space Administration (NASA) under Grant No. NAG2-1475, the National Institute of Health (NIH) under Grant No. 1 R01 EB000720, and the Environmental Protection Agency (EPA) under Grant No. R82960401.

References

1 S. NIE, R. N. ZARE, Optical detection of single molecules, *Annu. Rev. Biophys. Biomol. Struct.*, 26, 567–596 (**1997**).

2 S.-K. KRAEFT, A. LADANYI, K. GALIGER, A. HERLITZ, A. C. SHER, D. E. BERGSRUD, G. EVEN, S. BRUNELLE, L. HARRIS, R. SALGIA, T. DAHL, J. KESTERSON, L. B. CHEN, Reliable and sensitive identification of occult tumor cells using the improved rare event imaging system, *Clin. Cancer Res.*, 10, 3020 (**2004**).

3 http://optics.org/articles/news/8/3/27/1.

4 R. T. KRIVACIC, A. LADANYI, D. N. CURRY, H. B. HSIEH, P. KUHN, D. E. BERGSRUD, J. F. KEPROS, T. BARBERA, M. Y. HO, L. B. CHEN, R. A. LERNER, R. H. BRUCE, A rare-cell detector for cancer, *Proc. Natl. Acad. Sci. U.S.A.*, 101(29), 10 501–10 504 (**2004**).

5 M. MINUNNI, M. MASCINI, G. G. GUILBAULT, B. HOCK, The quartz crystal microbalance as biosensor, a status report on its future, *Anal. Lett.*, 28, 749–764 (**1995**).

6 A. G. GEHRING, D. L. PATTERSON, S.-I. TU, Use of a light-addressable potentiometric sensor for the detection of Escherichia coli O157:H7, *Anal. Biochem.*, 258, 293–298 (**1998**).

7 A. K. BEJ, M. H. MAHBUBANI, M. J. BOYCE, R. M. ATLAS, Detection of Salmonella sp. in oysters by PCR, *Appl. Environ. Microbiol.*, 60, 368–373 (**1994**).

8 B. A. ABACK, S. J. O'DAY, D. S. B. HOON, Quantification of circulating DNA in the plasma and serum of cancer patients, *Ann. New York Acad. Sci.*, 1022, 17 (**2004**).

9 F. OSHITA, A. SEKIYAMA, R. SUZUKI, M. IKEHARA, K. YAMADA, H. SAITO, K. NODA, Y. MIYAGI, Detection of occult tumor cells in peripheral blood from patients small cell lung cancer by promoter methylation and silencing of retinoic acid receptor-beta, *Oncol. Rep.* 10, 105 (**2003**).

10 S.-J. PARK, T. A. TATON, C. A. MIRKIN, Array-based electrical detection of DNA with nanoparticle probes, *Science*, 295, 1503–1506 (**2002**).

11 J. FRITZ, M. K. BALLER, H. P. LANG, H. ROTHUIZEN, P. VETTIGER, E. MEYER, H.-J. GUNTHERODT, Ch. GERBER, J. K. GIMZEWSKI, Translating biomolecular

recognition into nanomechanics, *Science*, 288, 316–318 (**2000**).

12 A. Schemmel, H. E. Gaub, Single molecule force spectrometer with magnetic force control and inductive detection, *Rev. Sci. Instrum.* 70(2), 1313–1317 (**1999**).

13 D. R. Baselt, G. U. Lee, R. J. Colton, Biosensor based on force microscope technology, *J. Vac. Sci. Technol.* B14(2), 798–793 (**1996**).

14 G. U. Lee, D. A. Kidwell, R. J. Colton, *U. S. Pat. Appl.* 38 (**1996**). NTIS Order No. PAT-APPL-8-505 547.

15 W. Han, S. M. Lindsay, T. Jing, A magnetically driven oscillating probe microscope for operation in liquids, *Appl. Phys. Lett.* 69(26), 4111–4113 (**1996**).

16 T. Thundat, E. A. Wachter, S. L. Sharp, R. J. Warmack, Detection of mercury vapor using resonating microcantilevers, *Appl. Phys. Lett.*, 66, 1695 (**1995**).

17 B. Ilic, D. Czaplewsli, H. G. Craighead, P. Neuzil, C. Campagnolo, C. Batt, Mechanical resonant immunospecific biological detector, *Appl. Phys. Lett.*, 77, 450 (**2000**).

18 P. I. Oden, G. Y. Chen, R. A. Steele, R. J. Warmack, T. Thundat, Viscous drag measurements utilizing microfabricated cantilevers, *Appl. Phys. Lett.*, 68, 3814 (**1996**).

19 Z. Lin, C. M. Yip, I. S. Josheph, M. D. Ward, Operation of an ultrasensitive 30-MHz quartz crystal microbalance in liquids, *Anal. Chem.*, 65, 1546–1551 (**1993**).

20 M. Basu, S. Seggerson, J. Henshaw, J. Jiang, R. del A Cordonal, C. Lefave, P. J. Boyle, A. Miller, M. Pugia, S. Basu, Nano-biosensor development for bacterial detection during human kidney infection: Use of glycoconjugate-specific antibody-bound gold nanowire arrays (GNWA), *J. Glycoconj.*, 21, 487–496 (**2004**).

21 B. Ilic, Y. Yang, K. Aubin, R. Reichenbach, S. Krylov, H. G. Craighead, Enumeration of DNA molecules bound to a nanomechanical oscillator, *Nano Lett.*, 5(5), 925 (**2005**).

22 W. Y. Shih, W.-H. Shih, Z. Shen, Piezoelectric cantilever sensor, US Patent Application No. PCT/US2004/036705, filed October 27, **2004**.

23 C. A. Rowe, L. M. Tender, M. J. Feldstein, J. P. Golden, S. B. Scruggs, B. D. MacCraith, J. J. Cras, F. S. Ligler, Array biosensor for simultaneous identification of bacterial, viral, and protein analytes, *Anal. Chem.* 71, 3846–3852 (**1999**).

24 S. P. Sakti, S. Rösler, R. Lucklum, P. Hauptmann, F. Bühling, S. Ansorge, Thick polystyrene-coated quartz crystal microbalance as a basis of a cost effective immunosensor, *Sens. Actuators A*, 76, 98 (**1999**).

25 M. D. Ward, D. A. Buttry, In situ interfacial mass detection with piezoelectric transducers, *Science*, 249, 1000–1007 (**1990**).

26 T. G. Thundat, J. Adams, The 2004 Scientific American 50 Award: Research Leaders, *Sci. Am.*, 291(6), 47 (**2004**).

27 J. W. Yi, W. Y. Shih, W.-H. Shih, Effect of length, width, and mode on the mass detection sensitivity of piezoelectric unimorph cantilevers, *J. Appl. Phys.*, 91(3), 1680 (**2002**).

28 W. Y. Shih, X. Li, H. Gu, W.-H. Shih, I. A. Aksay, Simultaneous liquid viscosity and density determination with piezoelectric unimorph cantilevers, *J. Appl. Phys.*, 89(2), 1497 (**2001**).

29 W. Y. Shih, X. Li, J. Vartuli, D. L. Milius, R. Prud'homme, I. A. Aksay, W.-H. Shih, Detection of water-ice transition using a lead zirconate titanate brass transducer, *J. Appl. Phys.* 92(1), 106 (**2002**).

30 J. Merhaut, *Theory of Electroacoustics*, McGraw-Hill, New York, **1981**.

31 C. de Silva, *Vibration, Fundamentals and Practice*, CRC Press, New York, **2000**, p. 302.

32 J. W. Yi, Wan Y. Shih, R. Mutharasan, Wei-Heng Shih, In situ cell detection using piezoelectric PZT-stainless steel cantilevers, *J. Appl. Phys.*, 93, 619 (**2003**).

33 R. G. Nuzzo, F. A. Fusco, D. L. Allara, *J. Am. Chem. Soc.*, 109, 2358 (**1987**).

34 R. G. Nuzzo, F. A. Fusco, D. L. Allara, *J. Am. Chem. Soc.*, 109, 733 (**1987**).

35 C. Steinem, A. Janshoff, H.-J. Galla, M. Sieber, *Bioelectrochem. Bioenerg.* 42, 213 (**1997**).

36 D. Clerc, W. Lukosz, Real-time analysis of avidin adsorption with an integrated-optical grating coupler: Adsorption kinetics and optical anisotropy of adsorbed monomolecular layers, *Biosensors Bioelectron.*, 12(3), 185–194 (**1997**).

37 B. L. Frey, C. E. Gordan, S. Jornguth, R. M. Corn, Control of the specific adsorption of proteins onto gold surfaces, *Anal. Chem.*, 67(24), 4452–4457 (**1995**).

38 Centers for Disease Control and Prevention, Division of Bacterial and Mycotic Diseases, Disease Information. http://www.cdc.gov/ncidod/dbmd/diseaseinfo/salmonellosis_t.htm. Accessed, 6 December **2004**.

39 P. J. Meehan, N. E. Rosenstein, M. Lillen, R. F. Meyer, M. J. Kiefer, S. Deitchman, R. E. Besser, R. L. Ehrenberg, K. M. Edwards, K. F. Martinez, Responding to detection of aerosolized Bacillus anthracis by autonomous detection systems in the workplace, *Morbidity Mortality Weekly Rep.*, **2004**, 30 Apr (53, early release), 1–11.

40 B. D. Ratner, A. S. Hoffman, F. J. Schoen, J. E. Lemons, *Biomaterials Science: An Introduction to Materials in Medicine*, Academic Press, San Diego, **1996**, p. 263.

41 V. Faust, Biofunctionalized biocompatible titania coatings for implants, *Euro. Ceramics, Key. Eng. Mat., VII.* part 1–3, 206-2, 1547–1550 (**2002**).

42 B. Arkles, Tailoring surfaces with silanes, *Chemtech*, 7, 766–778 (**1997**).

43 J. Yakovleva, R. Davidsson, A. Lobanova, M. Bengtsson, S. Eremin, T. Laurell, J. Emneus, Microfluidic enzyme immunoassay using silicon microchip with immobilized antibodies and chemiluminescence detection, *Anal Chem.*, 74(13), 2994–3004 (**2002**).

44 J. Piehler, A. Brecht, R. Valiokas, B. Liedberg, G. Gauglitz, A high-density poly(ethylene glycol) polymer brush for immobilization on glass-type surfaces, *Biosens. Bioelectron.*, 15(9–10), 473–481 (**2000**).

12

Toxicology of Nanoparticles in Environmental Air Pollution

Ken Donaldson, Nicholas Mills, David E. Newby, William MacNee, and Vicki Stone

12.1
Introduction

The toxicology of engineered nanoparticles is a topic of increasing interest. However the existing toxicology database on nanoparticles rests almost entirely on combustion-derived nanoparticle in environmental air. This research reached a peak in the mid to late 1990s, focused around the 'ultrafine hypothesis'[1,2]. This suggested that the combustion-derived nanoparticle component of PM was a key component of PM in causing adverse health effects, by virtue of its ability to cause oxidative stress and inflammation and translocate from the site of deposition[3]. This review puts forward the evidence that nanoparticles do play a role on the adverse health effects of environmental particles and what the mechanism may be, in the belief that this may illuminate the toxicology of engineered nanoparticles.

12.2
History of Air Pollution

The adverse health effects of air pollution have been recognized for centuries. In the UK, the burning of fossil fuels in towns and cities combined with periods of cold weather, where there is little mixing of air, have been associated with the generation of smogs. Due to the sulfurous nature of the coal, these smogs consisted mainly of sulfur dioxide and particles, the latter measured historically as "black smoke" (Table 12.1). The famous smog that occurred in London in December 1952 saw midday London appear more like midnight, with theatres closed due to the inability of the audience to see the stage! Interestingly, for the underlying theme of this article, analysis of the particles in the lungs of people dying during such episodes showed there to be a large proportion of carbon-centered combustion-derived nanoparticles [1], presumably from domestic coal combustion. The 1952 London smog episode, which was associated with thousands of deaths, had particle levels estimated at up to 4000 μg m^{-3} as compared to average current

Nanotechnologies for the Life Sciences Vol. 5
Nanomaterials – Toxicity, Health and Environmental Issues. Edited by Challa S. S. R. Kumar
Copyright © 2006 WILEY-VCH Verlag GmbH & Co. KGaA, Weinheim
ISBN: 3-527-31385-0

Tab. 12.1. Size fractions and description of the main size fractions of PM that are usually measured.

Size fraction	Unit	Description
Total suspended particulate (TSP)	$\mu g\ m^{-3}$	A TSP monitor measures by mass the atmospheric particulate smaller than about 40 μm in diameter
Black smoke	$\mu g\ m^{-3}$	This system was used in the UK and in other countries until the end of the 1980s. Air was drawn through a size-selective filter onto a white paper and the blackness of the "smudge" was measured; this method obviously is biased towards black, i.e., carbon-based, particles; there is a variable relationship between particles as measured by black smoke and PM_{10}
PM_{10}	$\mu g\ m^{-3}$	This size-selective sampling convention measures the mass per unit volume air of particles of aerodynamic diameter 10 μm with 50% efficiency; it roughly corresponds to the thoracic fraction of particles as defined by the International Standards Association (ISO) [3]
$PM_{2.5}$	$\mu g\ m^{-3}$	A size-selective sampling convention that measures the mass per unit volume air of particles of aerodynamic diameter 2.5 μm with 50% efficiency; it roughly corresponds to the respirable fraction of particles as defined by ISO [3]
$PM_{0.1}$/nanoparticles	$\mu g\ m^{-3}$	Also called ultrafine particles, these particles correspond to $PM_{0.1}$ and have a diameter of <0.1 μm (i.e., <100 nm).

London pollution levels of about 40 $\mu g\ m^{-3}$. As a result of such smogs, succeeding Governments introduced the Clean Air Acts that resulted in a steady decline in air pollution, leading to the present relatively clean air in UK cities.

However, in UK cities today there remains an air pollution problem, if lesser in magnitude than was seen in the past. The modern UK pollution environment is dominated by effluent from transport sources due to the increased number of cars on our roads and increased numbers of car journeys. In some cities, such as Los Angeles and Athens, air pollution is much worse than in the UK because of higher traffic density and the action of sunlight combining to produce "photochemical smog", which is qualitatively different, with ozone and particles as the dominant pollutants.

12.3
Introduction to Air Pollution Particles

Particles or particulate matter (PM) represent a part of the air pollution cocktail present in ambient air, which also consists of gases such as ozone, nitrogen dioxide etc. Particles have a special problem of nomenclature because they are measured by sampling conventions that collect only some fraction of the particulate material suspended in the air. These size fractions, including nanoparticles, are described in Table 12.1. Particles have received special attention because they are the most potent component of the air pollution mix in causing ill-health in the great majority of epidemiological studies. The adverse health effects of PM are seen at the levels that pertain in UK and other cities today and there is often no threshold. In other words, there is a background of ill-health caused by PM that increases when the ambient particle cloud increases in concentration and goes down when the amount of particles in the air decreases [2].

12.4
Adverse Effects of PM in Epidemiological Studies

The adverse effects of air pollution have been measured in thousands of studies. The levels of PM in any city vary both temporally – as a fluctuating hourly/daily level – and spatially, dependent on levels or traffic in an area or local industrial sources. The temporal and spatial variations in air pollution underlie the two main approaches to detecting and quantifying the adverse effects of PM (and indeed any air pollutant) in human populations. Time series studies utilize the temporal dimension and seek to relate the moving average of PM level to the moving average of a defined end-point, e.g., mortality [4]. Various lag-times are used to detect the relation between level of PM and the endpoint, e.g., 1 day. Environmental epidemiological studies seek to compare an endpoint such as the death rate in cities or suburbs characterized by high air pollution with the death rate in a city or suburb with lower air pollution [5]. Panel studies form a third category where small, well-characterized populations are followed for a defined endpoint that can be related to PM levels measured by personal or fixed point samplers [6].

When these different approaches are taken together and examined over hundreds of such studies there is good coherence between the acute effects, seen in time series and panel studies, and the chronic effects seen in environmental studies. More importantly, many of these studies show no evidence of a threshold, i.e., these adverse effects are occurring at the levels of PM that pertain in our cities today. Table 12.2 summarizes these adverse effects; the quantitative extent of mortality effects of PM in European studies is shown in Fig. 12.1.

Tab. 12.2. Adverse health effects due to PM.

Mortality from cardiovascular and respiratory causes
Admission to hospital for cardiovascular causes

Exacerbations of asthma in pre-existing asthmatics
Symptoms and use of asthma medication in asthmatics

Exacerbations of Chronic Obstructive Pulmonary Disease
Lung function decrease

Lung cancer

Within a short-lag time of one or two days following an increase in PM there are increases in the following (summarized in Ref. [8]) – (1) all-cause mortality; (2) attacks of asthma and increased usage of asthma medication; (3) deaths in COPD patients; (4) exacerbations of COPD; (5) deaths and hospitalizations for cardiovascular disease. The adverse cardiovascular effects associated with increases in PM are well-documented. Panel studies have documented associations between elevated levels of particles and

1. onset of myocardial infarction [6];
2. increased heart rate [9];
3. decreased heart rate variability [10].

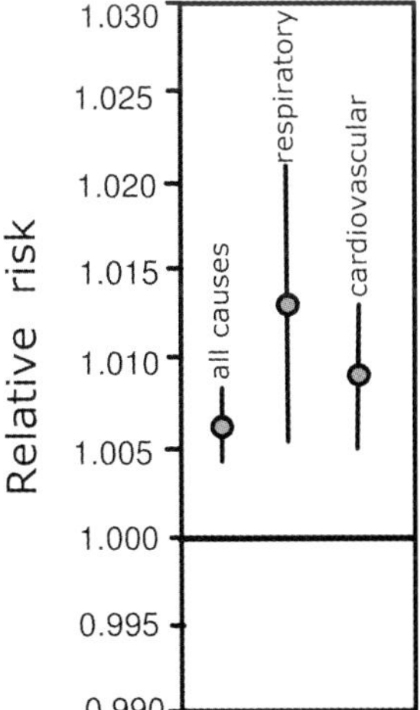

Fig. 12.1. Relative risk for mortality endpoints related to a 10 μg m^{-3} increase in PM$_{10}$, summarized from European studies [7].

Chamber studies with concentrated airborne particles (CAPs) have also shown increased lung inflammation [11] and altered brachial artery diameter in relation to increased exposure [12]. A recent epidemiological study of carotid intima-media thickness (CIMT), a measure of atherosclerosis, demonstrated evidence of an association between atherosclerosis and ambient air pollution level. Living in an area with a 10 μg m^{-3} higher level of PM$_{2.5}$ was associated with a CIMT increase of 5.9% (95% confidence interval, 1–11%); an even larger effect, 15.7%, was seen in older women.

12.5
Nanoparticles are an Important Component of PM

Urban PM is a complex mixture of particle types that depend on season, time of day, siting of sampler etc. Clearly, however, combustion-derived nanoparticles represent a major toxicologically important component (see below). Table 12.3 shows the common components of PM.

Whilst levels of PM, along with other pollutants, have gone down markedly over the last 35 years, the traffic-derived portion has gone up as the numbers of vehicles on the road has increased [13]. Release of nanoparticles from vehicle tailpipes during diesel and petrol combustion is the predominant source of nanoparticles (Fig. 12.2) although there are other sources, such as energy production (power stations) and industry [14]. The medians of ultrafine (NP) particle number concentrations across three European cities were recently (2005) found to range from 15 000 to 18 000 particles per cm^3 [15]. However, locally much higher exposures can be experienced. Using a nanoScanning Mobility Particle Sizer, multiple samples were

Tab. 12.3. Common components of PM and comments on their origin, nature and likely toxic potency.

PM$_{10}$ component	Comment	Toxic potency
Combustion-derived nanoparticles	Nanoparticles containing metals and organic volatiles; derived from combustion, e.g., vehicle exhaust particles	High
Sodium/magnesium compounds	Derived from sea spray	Low
Sulfate	Predominantly ammonium sulfate	Low
Nitrate	Predominantly ammonium nitrate	Low
Calcium/potassium compounds and insoluble minerals	Derived from the Earth's crust, e.g., clay	Low
Biologically-derived materials	For example, endotoxin	High

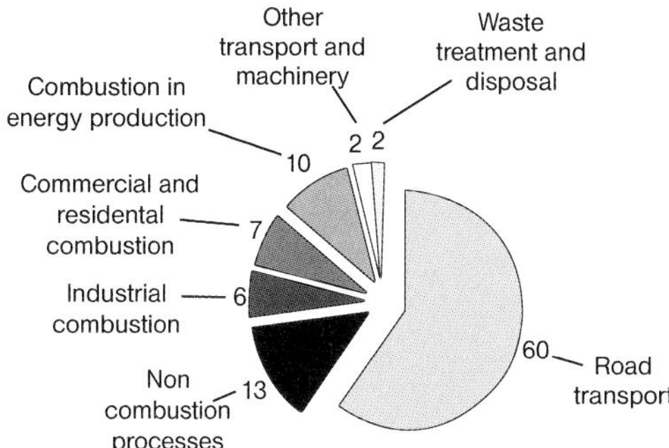

Fig. 12.2. Pie chart illustrating the source of $PM_{0.1}$ emissions in the UK in 1996 [21].

taken in Marylebone Road, a busy London street [16]. These showed that approximately 10 000 to 50 000 particles per cm^3 in the size range 30–100 nm were present. The daily pattern showed a typical one for traffic-derived nanoparticles [16]. A study on US highways suggested that the likely exposure in a vehicle traveling in busy traffic could be 200 to 560×10^3 particles (predominantly nanoparticles) per cm^3 [17, 18]. Indoors, there are also sources of nanoparticles such as cooking, vacuuming and burning wax candles [19]. Nanoparticles arise from combustion of domestic gas and in one study three gas rings produced a peak approaching 50 000 particles per cm^3 that underwent rapid aggregation, with increases in particle size and decrease in apparent number within a few minutes [20] (Fig. 12.3). Other sources of combustion, primarily industry, make a contribution to the nanoparticle cloud (Fig. 12.2). Secondary nanoparticles may also arise from environmental

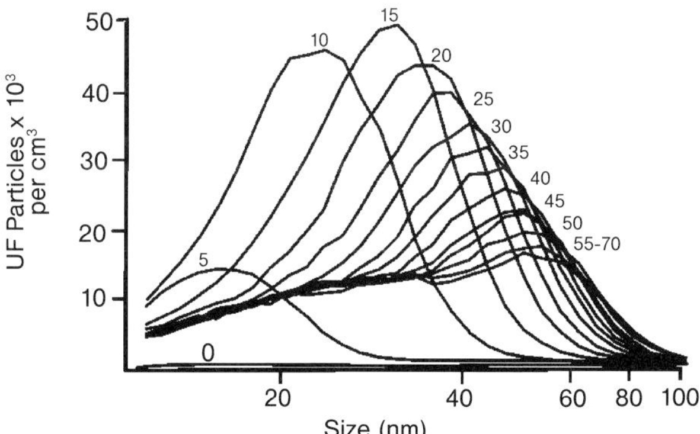

Fig. 12.3. Nanoparticle size distribution arising from burning of domestic gas in cooker rings. The family of curves shows the particle size distribution 5, 10, 15, 20 min etc. after lighting the rings. Note the decrease in number and increase in size with time, indicative of particle aggregation [20].

chemistry, e.g., nitrates, but these are unlikely to be as toxicologically potent as combustion-derived nanoparticles (see below).

12.6
Role of Nanoparticles in Mediating the Adverse Pulmonary Effects of PM

Epidemiological studies do not readily allow associations of adverse effects with sub-components of PM such as the nanoparticles, dependent as they usually are on a simple mass measure such as PM_{10} or $PM_{2.5}$. However, a few epidemiological studies have been able to identify combustion-derived particles as an important component in driving adverse effects of PM [22–24].

Toxicology can more readily address the relative toxic potency of the components of PM. Examining the components of PM toxicologically allows them to be divided into those with a high toxic potency and those with low toxic potency (Table 12.3). Combustion-derived nanoparticles (CDNP) stand out as common components of PM with high potency and several such nanoparticles have been utilized in toxicological studies, including carbon black and welding fume. As the predominant CDNP in PM, diesel soot has been specially studied.

Diesel fuel is a distillate of petroleum that contains paraffins, alkenes and aromatics [25]. On combustion in automobile engines it is transformed into low solubility carbon-centered nanoparticles of complex chemical and physical structure. Sulfates and organics, consisting of unburnt fuel, lube oil and polycyclic aromatic hydrocarbons (PAHs), also condense on the particles [25, 26]. Single diesel nanoparticles are 60 to 100 nm but these readily form complex chains and aggregates [27]. Diesel exhaust particles (DEP) are usually the most common CDNP in urban environmental air and in environmental particulate air pollution (PM_{10}) in conurbations generally. In the general environment the concentration of DEP in PM_{10} is likely to range from 5 to 30 $\mu g\ m^{-3}$.

Exposure to DEP is also highly inflammatory in rats and mice in non-overload conditions [28–30] and induces pro-inflammatory effects on cells *in vitro* [31–33]. The well-documented link between inflammation and lung cancer [34–36] supports the idea that diesel exhaust may indeed be carcinogenic via an inflammatory pathway. The inflammatory effects of DEP appear to be driven by the particulate component, i.e., the surface area effect [37]. However, the organic [38] and metal components [39] also appear to play a role in pro-inflammatory effects and thereby affect pathogenicity.

CDNP, such as carbon black and diesel soot, show several pro-inflammatory effects that are relevant to pulmonary inflammation (summarized in Table 12.4). Inflammation caused by CDNP in PM could be important in mediating the observed adverse effects (Fig. 12.4).

Studies with several CDNP suggest that various different ones are a hazard to the lungs through the pathways of oxidative stress and inflammation [45] (Fig. 12.5). Heterogeneity in composition and solubility means that the key initiating

Tab. 12.4. Some pro-inflammatory effects of combustion-derived nanoparticles.

Pulmonary-related effects of combustion-derived nanoparticles	Ref.
Generate oxidative stress in cell-free systems	40
Cause oxidative stress and activation of oxidative stress-responsive signaling pathways in epithelial cells	41, 42
Cause synthesis and release of pro-inflammatory cytokines	43
Causes pulmonary inflammation in laboratory animals	44

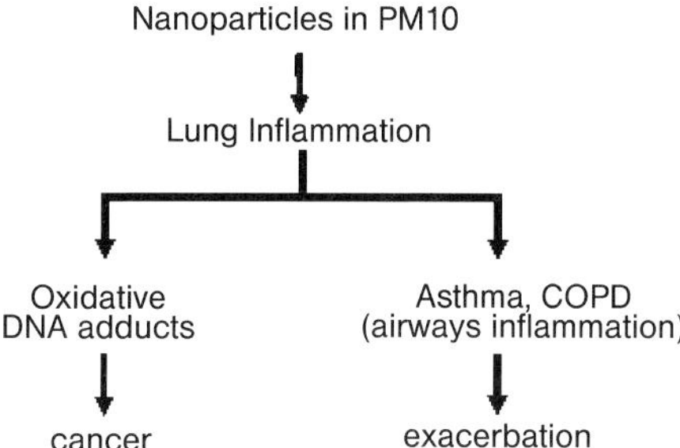

Fig. 12.4. Role of inflammation in the pulmonary effects of combustion-derived nanoparticles (CDNP).

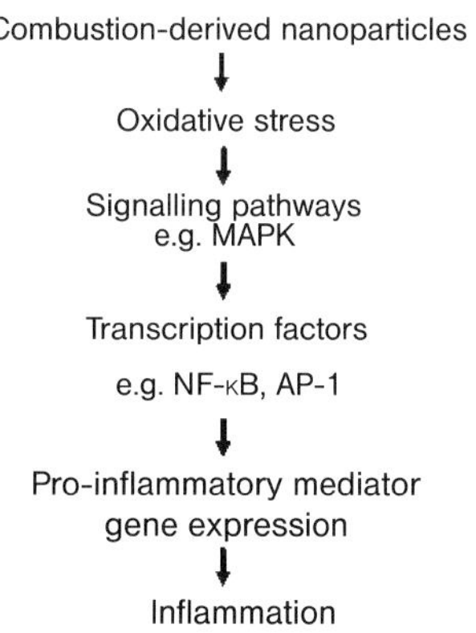

Fig. 12.5. Mechanism of the pro-inflammatory effects of combustion-derived nanoparticles (CDNP).

event of oxidative stress may originate from different components such as surfaces, metals or organics. These may even interact to produce oxidative stress as in the case of metals and organics interacting in the redox-cycling of quinoids [46] or nanoparticle surfaces and transition metals interacting additively in their ability to cause inflammation [47]. Inside the lung cells, oxidative stress can trigger inflammation through well-documented oxidative stress–response signaling pathways, including the MAPK and NF-κB. This culminates in the expression of genes that are involved in the recruitment and activation of leukocytes that are characteristic of inflammation. In addition, CDNP may exert genotoxic effects via well-documented pathways, including both oxidative stress adducts, such as 8-hydroxydeoxyguanosine, and bulky PAH-type adducts.

Lifetime animal exposure studies generally demonstrate that exposures to DEP and other nanoparticulate forms of carbon are carcinogenic [37] but these findings are complicated by the issue of rat lung overload [48]. Rat lung overload is a condition when very high lung surface area burden [49] of low toxicity, low soluble particles leads to failure of clearance. This leads to rapid accumulation of dose with concomitant inflammation and proliferation, which culminates in fibrosis and cancer [50]. Humans are unlikely to experience overload levels of diesel soot, even in occupational settings, and there is a question over whether overload can occur at all in humans. It is, therefore, unlikely that cancer associated with DEP exposure in humans results from a mechanism similar to rat lung overload.

12.7
Effects of Nanoparticles on the Cardiovascular System

Adverse cardiovascular effects associated with increases in PM are well-documented and have been reviewed extensively [51]. Exposure to particulate matter is associated with increases in mortality from ischemic heart disease and hospital admissions with acute myocardial infarction, heart failure and arrhythmia in time-series studies and population studies. Cohort studies have provided insight into the likely mechanisms responsible for these clinical events, and have documented associations between elevated levels of particles and heart rate, heart rate variability, blood pressure, myocardial ischemia and sub-clinical atherosclerosis.

Epidemiological studies cannot prove a causative biological effect of particulate exposure. Systems designed to deliver controlled amounts of ambient and combustion derived particulate now exist to allow a mechanistic approach to determining the effect of inhaled PM. Several plausible interlinking mechanisms have been proposed to explain these observations, and multiple pathways are likely to be involved (Fig. 12.6). These putative biological mechanisms involve direct effects of particles on the cardiovascular system following translocation across the alveolar-epithelial barrier, and indirect effects mediated by pulmonary oxidative stress and inflammation.

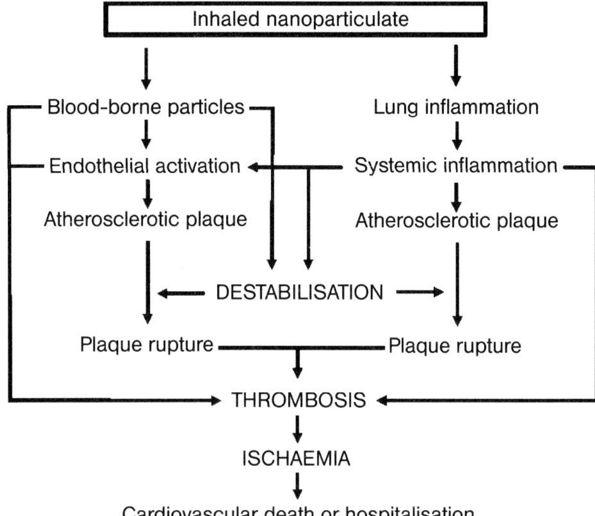

Fig. 12.6. Hypothetical pathways for the impact of inhaled nanoparticles on cardiovascular endpoints.

12.8
Inflammation, Atherosclerosis and Plaque Rupture

Markers of systemic inflammation are elevated in patients with cardiovascular disease [52]. In previously healthy individuals, elevated plasma concentrations of the acute phase reactant C-reactive protein have been shown to predict the development of ischemic heart disease [53], and in particular the risk of a first myocardial or cerebral infarction, independent of other risk factors [54].

Atherosclerosis is widely recognized to be an inflammatory process that is initiated through an injury to the vascular endothelium [55]. The resulting endothelial dysfunction leads to increased expression of leukocyte adhesion proteins, reduced anticoagulant activity and the release of growth factors, inflammatory mediators and cytokines. Continued inflammation results in leukocyte and monocyte recruitment, induction of atheroma formation and further arterial damage. Cycles of further damage cause plaque expansion and disruption that can lead to angina, crescendo angina and acute coronary syndromes, including myocardial infarction.

Inhalation of PM induces changes that are indicative of systemic inflammation such as increases in white blood cells and platelets [56], rises in C-reactive protein [57] and fibrinogen [58]. Experimental exposures mirror these clinical findings [59] and demonstrate evidence of combined systemic inflammation and endothelial dysfunction [60]. Repeated exposure to PM_{10} may, therefore, induce or exacerbate the vascular inflammation of atherosclerosis and promote plaque expansion or rupture. Indeed, using a Watanabe hereditary hyper-lipidemic rabbit model, Suwa and coworkers [61] have described plaque progression and destabilization following instillation of high doses of PM_{10}.

12.9
Nanoparticle Translocation and Direct Vascular Effects

It has been proposed that environmental nanoparticulate may translocate from the lung into the circulation and exert a direct effect on the vasculature. Once circulating, particles could interact with the vascular endothelium, or have direct effects on atherosclerotic plaque. Local inflammation could destabilize a coronary plaque, resulting in rupture, thrombosis, and an acute coronary syndrome.

Several factors influence the likelihood of particle translocation, including the properties of the particles themselves, size, affinity for water, rate of aggregation, surface charge, and the integrity or permeability of the alveolar-blood barrier. Inhaled nanoparticles deposit in the nasopharyngeal, tracheobronchial or alveolar regions by diffusion as discussed previously. Particles can evade phagocytosis by the alveolar macrophage and gain access to the interstitium whereupon they may be cleared directly into the circulation or by the lymphatic system. Environmental nanoparticulate may, possibly, gain access to the interstitium through pores in the blood-alveolar barrier. A three-pore model has been proposed to explain the ability of macromolecules to cross the alveolar-airway barrier, where the alveolar epithelial barrier contains many pores of 1 nm (68%) and 40 nm (30%) diameter and a few larger, 400 nm (2%), ones [62]. Alternatively, the mechanism may be analogous to the receptor-mediated transcytosis used by virus particles to enter and transfer between epithelial cells.

The first description of nanoparticulate translocation across the alveolar epithelium identified 30 nm gold particles in pulmonary capillary platelets following intra-tracheal instillation [63]. As well as identifying a novel pathway for the clearance of inhaled particulate, platelet uptake of particulate could result in aggregation and predispose to thrombus formation and atherothrombosis. Since then several studies have used radiolabeling to track the distribution of inhaled particles. Following intra-tracheal instillation of 99^{m}-technetium-labeled albumin nanocolloid (particles < 80 nm diameter) radioactivity was detected in the blood stream within 5 min, suggesting that particles underwent rapid translocation rather than being phagocytosed [64]. The mass of activity instilled did not alter the proportion of blood-borne radioactivity, suggesting the translocation is passive. Quantification of this process is hampered by methodological difficulties such as leaching of radiolabel and partial solubility of some particles. However, rats inhaling elemental carbon (^{13}C; particles 20–29 nm diameter) showed a substantial accumulation of ^{13}C in the liver 24 h after exposure [65]. The lung is not the only portal where airborne nanoparticles may enter the body and low levels of pulmonary deposition of ^{13}C suggested that ingestion or up-take through the skin could also be important.

Evidence that inhaled nanoparticulate can translocate in humans is less well established and the only two published studies to date are contradictory. In Nemmar et al. inhaled 99^mTc-labeled ultrafine carbon particles (5 nm diameter) were detected almost immediately in whole blood by thin-layer chromatography and accumulated in the liver and other extra-pulmonary organs [66]. In the study by Brown et al. using the same radiolabeled particles no such accumulation of activity outwith the lungs was reported [67]. Leaching of soluble radiolabel from the particles prevents any form of quantification of the extent of particle translocation using these methods. Further studies are required to determine whether alveolar translocation of nanoparticulate is an important pathway for the clearance of inhaled nanoparticles in man, and whether it can potentially explain the extra-pulmonary effects of particulate air pollution.

Oxidative stress increases the permeability of lung epithelium [68], potentially allowing particles, and macrophage laden with particles, to pass into the pulmonary interstitium and enter the circulation. Transfer of particles into the blood stream may be a more important pathway in patients with reduced antioxidant defenses, such as cigarette smokers or patients with underlying lung disease. In support of this, in an *ex vivo* lung perfusion model histamine and hydrogen peroxide perfusion increased alveolar permeability and the proportion of instilled iridium particles detected in the pulmonary venous effluent [69]. Likewise, increased epithelial permeability may encourage molecules produced in the lungs response to particles, e.g., IL-6 or oxidized LDL, to enter the interstitium and diffuse into the circulation.

Once circulating, nanoparticles could incorporate into sites of arterial injury and contribute to the development and progression of arterial atherosclerotic plaque formation. Oxidized LDL particles are typically around 200 nm in diameter and incorporation of these particles into the arterial wall is the central pathological process in the formation of atheroma. There is evidence that non-biological nanoparticles can be incorporated into the arterial wall. Following the infusion of ultrasmall paramagnetic iron oxide particles (USPIOs, 18 nm mean diameter) into hyperlipidaemic rabbits, the particles are phagocytosed by macrophages in atherosclerotic plaques of the aortic wall in quantities sufficient to be detectable by MRI [70]. Fluorescent-labeled nanoparticles, infused intra-luminally after balloon injury to the carotid artery in an animal model, successfully penetrated into the vessel wall and persisted for up to 14 days [71]. In humans, infusion of USPIOs before elective carotid endarterectomy revealed uptake of particles in 75% of ruptured, 54% of rupture-prone and only 7% of stable atheromatous plaques, suggesting that nanoparticles may incorporate into unstable plaques preferentially and potentially precipitate clinical events [72].

12.10
Endothelial Dysfunction and Endogenous Fibrinolysis

The endothelium plays a vital role in the control of blood flow, coagulation, fibrinolysis and inflammation. Following the seminal work of Furchgott and Zawadzki

[73], it is widely recognized that an array of mediators, including cigarette smoking, can influence vascular tone through endothelium-dependent actions, and there is now extensive evidence of abnormal endothelium-dependent vasomotion in patients with atherosclerosis [74–76]. Endothelial dysfunction is one of the earliest pathological processes in atherosclerosis and can predict the likelihood of future cardiovascular events and death in both healthy individuals [77] and those with coronary artery disease [78].

Endothelial dysfunction has been described in both the peripheral and coronary circulation of cigarette smokers [76, 79]. As combustion products and particulate matter are common to both air pollution and cigarette smoke, it is likely that air pollution is likely to have similar detrimental vascular effects. The tar phase of cigarette smoke and combustion-derived particulate may contribute to the oxidative burden and generate equivalent numbers of oxidative radicals. *In vitro* studies provide support for a mechanism of oxidative stress induced vascular dysfunction, with Ikeda et al. demonstrating that the incubation of aortic ring preparations with diesel exhaust particles results in a dose-dependent inhibition of acetylcholine-mediated relaxation, an effect abolished by co-incubation with superoxide dismutase [80].

Whilst endothelium-dependent vasomotion is important, it may not be representative of other aspects of endothelial function, such as the regulation of fibrinolysis. The fibrinolytic factor tissue plasminogen activator (t-PA) regulates the degradation of intravascular fibrin and is released from the endothelium through the translocation of a dynamic intracellular storage pool. Acute endogenous t-PA release from the endothelium regulates the dissolution of intravascular thrombosis and is a critical determinant of cardiovascular outcome. This is exemplified by the clinical observation that in approximately 30% of patients with acute myocardial infarction, spontaneous reperfusion occurs within 12 h of vessel occlusion. Culture of human umbilical vein endothelial cells with particulate matter for 6 h inhibits both the synthesis and release of t-PA in a dose-dependent manner [81].

Direct adverse effects of particulates on the vascular endothelial are important only if sufficient quantities of inhaled particulate reach the circulation. Alternatively, it is possible that the local pulmonary inflammatory response to particulate is sufficient to release inflammatory cells and cytokines into the circulation and induce a mild systemic inflammation. Healthy adults exposed to dilute diesel exhaust for one hour had a mild systemic inflammatory response with increased blood neutrophil levels [82], and mild systemic inflammation also causes a profound, but temporary, suppression of endothelium-dependent vasodilatation [83].

Our group has recently shown that these pathways are adversely affected *in vivo* following controlled human exposures to dilute diesel exhaust [84]. Exposures to particulate at levels encountered in the urban environment for one hour markedly impaired the regulation of vascular tone and endogenous fibrinolysis. This study confirms *in vitro* findings and provides a plausible mechanism linking air pollution to the pathogenesis of atherothrombosis.

The acute effects of exposure to particulate are consistent with recent epidemiological studies that report a significant increase in risk of acute myocardial infarction as little as two hours after exposure to road traffic [85] or an increase in $PM_{2.5}$ [6]. Our studies add to those of Brook et al. who demonstrated a reduction in brachial artery diameter immediately after exposure to a mixture of concentrated ambient particles and ozone [12]. In contrast, they found no effect on endothelium-dependent or -independent vasodilatation using flow-mediated and nitro-glycerine induced dilatation. This may reflect differences in the potency of the pollution models used or the technique used to assess vascular function. Exposures to concentrated ambient particulate are inherently variable in magnitude and composition, whereas in our study each volunteer received a standard exposure to combustion-derived particulate of known toxicity. Alternatively, the vascular effects of particulate matter are, possibly, mediated primarily in the resistance vessels assessed by plethysmography rather than in the conduit arteries assessed by ultrasound of the brachial artery.

12.11
Coagulation and Thrombosis

Atherothrombosis is characterized by disruption of an atherosclerotic plaque with thrombus formation and is the major cause of acute coronary syndromes and cardiovascular death worldwide. To explain the association between environmental air pollution and acute cardiovascular events we proposed a hypothetical mechanism whereby particles reaching the alveolar epithelium could influence blood coagulability, resulting in a prothrombotic state and potentially triggering acute thrombus formation in susceptible individuals [86]. Over the last 10 years, experimental evidence has provided support for this mechanism.

Exposure to particulate matter increases the levels of blood fibrinogen, a key component in coagulation cascade and determinant of blood viscosity [11]. Fibrinogen levels are associated with increased blood thrombogenicity, and can independently predict outcome in patients with existing cardiovascular disease [87]. Blood viscosity also increased in association with levels of ambient particulate during a prolonged air pollution episode [88]. In human challenge studies, exposure to diesel exhaust increases the number of circulating platelets [56], and in animal studies exposure to diesel exhaust and ultrafine particles rapidly activates blood platelets and enhances venous thrombosis in a model of endothelial injury [89, 90]. In a subsequent study, the pro-thrombotic effect of diesel exhaust particles was abrogated by intra-peritoneal dexamethasone, suggesting this effect is mediated by either local or systemic inflammation [91].

PM can induce various prothrombotic effects, including enhanced tissue factor expression on endothelial cells [92]. Tissue factor is the main stimulus for thrombin generation, initiates the extrinsic coagulation pathway, and circulating levels are present in patients with coronary artery disease. In an *in vivo* perfusion study, intra-arterial carbon ultrafine particles enhanced endothelial von-Willebrand factor expression and accumulation of fibrin and platelets on the endothelial surface [93]. In addition to altering the properties of endothelial cells and platelets to promote thrombus formation, nanoparticles may themselves act as a focus for thrombus formation. Scanning electron micrographic evaluation of thrombus on the surface of explanted temporary vena cava filters revealed the presence of foreign nanoparticles within thrombus [94]. These findings taken together support the notion that particulate matter causes inflammation, endothelial activation and can acutely increase the risk of thrombosis, thus predisposing to acute cardiovascular events.

12.12
Cardiac Autonomic Dysfunction

There is an important relationship between autonomic regulation of the cardiac cycle and cardiovascular mortality [95]. Variation in the interval between consecutive heart beats or heart rate variability (HRV) is controlled by the contrasting effects of the sympathetic and parasympathetic nervous systems. Reduction in HRV reflects either an increase in sympathetic drive or a decrease in vagal parasympathetic tone. Reduced HRV increases the risk of cardiovascular morbidity and mortality in both healthy individuals [96] and patients following myocardial infarction [97].

Several panel studies have reported a consistent association between reduced HRV and high ambient PM [98, 99]. The finding of altered HRV in an elderly cohort exposed to concentrated ambient particles (CAPs) provides direct evidence of the effects of PM on autonomic activity [10]. The importance of this effect in the presence of cardiovascular pathology was demonstrated in a canine model of myocardial infarction, where exposure to residual oil fly ash (ROFA) reduced HRV and increased cardiac arrhythmia following infarction [100]. Furthermore, in patients with implanted cardiac defibrillators there appears to be a relationship between ambient PM and the incidence of ventricular fibrillation [101].

The influence of PM on the autonomic nervous system may result in hospitalization or cardiac death by triggering tachyarrhythmia or altering coronary vascular tone. How inhaled PM interacts with the nervous system is unclear. The effects could be mediated by interstitialized particles directly irritating the nerve endings or through the local release of inflammatory cytokines in response to the particles.

12.13
Effects of Nanoparticles on the Liver and Gastrointestinal Tract

As described previously, particles depositing in the lung may gain access to the blood, allowing them to be transported around the body. For many years it has been recognized that the reticulo-endothelial system, a concentration of phagocytic cells in organs such as the liver and spleen, quickly accumulate particles injected into the blood [102]. For example, in 1970 Stuart reported that colloidal carbon injected into the tail vein accumulates in the liver and spleen [103]. NP designed as drug carriers are also sequestered by these phagocytic cells, with negatively charged carboxylated polystyrene nanoparticles being taken up into the liver via scavenger receptors [104]. The effects of interaction between such cells and nanoparticles is as yet unclear, but some studies suggest that polystyrene nanoparticles can induce mild oxidative stress in the liver via activation of the macrophage phagocytic burst [105]. *In vitro*, both nanoparticles [106, 107] and PM_{10} [108] stimulate calcium signaling via oxidative stress, leading to the production of pro-inflammatory cytokines such as tumor necrosis factor alpha. On the whole this work has been discussed in relation to the effects of PM and NP on the lung, but it is also feasible that NP translocated into the blood could initiate an inflammatory reaction in other organs such as the liver via their effects on macrophages.

There have been very few studies on the potentially adverse effects of NP directly on hepatocytes. Ref. [109] investigated the effects of various nanoparticles, differing in size and chemical composition, on hepatocyte viability as assessed by standard assays, including a measure of mitochondrial function (MTT assay) and enzyme leak through dysfunctional membranes (LDH assay). The silver particles studied ($10–50$ μg mL^{-1}) were found to be highly toxic via oxidative stress (glutathione depletion), whereas molybdenum, aluminum, iron oxide and titanium dioxide only exhibited mild toxicity at doses greater than 100 μg mL^{-1}. Further studies are required to establish how nanoparticles interact with liver cells and to elucidate the consequences.

The main route of particle clearance from the lung is via the mucociliary escalator, some of which is cleared via the nose, but most of which is swallowed [110]. This means that a significant proportion of the particle dose enters the gastrointestinal system. In relation to drug delivery, translocation of particulate matter across the gastrointestinal tract is now well documented [111]. There is histological, radiological and chemical evidence that 50–100 nm polystyrene nanoparticles cross the gastrointestinal tract and pass via the mesentery lymph supply and lymph nodes to the liver and spleen [112]. In addition, gastrointestinal uptake of particles of 100 nm diameter is 15–250-fold higher than for larger microparticles [113].

Furthermore, the uptake efficiency of the particles was dependent upon the tissue type, with Peyer's patches exhibiting a 2–200-fold higher uptake than non-patch tissue. The 100 nm particles were diffuse throughout the submucosal layers of the gastrointestinal tract while the larger particles were predominantly localized in the epithelium. In fact, histological studies have revealed a build-up of pigment particles accumulated from the diet in Peyer's patches of humans [114]. The ability of NP to cross the gut wall suggests that they would enter the body by the hepatic portal vein that drains the intestine, delivering blood directly to the liver. Hence, this would suggest that the liver would be exposed to particles translocated into the blood from both the lungs and the gastrointestinal tract.

The impact of nanoparticles on the gastrointestinal tract is relatively unknown. Some studies have suggested that exposure to nano- and microparticles in our diet could be associated with worsening of conditions such as Crohn's disease [115] through their ability to act as adjuvants in antigen-mediated immune responses, in tissues such as the Peyer's patches [116]. More work is required to verify these results and to elucidate the mechanisms involved.

12.14
Effects of NP on the Nervous System

In the past, in relation to the pulmonary toxicology of inhaled particles, most emphasis has been placed on the particles ($<$2.5 µm diameter) that gain access to the respiratory regions of the lung. This is because those particles depositing in the upper airways are easily cleared via the mucociliary escalator, whereas those depositing in the deeper lung depend upon the phagocytic activity of macrophages to minimize exposure and hence toxicity. However, more recently, Oberdörster et al. have demonstrated the ability of ^{13}C nanoparticles depositing on the nasal epithelium to gain access to the brain via the olfactory nerve [117]. They pointed out that neurophysiologists have long recognized that polio virus particles can be transported along nerve axons.

12.15
Summary

We have reviewed the literature pertaining to the role of nanoparticles in the adverse health effects of air pollution particles. Nanoparticles from combustion processes, predominantly automobile engines, are ubiquitous in ambient air and are strongly suspected of driving inflammatory and oxidative stress responses in the lungs. This is a factor in the exacerbations of airways disease seen in association with increases in PM. In addition to effects in the lungs, marked effects are seen in patients with cardiovascular disease. These may be explained by the ability of nanoparticles to gain access to the blood or indirectly through pro-inflammatory effects in the lungs, causing activation and destabilization of atheromatous plaques and leading to atherothrombosis and its associated morbidity and mortality. Nanoparticles may also gain access to the blood and directly affect the endothelium and the coagulation system. Nanoparticles in blood could affect the liver and potentially cross the brain barrier although nanoparticles could also enter the brain via olfactory neurones after deposition in the nose. Nanoparticles in the air are cleared to the gut via the mucociliary escalator and so their potential effects on the gut were considered. Once combustion-derived nanoparticles are resident in tissue they can deliver oxidative stress and affect inflammatory pathways and have genotoxic effects. Combustion-derived nanoparticles are a main driver of pulmonary and cardiovascular mortality and morbidity and, because of their potential to translocate around the body, have the potential to affect a number of extra-pulmonary target organs and systems.

References

1 HUNT, A., J. L. ABRAHAM, B. JUDSON, C. L. BERRY. 2003. Toxicologic and epidemiologic clues from the characterization of the 1952 London

smog fine particulate matter in archival autopsy lung tissues. *Environ. Health Perspect.* 111, 1209–1214.

2 BRUNEKREEF, B., S. T. HOLGATE. **2002.** Air pollution and health. *Lancet* 360, 1233–1242.

3 International Standards Organisation. **1994.** *Air Quality: Particle Size Fraction definitions for Health-related Sampling.* IS 7708, ISO, Geneva.

4 SCHWARTZ, J., D. W. DOCKERY, L. M. NEAS. **1996.** Is daily mortality associated specifically with fine particles? *J. Air Waste Manage. Assoc.* 46, 927–939.

5 DOCKERY, D. W., C. A. POPE, X. P. XU, J. D. SPENGLER, J. H. WARE, M. E. FAY, B. G. FERRIS, F. E. SPEIZER. **1993.** An association between air-pollution and mortality in 6 United-States cities. *N. Eng. J. Med.* 329, 1753–1759.

6 PETERS, A., D. W. DOCKERY, J. E. MULLER, M. A. MITTLEMAN. **2001.** Increased particulate air pollution and the triggering of myocardial infarction. *Circulation* 103, 2810–2815.

7 World Health Organization Europe. **2004.** *Health Aspects of Air Pollution.* World Health Organization Copenhagen, Denmark.

8 POPE, C. A., D. W. DOCKERY. **1999.** Epidemiology of particle effects, in *Air Pollution and Health.* ed. HOLGATE S. T., SAMET J. M., KOREN H. S., MAYNARD R. L., Academic Press, San Diego, pp. 673–705.

9 GONG, H., JR., W. S. LINN, S. L. TERRELL, K. W. CLARK, M. D. GELLER, K. R. ANDERSON, W. E. CASCIO, C. SIOUTAS. **2004.** Altered heart-rate variability in asthmatic and healthy volunteers exposed to concentrated ambient coarse particles. *Inhal. Toxicol.* 16, 335–343.

10 DEVLIN, R. B., A. J. GHIO, H. KEHRL, G. SANDERS, W. CASCIO. **2003.** Elderly humans exposed to concentrated air pollution particles have decreased heart rate variability. *Eur. Respir. J. Suppl.* 40, 76s–80s.

11 GHIO, A. J., C. KIM, R. B. DEVLIN. **2000.** Concentrated ambient air particles induce mild pulmonary inflammation in healthy human volunteers. *Am. J. Respir. Crit Care Med.* 162, 981–988.

12 BROOK, R. D., J. R. BROOK, B. URCH, R. VINCENT, S. RAJAGOPALAN, F. SILVERMAN. **2002.** Inhalation of fine particulate air pollution and ozone causes acute arterial vasoconstriction in healthy adults. *Circulation* 105, 1534–1536.

13 Airborne Particles Expert Group. **1999.** *Source Apportionment of Airborne Patriculate Matter in the United Kingdom.* Prepared on behalf of the Department of Environment, Transport and the Regions, the Welsh Office, the Scottish Office and the Department of the Environment (Northern Ireland), Crown copyright 158 pp.

14 WICHMANN, H. E. **2004.** 20 years after the winter smog episode 1985 – the particle problem then and today. *Int. J. Hyg. Environ. Health* 207, 489–491.

15 DE HARTOG, J. J., G. HOEK, A. MIRME, T. TUCH, G. P. KOS, H. M. TEN BRINK, B. BRUNEKREEF, J. CYRYS, J. HEINRICH, M. PITZ, T. LANKI, M. VALLIUS, J. PEKKANEN, W. G. KREYLING. **2005.** Relationship between different size classes of particulate matter and meteorology in three European cities. *J. Environ. Monit.* 7, 302–310.

16 Air Quality Expert Group. **2005.** *Particulate Matter in the United Kingdom.* Department of Environment, Food and Rural Affairs, London.

17 ELDER, A., R. GELEIN, J. FINKELSTEIN, R. PHIPPS, M. FRAMPTON, M. UTELL, D. B. KITTELSON, W. F. WATTS, P. HOPKE, C. H. JEONG, E. KIM, W. LIU, W. ZHAO, L. ZHUO, R. VINCENT, P. KUMARATHASAN, G. OBERDÖRSTER. **2004.** On-road exposure to highway aerosols. 2. Exposures of aged, compromised rats. *Inhal. Toxicol.* 16(Suppl. 1), 41–53.

18 KITTELSON, D. B., W. F. WATTS, J. P. JOHNSON, M. L. REMEROWKI, E. E. ISCHE, G. OBERDÖRSTER, R. M. GELEIN, A. ELDER, P. K. HOPKE, E. KIM, W. ZHAO, L. ZHOU, C. H. JEONG. **2004.** On-road exposure to highway aerosols. 1. Aerosol and gas measurements. *Inhal. Toxicol.* 16(Suppl. 1), 31–39.

19 AFSHARI, A., U. MATSON, L. E. EKBERG. **2005**. Characterization of indoor sources of fine and ultrafine particles: A study conducted in a full-scale chamber. *Indoor Air* 15, 141–150.

20 DENNEKAMP, M., S. HOWARTH, C. A. DICK, J. W. CHERRIE, K. DONALDSON, A. SEATON. **2001**. Ultrafine particles and nitrogen oxides generated by gas and electric cooking. *Occup. Environ. Med.* 58, 511–516.

21 HARRISON, R. M., J. P. SHI, S. XI, A. KHAN, D. MARK, R. KINNERSLEY, J. YIN. **2000**. Measurment of number, mass and size distribution of particles in the atmosphere. *Phil. Trans. R. Soc. London A* 358, 2567–2580.

22 POPE, C. A., III, R. W. HILL, G. M. VILLEGAS. **1999**. Particulate air pollution and daily mortality on Utah's wasatch front. *Environ. Health Perspect.* 107, 567–573.

23 LADEN, F., L. M. NEAS, D. W. DOCKERY, J. SCHWARTZ. **2000**. Association of fine particulate matter from different sources with daily mortality in six U.S. cities. *Environ. Health Perspect.* 108, 941–947.

24 SCHWARTZ, J. **1994**. What are people dying of on high air-pollution days. *Environ. Res.* 64, 26–35.

25 HEI. **1995**. *Diesel Exhaust: A Critical Analysis of Emissions, Exposure and Health Effects. A Special Report of the Insititute's Diesel Working Group.* HEI Research Report, HEI Cambridge, MA, USA.

26 TOBIAS, H. J., D. E. BEVING, P. J. ZIEMANN, H. SAKURAI, M. ZUK, P. H. MCMURRY, D. ZARLING, R. WAYTULONIS, D. B. KITTELSON. **2001**. Chemical analysis of diesel engine nanoparticles using a nano-DMA/thermal desorption particle beam mass spectrometer. *Environ. Sci. Technol.* 35, 2233–2243.

27 KITTELSON, DB. **1998**. Engines and nanoparticles: A review. *J. Aerosol Sci.* 29, 575–588.

28 ISHIHARA, Y., J. KAGAWA. **2003**. Chronic diesel exhaust exposures of rats demonstrate concentration and time-dependent effects on pulmonary inflammation. *Inhal. Toxicol.* 15, 473–492.

29 MAUDERLY, J. L., M. B. SNIPES, E. B. BARR, S. A. BELINSKY, J. A. BOND, A. L. BROOKS, I. Y. CHANG, Y. S. CHENG, N. A. GILLETT, W. C. GRIFFITH. **1994**. Pulmonary toxicity of inhaled diesel exhaust and carbon black in chronically exposed rats. Part I: Neoplastic and nonneoplastic lung lesions. *Res. Rep. Health Eff. Inst.*, 1–75.

30 MIYABARA, Y., R. YANAGISAWA, N. SHIMOJO, H. TAKANO, H. B. LIM, T. ICHINOSE, M. SAGAI. **1998**. Murine strain differences in airway inflammation caused by diesel exhaust particles. *Eur. Respir. J.* 11, 291–298.

31 ABE, S., H. TAKIZAWA, I. SUGAWARA, S. KUDOH. **2000**. Diesel exhaust (DE)-induced cytokine expression in human bronchial epithelial cells: A study with a new cell exposure system to freshly generated DE in vitro. *Am. J. Respir. Cell Mol. Biol.* 22, 296–303.

32 BONVALLOT, V., A. BAEZA-SQUIBAN, A. BAULIG, S. BRULANT, S. BOLAND, F. MUZEAU, R. BAROUKI, F. MARANO. **2001**. Organic compounds from diesel exhaust particles elicit a proinflammatory response in human airway epithelial cells and induce cytochrome p450 1A1 expression. *Am. J. Respir. Cell Mol. Biol.* 25, 515–521.

33 HENDERSON, R. F., H. W. LEUNG, A. G. HARMSEN, R. O. MCCLELLAN. **1988**. Species differences in release of arachidonate metabolites in response to inhaled diluted diesel exhaust. *Toxicol. Lett.* 42, 325–332.

34 KNAAPEN, A. M., F. SEILER, P. A. SCHILDERMAN, P. NEHLS, J. BRUCH, R. P. SCHINS, P. J. BORM. **1999**. Neutrophils cause oxidative DNA damage in alveolar epithelial cells. *Free Radic. Biol. Med.* 27, 234–240.

35 PARKE, D. V., A. L. PARKE. **1996**. Chemical-induced inflammation and inflammatory diseases. (Review; 24 refs). *Int. J. Occup. Med. Environ. Health* 9, 211–217.

36 SCHINS, R. P. **2002**. Mechanisms of genotoxicity of particles and fibers. *Inhal. Toxicol.* 14, 57–78.

37 Heinrich, U., R. Fuhst, S. Rittinghausen, O. Creutzenberg, B. Bellmann, W. Koch, K. Levsen. **1995**. Chronic inhalation exposure of Wistar rats and 2 different strains of mice to diesel-engine exhaust, carbon-black, and titanium-dioxide. *Inhal. Toxicol.* 7, 533–556.

38 Baeza-Squiban, A., V. Bonvallot, S. Boland, F. Marano. **1999**. Airborne particles evoke an inflammatory response in human airway epithelium. Activation of transcription factors. *Cell Biol. Toxicol.* 15, 375–380.

39 Ball, J. C., A. M. Straccia, W. C. Young, A. E. Aust. **2000**. The formation of reactive oxygen species catalyzed by neutral, aqueous extracts of NIST ambient particulate matter and diesel engine particles. *J. Air Waste Manag. Assoc.* 50, 1897–1903.

40 Wilson, M., J. H. Lightbody, K. Donaldson, V. Stone. **2002**. Interactions between ultrafine particles and metals *in vitro* and *in vivo*. *Toxicol. Appl. Pharmacol.* 184, 172–179.

41 Beck-Speier, I., N. Dayal, E. Karg, K. L. Maier, G. Schumann, H. Schulz, M. Semmler, S. Takenaka, K. Stettmaier, W. Bors, A. Ghio, J. M. Samet, J. Heyder. **2005**. Oxidative stress and lipid mediators induced in alveolar macrophages by ultrafine particles. *Free Radic. Biol. Med.* 38, 1080–1092.

42 Tamaoki, J., K. Isono, K. Takeyama, E. Tagaya, J. Nakata, A. Nagai. **2004**. Ultrafine carbon black particles stimulate proliferation of human airway epithelium via EGF receptor-mediated signaling pathway. *Am. J. Physiol. Lung Cell Mol. Physiol.* 287, L1127–L1133.

43 Brown, D. M., K. Donaldson, P. J. Borm, R. P. Schins, M. Dehnhardt, P. Gilmour, L. A. Jimenez, V. Stone. **2004**. Calcium and ROS-mediated activation of transcription factors and TNF-alpha cytokine gene expression in macrophages exposed to ultrafine particles. *Am. J. Physiol. Lung Cell Mol. Physiol.* 286, L344–L353.

44 Gilmour, P. S., A. Ziesenis, E. R. Morrison, M. A. Vickers, E. M. Drost, I. Ford, E. Karg, C. Mossa, A. Schroeppel, G. A. Ferron, J. Heyder, M. Greaves, W. MacNee, K. Donaldson. **2004**. Pulmonary and systemic effects of short-term inhalation exposure to ultrafine carbon black particles. *Toxicol. Appl. Pharmacol.* 195, 35–44.

45 Donaldson, K., C. L. Tran, L. A. Jimenez, R. Duffin, D. E. Newby, N. Mills, W. MacNee, V. Stone. **2005**. Combustion-derived nanoparticles: A review of their toxicology following inhalation exposure. *Particle Fibre Toxicol.* 2, 10.

46 Squadrito, G. L., R. Cueto, B. Dellinger, W. A. Pryor. **2001**. Quinoid redox cycling as a mechanism for sustained free radical generation by inhaled airborne particulate matter. *Free Radic. Biol. Med.* 31, 1132–1138.

47 Wilson, M., J. H. Lightbody, K. Donaldson, J. H. Sales, V. Stone. **2001**. Oxidative Interactions between ultrafine particles and transition metals *in vivo* and *in vitro*. *Toxicol. Appl. Pharmacol.* 163, A495.

48 ILSI Workshop. **2000**. The relevance of the rat lung response to particle overload for human risk assessment. *Inhal. Toxicol.* 12, 1–148.

49 Tran, C. L., D. Buchanan, R. T. Cullen, A. Searl, A. D. Jones, K. Donaldson. **2000**. Inhalation of poorly soluble particles. II. Influence of particle surface area on inflammation and clearance. *Inhal. Toxicol.* 12, 1113–1126.

50 Mauderly, J. L. **1997**. Relevance of particle-induced rat lung tumors for assessing lung carcinogenic hazard and human lung cancer risk. *Environ. Health Perspect.* 105, 1337–1346.

51 Brook, R. D., B. Franklin, W. Cascio, Y. Hong, G. Howard, M. Lipsett, R. Luepker, M. Mittleman, J. Samet, S. C. Smith, Jr., I. Tager. **2004**. Air pollution and cardiovascular disease: A statement for healthcare professionals from the Expert Panel on Population and Prevention Science of the American Heart Association. *Circulation* 109, 2655–2671.

52 LIUZZO, G., L. M. BIASUCCI, J. R. GALLIMORE, R. L. GRILLO, A. G. REBUZZI, M. B. PEPYS, A. MASERI. **1994**. The prognostic value of C-reactive protein and serum amyloid a protein in severe unstable angina. *N. Engl. J. Med.* 331, 417–424.

53 RIDKER, P. M., P. HAUGHIE. **1998**. Prospective studies of C-reactive protein as a risk factor for cardiovascular disease. *J. Investig. Med.* 46, 391–395.

54 RIDKER, P. M., M. CUSHMAN, M. J. STAMPFER, R. P. TRACY, C. H. HENNEKENS. **1997**. Inflammation, aspirin, and the risk of cardiovascular disease in apparently healthy men. *N. Engl. J. Med.* 336, 973–979.

55 LIBBY, P., P. M. RIDKER, A. MASERI. **2002**. Inflammation and athero-sclerosis. *Circulation* 105, 1135–1143.

56 SALVI, S., A. BLOMBERG, B. RUDELL, F. KELLY, T. SANDSTROM, S. T. HOLGATE, A. FREW. **1999**. Acute inflammatory responses in the airways and peripheral blood after short-term exposure to diesel exhaust in healthy human volunteers. *Am. J. Respir. Crit Care Med.* 159, 702–709.

57 POPE, C. A., III. **2001**. Particulate air pollution, C-reactive protein, and cardiac risk. *Eur. Heart J.* 22, 1149–1150.

58 PEKKANEN, J., E. J. BRUNNER, H. R. ANDERSON, P. TIITTANEN, R. W. ATKINSON. **2000**. Daily concentrations of air pollution and plasma fibrinogen in London. *Occup. Environ. Med.* 57, 818–822.

59 VAN EEDEN, S. F., W. C. TAN, T. SUWA, H. MUKAE, T. TERASHIMA, T. FUJII, D. QUI, R. VINCENT, J. C. HOGG. **2001**. Cytokines involved in the systemic inflammatory response induced by exposure to particulate matter air pollutants (pm(10)). *Am. J. Respir. Crit Care Med.* 164, 826–830.

60 VINCENT, R., P. KUMARATHASAN, P. GEOGAN, S. G. BJARNASON, J. GUENETTE, D. BERUBE, I. Y. ADAMSON, S. DESJARDINS, R. BURNETT, F. J. MILLER, B. BATTISTINI. **2001**. Inhalation toxicology of urban ambient particulate matter: Acute cardiovascular effects in rats. *Health Effects Res. Rep.* No 104.

61 SUWA, T., J. C. HOGG, K. B. QUINLAN, A. OHGAMI, R. VINCENT, S. F. VAN EEDEN. **2002**. Particulate air pollution induces progression of atherosclerosis. *J. Am. Coll. Cardiol.* 39, 935–942.

62 CONHAIM, R. L., A. EATON, N. C. STAUB, T. D. HEATH. **1988**. Equivalent pore estimate for the alveolar-airway barrier in isolated dog lung. *J. Appl. Physiol.* 64, 1134–1142.

63 BERRY, J. P., B. ARNOUX, G. STANISLAS, P. GALLE, J. CHRETIEN. **1977**. A microanalytic study of particles transport across the alveoli: Role of blood platelets 25. *Biomedicine.* 27, 354–357.

64 NEMMAR, A., H. VANBILLOEN, M. F. HOYLAERTS, P. H. HOET, A. VERBRUGGEN, B. NEMERY. **2001**. Passage of intratracheally instilled ultrafine particles from the lung into the systemic circulation in hamster. *Am. J. Respir. Crit Care Med.* 164, 1665–1668.

65 OBERDÖRSTER, G., Z. SHARP, V. ATUDOREI, A. P. ELDER, R. GELEIN, A. K. LUNTS, W. KREYLING, C. COX. **2002**. Extrapulmonary translocation of ultrafine carbon particles following inhalation exposure. *J. Toxicol. Environ. Health.* 65, 1531–1545.

66 NEMMAR, A., P. H. HOET, B. VANQUICKENBORNE, D. DINSDALE, M. THOMEER, M. F. HOYLAERTS, H. VANBILLOEN, L. MORTELMANS, B. NEMERY. **2002**. Passage of inhaled particles into the blood circulation in humans. *Circulation* 105, 411–414.

67 BROWN, J. S., K. L. ZEMAN, W. D. BENNETT. **2002**. Ultrafine particle deposition and clearance in the healthy and obstructed lung. *Am. J. Respir. Crit. Care Med.* 166, 1240–1247.

68 LI, X. Y., K. DONALDSON, I. RAHMAN, W. MacNEE. **1994**. An investigation of the role of glutathione in increased epithelial permeability induced by cigarette smoke in vivo and in vitro. *Am. J. Respir. Crit. Care Med.* 149, 1518–1525.

69 Meiring, J. J., P. J. Borm, K. Bagate, M. Semmler, J. Seitz, S. Takenaka, W. G. Kreyling. **2005**. The influence of hydrogen peroxide and histamine on lung permeability and translocation of iridium nanoparticles in the isolated perfused rat lung 2. *Part Fibre Toxicol.* 2, 3.

70 Ruehm, S. G., M. Goyen, J. Barkhausen, K. Kroger, S. Bosk, M. E. Ladd, J. F. Debatin. **2001**. Rapid magnetic resonance angiography for detection of athero-sclerosis. *Lancet* 357, 1086–1091.

71 Guzman, L. A., V. Labhasetwar, C. Song, Y. Jang, A. M. Lincoff, R. Levy, E. J. Topol. **1996**. Local intraluminal infusion of biodegradable polymeric nanoparticles. A novel approach for prolonged drug delivery after balloon angioplasty. *Circulation* 94, 1441–1448.

72 Kooi, M. E., V. C. Cappendijk, K. B. Cleutjens, A. G. Kessels, P. J. Kitslaar, M. Borgers, P. M. Frederik, M. J. Daemen, J. M. van Engelshoven. **2003**. Accumulation of ultrasmall superparamagnetic particles of iron oxide in human atherosclerotic plaques can be detected by in vivo magnetic resonance imaging. *Circulation* 107, 2453–2458.

73 Furchgott, R. F., J. V. Zawadzki. **1980**. The obligatory role of endo-thelial cells in the relaxation of arterial smooth muscle by acetylcholine. *Nature* 288, 373–376.

74 Ludmer, P. L., A. P. Selwyn, T. L. Shook, R. R. Wayne, G. H. Mudge, R. W. Alexander, P. Ganz. **1986**. Paradoxical vasoconstriction induced by acetylcholine in atherosclerotic coronary arteries. *N. Engl. J. Med.* 315, 1046–1051.

75 Celermajer, D. S., M. R. Adams, P. Clarkson, J. Robinson, R. McCredie, A. Donald, J. E. Deanfield. **1996**. Passive smoking and impaired endothelium-dependent arterial dilatation in healthy young adults. *N. Engl. J. Med.* 334, 150–154.

76 Newby, D. E., A. L. McLeod, N. G. Uren, L. Flint, C. A. Ludlam, D. J. Webb, K. A. Fox, N. A. Boon. **2001**. Impaired coronary tissue plasminogen activator release is associated with coronary atherosclerosis and cigarette smoking: Direct link between endothelial dysfunction and atherothrombosis. *Circulation* 103, 1936–1941.

77 Halcox, J. P., W. H. Schenke, G. Zalos, R. Mincemoyer, A. Prasad, M. A. Waclawiw, K. R. Nour, A. A. Quyyumi. **2002**. Prognostic value of coronary vascular endothelial dysfunction. *Circulation* 106, 653–658.

78 Heitzer, T., T. Schlinzig, K. Krohn, T. Meinertz, T. Munzel. **2001**. Endothelial dysfunction, oxidative stress, and risk of cardiovascular events in patients with coronary artery disease. *Circulation* 104, 2673–2678.

79 Newby, D. E., R. A. Wright, C. Labinjoh, C. A. Ludlam, K. A. Fox, N. A. Boon, D. J. Webb. **1999**. Endothelial dysfunction, impaired endogenous fibrinolysis, and cigarette smoking: A mechanism for arterial thrombosis and myocardial infarction. *Circulation* 99, 1411–1415.

80 Ikeda, M., M. Shitashige, H. Yamasaki, M. Sagai, T. Tomita. **1995**. Oxidative modification of low density lipoprotein by diesel exhaust particles. *Biol. Pharm. Bull.* 18, 866–871.

81 Gilmour, P. S., E. R. Morrison, M. A. Vickers, I. Ford, C. A. Ludlam, M. Greaves, K. Donaldson, W. MacNee. **2005**. The procoagulant potential of environmental particles (PM10). *Occup. Environ. Med.* 62, 164–171.

82 Salvi, S., A. Blomberg, B. Rudell, F. Kelly, T. Sandstrom, S. T. Holgate, A. Frew. **1999**. Acute inflammatory responses in the airways and peripheral blood after short-term exposure to diesel exhaust in healthy human volunteers. *Am. J. Respir. Crit. Care Med.* 159, 702–709.

83 Hingorani, A. D., J. Cross, R. K. Kharbanda, M. J. Mullen, K. Bhagat, M. Taylor, A. E. Donald, M. Palacios, G. E. Griffin, J. E. Deanfield, R. J. MacAllister, P.

VALLANCE. **2000**. Acute systemic inflammation impairs endothelium-dependent dilatation in humans. *Circulation* 102, 994–999.

84 MILLS, N., S. TORNQUIST, S. ROBINSON, K. DARNLEY, N. A. BOON, W. MACNEE, K. DONALDSON, A. BLOMBERG, T. SANDSTROM, D. NEWBY. **2005**. Diesel exhaust inhalation causes vascular dysfunction and impaired endogenous fibrinolysis. *Circulation* 112, 3930–3936.

85 PETERS, A., K. S. VON M. HEIER, I. TRENTINAGLIA, A. HORMANN, H. E. WICHMANN, H. LOWEL. **2004**. Exposure to traffic and the onset of myocardial infarction. *N. Engl. J. Med.* 351, 1721–1730.

86 SEATON, A., W. MACNEE, K. DONALDSON, D. GODDEN. **1995**. Particulate air-pollution and acute health-effects. *Lancet* 345, 176–178.

87 THOMPSON, S. G., J. KIENAST, S. D. PYKE, F. HAVERKATE, J. C. VAN DE LOO. **1995**. Hemostatic factors and the risk of myocardial infarction or sudden death in patients with angina pectoris. European Concerted Action on Thrombosis and Disabilities Angina Pectoris Study Group. *N. Engl. J. Med.* 332, 635–641.

88 PETERS, A., A. DORING, H. E. WICHMANN, W. KOENIG. **1997**. Increased plasma viscosity during an air pollution episode: A link to mortality? *Lancet* 349, 1582–1587.

89 NEMMAR, A., P. H. HOET, D. DINSDALE, J. VERMYLEN, M. F. HOYLAERTS, B. NEMERY. **2003**. Diesel exhaust particles in lung acutely enhance experimental peripheral thrombosis. *Circulation* 107, 1202–1208.

90 NEMMAR, A., M. F. HOYLAERTS, P. H. HOET, J. VERMYLEN, B. NEMERY. **2003**. Size effect of intratracheally instilled particles on pulmonary inflammation and vascular thrombosis. *Toxicol. Appl. Pharmacol.* 186, 38–45.

91 NEMMAR, A., M. F. HOYLAERTS, P. H. HOET, B. NEMERY. **2004**. Possible mechanisms of the cardiovascular effects of inhaled particles: Systemic translocation and prothrombotic effects. *Toxicol. Lett.* 149, 243–253.

92 GILMOUR, P. S., E. R. MORRISON, M. A. VICKERS, I. FORD, C. A. LUDLAM, M. GREAVES, K. DONALDSON, W. MACNEE. **2005**. The procoagulant potential of environmental particles (PM10). *Occup. Environ. Med.* 62, 164–171.

93 KHANDOGA, A., A. STAMPFL, S. TAKENAKA, H. SCHULZ, R. RADYKEWICZ, W. KREYLING, F. KROMBACH. **2004**. Ultrafine particles exert prothrombotic but not inflammatory effects on the hepatic microcirculation in healthy mice in vivo. *Circulation* 109, 1320–1325.

94 GATTI, A. M., S. MONTANARI, E. MONARI, A. GAMBARELLI, F. CAPITANI, B. PARISINI. **2004**. Detection of micro- and nano-sized biocompatible particles in the blood. *J. Mater. Sci. Mater. Med.* 15, 469–472.

95 BECKETT, W. S., D. F. CHALUPA, A. PAULY-BROWN, D. M. SPEERS, J. C. STEWART, M. W. FRAMPTON, M. J. UTELL, L. S. HUANG, C. COX, W. ZAREBA, G. OBERDÖRSTER. **2005**. Comparing inhaled ultrafine versus fine zinc oxide particles in healthy adults: A human inhalation study. *Am. J. Respir. Crit Care Med.* 171, 1129–1135.

96 TSUJI, H., F. J. VENDITTI, JR., E. S. MANDERS, J. C. EVANS, M. G. LARSON, C. L. FELDMAN, D. LEVY. **1994**. Reduced heart rate variability and mortality risk in an elderly cohort. The Framingham Heart Study. *Circulation* 90, 878–883.

97 KLEIGER, R. E., P. K. STEIN, J. T. BIGGER, JR. **2005**. Heart rate variability: Measurement and clinical utility. *Ann. Noninvasive. Electrocardiol.* 10, 88–101.

98 POPE, C. A., III, D. J. EATOUGH, D. R. GOLD, Y. PANG, K. R. NIELSEN, P. NATH, R. L. VERRIER, R. E. KANNER. **2001**. Acute exposure to environmental tobacco smoke and heart rate variability. *Environ. Health Perspect.* 109, 711–716.

99 MAGARI, S. R., J. SCHWARTZ, P. L.

WILLIAMS, R. HAUSER, T. J. SMITH, D. C. CHRISTIANI. **2002**. The association between personal measurements of environmental exposure to particulates and heart rate variability. *Epidemiology* 13, 305–310.

100 WELLENIUS, G. A., P. H. SALDIVA, J. R. BATALHA, G. G. KRISHNA MURTHY, B. A. COULL, R. L. VERRIER, J. J. GODLESKI. **2002**. Electrocardiographic changes during exposure to residual oil fly ash (ROFA) particles in a rat model of myocardial infarction. *Toxicol. Sci.* 66, 327–335.

101 DOCKERY, D. W., H. LUTTMANN-GIBSON, D. Q. RICH, M. S. LINK, M. A. MITTLEMAN, D. R. GOLD, P. KOUTRAKIS, J. D. SCHWARTZ, R. L. VERRIER. **2005**. Association of air pollution with increased incidence of ventricular tachyarrhythmias recorded by implanted cardioverter defibrillators. *Environ. Health Perspect.* 113, 670–674.

102 MOGHIMI, S. M., A. E. HAWLEY, N. M. CHRISTY, T. GRAY, L. ILLUM, S. S. DAVIS. **1994**. Surface engineered nanospheres with enhanced drainage into lymphatics and uptake by macrophages of the regional lymph nodes. *FEBS Lett.* 344, 25–30.

103 STUART, A. E. **1970**. *The Reticuloendothelial System.* E and S Livingstone, Edinburgh.

104 FURUMOTO, K., S. NAGAYAMA, K. OGAWARA, Y. TAKAKURA, M. HASHIDA, K. HIGAKI, T. KIMURA. **2004**. Hepatic uptake of negatively charged particles in rats: Possible involvement of serum proteins in recognition by scavenger receptor. *J. Controlled Release* 97, 133–141.

105 FERNANDEZ-URRUSUNO, R., E. FATTAL, J. FEGER, P. COUVREUR, P. THEROND. **1997**. Evaluation of hepatic antioxidant systems after intravenous administration of polymeric nanoparticles. *Biomaterials* 18, 511–517.

106 BROWN, D. M., K. DONALDSON, P. J. BORM, R. P. SCHINS, M. DEHNHARDT, P. GILMOUR, L. A. JIMENEZ, V. STONE. **2004**. Calcium and ROS-mediated activation of transcription factors and TNF-alpha cytokine gene expression in macrophages exposed to ultrafine particles. *Am. J. Physiol. Lung Cell Mol. Physiol.* 286, L344–L353.

107 STONE, V., M. TUINMAN, J. E. VAMVAKOPOULOS, J. SHAW, D. BROWN, S. PETTERSON, S. P. FAUX, P. BORM, W. MACNEE, F. MICHAELANGELI, K. DONALDSON. **2000**. Increased calcium influx in a monocytic cell line on exposure to ultrafine carbon black. *Eur. Respir. J.* 15, 297–303.

108 BROWN, D. M., K. DONALDSON, V. STONE. **2004**. Effects of PM10 in human peripheral blood monocytes and J774 macrophages. *Respir. Res.* 5, 29.

109 HUSSAIN, S. M., K. L. HESS, J. M. GEARHART, K. T. GEISS, J. J. SCHLAGER. **2005**. In vitro toxicity of nanoparticles in BRL 3A rat liver cells. *Toxicol. in Vitro.* 19, 975–983.

110 DONALDSON, K., D. BROWN, A. CLOUTER, R. DUFFIN, W. MACNEE, L. RENWICK, L. TRAN, V. STONE. **2002**. The pulmonary toxicology of ultrafine particles. *J. Aerosol Med.* 15, 213–220.

111 FLORENCE, A. T., N. HUSSAIN. **2001**. Transcytosis of nanoparticle and dendrimer delivery systems: Evolving vistas. *Adv. Drug Deliv. Rev.* 50(Suppl. 1), S69–S89.

112 JANI, P., G. W. HALBERT, J. LANGRIDGE, A. T. FLORENCE. **1990**. Nanoparticle uptake by the rat gastrointestinal mucosa: Quantitation and particle size dependency. *J. Pharm. Pharmacol.* 42, 821–826.

113 DESAI, M. P., V. LABHASETWAR, G. L. AMIDON, R. J. LEVY. **1996**. Gastrointestinal uptake of biodegradable microparticles: Effect of particle size. *Pharm. Res.* 13, 1838–1845.

114 POWELL, J. J., C. C. AINLEY, R. S. HARVEY, I. M. MASON, M. D. KENDALL, E. A. SANKEY, A. P. DHILLON, R. P. THOMPSON. **1996**. Characterisation of inorganic microparticles in pigment cells of human gut associated lymphoid tissue. *Gut* 38, 390–395.

115 POWELL, J. J., R. S. HARVEY, P. ASHWOOD, R. WOLSTENCROFT, M. E. GERSHWIN, R. P. THOMPSON. **2000**.

Immune potentiation of ultrafine dietary particles in normal subjects and patients with inflammatory bowel disease. *J. Autoimmun.* **14**, 99–105.

116 LOMER, M. C., R. P. THOMPSON, J. J. POWELL. **2002**. Fine and ultrafine particles of the diet: Influence on the mucosal immune response and association with Crohn's disease. *Proc. Nutr. Soc.* **61**, 123–130.

117 OBERDÖRSTER, G., Z. SHARP, V. ATUDOREI, A. ELDER, R. GELEIN, W. KREYLING, C. COX. **2004**. Translocation of inhaled ultrafine particles to the brain. *Inhal. Toxicol.* **16**, 437–445.

Index

Nanotechnologies for the Life Sciences Vol. 5
Nanomaterials – Toxicity, Health and Environmental Issues. Edited by Challa S. S. R. Kumar
Copyright © 2006 WILEY-VCH Verlag GmbH & Co. KGaA, Weinheim
ISBN: 3-527-31385-0